Foundation and Anchor Design Guide for Metal Building Systems

Foundation and Anchor Design Guide for Metal Building Systems

Alexander Newman, P.E., F.ASCE

New York Chicago San Francisco
Lisbon London Madrid Mexico City
Milan New Delhi San Juan
Seoul Singapore Sydney Toronto

The McGraw·Hill Companies

Cataloging-in-Publication Data is on file with the Library of Congress

Copyright © 2013 by The McGraw-Hill Companies, Inc. All rights reserved. Printed in the United States of America. Except as permitted under the United States Copyright Act of 1976, no part of this publication may be reproduced or distributed in any form or by any means, or stored in a data base or retrieval system, without the prior written permission of the publisher.

1 2 3 4 5 6 7 8 9 0   DOC/DOC   1 8 7 6 5 4 3 2

ISBN 978-0-07-176635-7
MHID 0-07-176635-9

Sponsoring Editor
Larry S. Hager

Editing Supervisor
Stephen M. Smith

Production Supervisor
Richard C. Ruzycka

Acquisitions Coordinator
Bridget L. Thoreson

Project Manager
Rohini Deb,
Cenveo Publisher Services

Copy Editor
Rachel Hockett

Proofreader
Surendra Nath Shivam,
Cenveo Publisher Services

Art Director, Cover
Jeff Weeks

Composition
Cenveo Publisher Services

Printed and bound by RR Donnelley.

McGraw-Hill books are available at special quantity discounts to use as premiums and sales promotions, or for use in corporate training programs. To contact a representative, please e-mail us at bulksales@mcgraw-hill.com.

This book is printed on acid-free paper.

Information contained in this work has been obtained by The McGraw-Hill Companies, Inc. ("McGraw-Hill") from sources believed to be reliable. However, neither McGraw-Hill nor its authors guarantee the accuracy or completeness of any information published herein, and neither McGraw-Hill nor its authors shall be responsible for any errors, omissions, or damages arising out of use of this information. This work is published with the understanding that McGraw-Hill and its authors are supplying information but are not attempting to render engineering or other professional services. If such services are required, the assistance of an appropriate professional should be sought.

About the Author

Alexander Newman, P.E., F.ASCE, is a forensic and structural consultant in Needham, Massachusetts. During more than 30 years of professional practice, he has been involved with the structural design, renovation, and failure investigation of numerous structures around the country. His areas of expertise include design and failure analysis of pre-engineered metal buildings, metal roofs, and other structures. He is one of the country's foremost experts on metal building systems. He has served as an expert consultant for litigation involving metal building systems and other structures.

Mr. Newman's diverse engineering and managerial experience includes positions as principal structural engineer with a large architectural and engineering firm, manager of a steel fabrication shop, and engineer with light-gage metal and precast-concrete manufacturers. Most recently, he has been a managing engineer with Exponent Failure Analysis Associates, the country's premier construction failure and accident investigation firm.

Mr. Newman's many publications include a number of award-winning articles that appeared in leading engineering magazines. His definitive reference book *Structural Renovation of Buildings: Methods, Details, and Design Examples* was published by McGraw-Hill in 2001. He is the author of another authoritative book from McGraw-Hill, *Metal Building Systems: Design and Specifications*, now in its Second Edition and translated into Chinese.

Mr. Newman has led numerous educational seminars around the country for design professionals, building officials, owners, and contractors. He has conducted training programs for the employees of the U.S. State Department, NASA, the U.S. Air Force, the Iraq Reconstruction Team, and many educational and professional groups. He has taught a number of courses at Northeastern University.

Contents

Preface xiii

1 Introduction to Metal Building Systems 1
1.1 Two Main Classes of Metal Building Systems 1
1.2 Frame-and-Purlin Buildings: Primary and Secondary Framing . . . 1
    1.2.1 Primary Frames: Usage and Terminology 3
    1.2.2 Single-Span Rigid Frames 3
    1.2.3 Multiple-Span Rigid Frames 4
    1.2.4 Tapered Beam 5
    1.2.5 Trusses 6
    1.2.6 Other Primary Framing Systems 7
    1.2.7 Endwall and Sidewall Framing 7
1.3 Frame-and-Purlin Buildings: Lateral-Force-Resisting Systems . . . 9
1.4 Quonset Hut–Type Buildings 13
References 14

2 Foundation Design Basics 15
2.1 Soil Types and Properties 15
    2.1.1 Introduction 15
    2.1.2 Some Relevant Soil Properties 15
    2.1.3 Soil Classification 16
    2.1.4 Characteristics of Coarse-Grained Soils 17
    2.1.5 Characteristics of Fine-Grained Soils 17
    2.1.6 The Atterberg Limits 19
    2.1.7 Soil Mixtures 20
    2.1.8 Structural Fill 21
    2.1.9 Rock 21
2.2 Problem Soils 22
    2.2.1 Expansive Soils: The Main Issues 22
    2.2.2 Measuring Expansive Potential of Soil 22
    2.2.3 Organics 23
    2.2.4 Collapsing Soils and Karst 24
2.3 Soil Investigation 24
    2.3.1 Types of Investigation 24
    2.3.2 Preliminary Exploration 25
    2.3.3 Detailed Exploration: Soil Borings and
            Other Methods 26
    2.3.4 Laboratory Testing 28
2.4 Settlement and Heave Issues 29
    2.4.1 What Causes Settlement? 29
    2.4.2 Settlement in Sands and Gravels 29

		2.4.3	Settlement in Silts and Clays	30
		2.4.4	Differential Settlement	31
		2.4.5	Some Criteria for Tolerable Differential Settlement	32
	2.5	Determination of Allowable Bearing Value	33
		2.5.1	Why Not Simply Use the Code Tables?	33
		2.5.2	Special Provisions for Seismic Areas	34
		2.5.3	What Constitutes a Foundation Failure?	34
		2.5.4	Summary	35
	2.6	Shallow vs. Deep Foundations	35
	References	36
3	**Foundations for Metal Building Systems: The Main Issues**	37
	3.1	The Differences between Foundations for Conventional Buildings and Metal Building Systems	37
		3.1.1	Light Weight Means Large Net Uplift	37
		3.1.2	Large Lateral Reactions	40
		3.1.3	Factors of Safety and One-Third Stress Increase	41
		3.1.4	In Some Circumstances, Uncertainty of Reactions	42
	3.2	Estimating Column Reactions	43
		3.2.1	Methods of Estimating Reactions	43
		3.2.2	How Accurate Are the Estimates?	44
	3.3	Effects of Column Fixity on Foundations	45
		3.3.1	Is There a Cost Advantage?	45
		3.3.2	Feasibility of Fixed-Base Columns in MBS	45
		3.3.3	Communication Breakdown	46
	3.4	General Procedure for Foundation Design	46
		3.4.1	Assign Responsibilities	46
		3.4.2	Collect Design Information	47
		3.4.3	Research Relevant Code Provisions and Determine Reactions	47
		3.4.4	Determine Controlling Load Combinations	47
		3.4.5	Choose Shallow or Deep Foundations	49
		3.4.6	Establish Minimum Foundation Depth	49
		3.4.7	Design the Foundation	49
	3.5	Reliability, Versatility, and Cost	50
		3.5.1	Definitions	50
		3.5.2	Some Examples	50
	3.6	Column Pedestals (Piers)	52
		3.6.1	The Area Inviting Controversy	52
		3.6.2	Two Methods of Supporting Steel Columns in Shallow Foundations	52
		3.6.3	Establishing Sizes of Column Pedestals (Piers)	54
		3.6.4	Minimum Reinforcement of Piers	54
	References	57

4 Design of Isolated Column Footings ... 59
4.1 The Basics of Footing Design and Construction ... 59
- 4.1.1 Basic Design Requirements ... 59
- 4.1.2 Construction Requirements ... 60
- 4.1.3 Seismic Ties ... 60
- 4.1.4 Reinforced-Concrete Footings ... 60
- 4.1.5 Plain-Concrete and Other Footings ... 60
- 4.1.6 Nominal vs. Factored Loading ... 61

4.2 The Design Process ... 62
- 4.2.1 General Design Procedure ... 62
- 4.2.2 Using ASD Load Combinations ... 62
- 4.2.3 Using Load Combinations for Strength Design ... 63
- 4.2.4 What Is Included in the Dead Load? ... 63
- 4.2.5 Designing for Moment ... 64
- 4.2.6 Designing for Shear ... 65
- 4.2.7 Minimum Footing Reinforcement ... 68
- 4.2.8 Distribution of Reinforcement in Rectangular Footings ... 68
- 4.2.9 Designing for Uplift ... 69
- 4.2.10 Reinforcement at Top of Footings ... 70

References ... 77

5 Foundation Walls and Wall Footings ... 79
5.1 The Basics of Design and Construction ... 79
- 5.1.1 Foundation Options for Support of Exterior Walls ... 79
- 5.1.2 Design and Construction Requirements for Foundation Walls ... 80
- 5.1.3 Construction of Wall Footings ... 83
- 5.1.4 Design of Wall Footings ... 84

References ... 87

6 Tie Rods, Hairpins, and Slab Ties ... 89
6.1 Tie Rods ... 89
- 6.1.1 The Main Issues ... 89
- 6.1.2 Some Basic Tie-Rod Systems ... 90
- 6.1.3 A Reliable Tie-Rod Design ... 92
- 6.1.4 Development of Tie Rods by Standard Hooks ... 95
- 6.1.5 Design of Tie Rods Considering Elastic Elongation ... 96
- 6.1.6 Post-Tensioned Tie Rods ... 97
- 6.1.7 Tie-Rod Grid ... 99
- 6.1.8 Which Tie-Rod Design Is Best? ... 100

6.2 Hairpins and Slab Ties ... 103
- 6.2.1 Hairpins: The Essence of the System ... 103
- 6.2.2 Hairpins in Slabs on Grade ... 104

		6.2.3	Hairpins: The Design Process	105
		6.2.4	Development of Straight Bars in Slabs	107
		6.2.5	Slab Ties (Dowels)	109
		6.2.6	Using Foundation Seats	111
	References			111

7 Moment-Resisting Foundations — 113

- 7.1 The Basic Concept — 113
  - 7.1.1 A Close Relative: Cantilevered Retaining Wall — 113
  - 7.1.2 Advantages and Disadvantages — 115
- 7.2 Active, Passive, and At-Rest Soil Pressures — 115
  - 7.2.1 The Nature of Active, Passive, and At-Rest Pressures — 115
  - 7.2.2 How to Compute Active, Passive, and At-Rest Pressure — 117
  - 7.2.3 Typical Values of Active, Passive, and At-Rest Coefficients — 117
- 7.3 Lateral Sliding Resistance — 119
  - 7.3.1 The Nature of Lateral Sliding Resistance — 119
  - 7.3.2 Combining Lateral Sliding Resistance and Passive Pressure Resistance — 120
- 7.4 Factors of Safety against Overturning and Sliding — 121
  - 7.4.1 No Explicit Factors of Safety in IBC Load Combinations — 121
  - 7.4.2 Explicit Factors of Safety for Retaining Walls — 121
  - 7.4.3 How to Increase Lateral Sliding Resistance — 122
- 7.5 The Design Procedures — 122
  - 7.5.1 Design Input — 122
  - 7.5.2 Design Using Combined Stresses Acting on Soil — 123
  - 7.5.3 The Pressure Wedge Method — 126
  - 7.5.4 General Design Process — 127
  - 7.5.5 Moment-Resisting Foundations in Combination with Slab Dowels — 127
- References — 141

8 Slab with Haunch, Trench Footings, and Mats — 143

- 8.1 Slab with Haunch — 143
  - 8.1.1 General Issues — 143
  - 8.1.2 The Role of Girt Inset — 144
  - 8.1.3 Resisting the Column Reactions — 144
- 8.2 Trench Footings — 164
- 8.3 Mats — 165
  - 8.3.1 Common Uses — 165
  - 8.3.2 The Basics of Design — 167
  - 8.3.3 Typical Construction in Cold Climates — 168
  - 8.3.4 Using Anchor Bolts in Mats — 171
- References — 172

Contents

9 Deep Foundations 173
   9.1 Introduction 173
   9.2 Deep Piers 173
      9.2.1 The Basics of Design and Construction 173
      9.2.2 Resisting Uplift and Lateral Column Reactions with Deep Piers 174
   9.3 Piles 176
      9.3.1 The Basic Options 176
      9.3.2 The Minimum Number of Piles 177
      9.3.3 Using Structural Slab in Combination with Deep Foundations 178
      9.3.4 Resisting Uplift with Piles 181
      9.3.5 Resisting Lateral Column Reactions with Piles 181
   References 183

10 Anchors in Metal Building Systems 185
   10.1 General Issues 185
      10.1.1 Terminology and Purpose 185
      10.1.2 The Minimum Number of Anchor Bolts 186
   10.2 Anchor Bolts: Construction and Installation 186
      10.2.1 Typical Construction 186
      10.2.2 Field Installation 187
      10.2.3 Placement Tolerances vs. Oversized Holes in Column Base Plates 188
      10.2.4 Using Anchor Bolts for Column Leveling 190
      10.2.5 Should Anchor Bolts Be Used to Transfer Horizontal Column Reactions? 191
   10.3 Design of Anchor Bolts: General Provisions 193
      10.3.1 Provisions of the International Building Code 193
      10.3.2 ACI 318-08 Appendix D 196
   10.4 Design of Anchor Bolts for Tension per ACI 318-08 Appendix D 198
      10.4.1 Tensile Strength of Anchor Bolt vs. Tensile Strength of Concrete for a Single Anchor 198
      10.4.2 Tensile Strength of an Anchor Group 198
      10.4.3 Tensile Strength of Steel Anchors 200
      10.4.4 Pullout Strength of Anchor in Tension 201
      10.4.5 Concrete Side-Face Blowout Strength of Headed Anchors in Tension 202
      10.4.6 Concrete Breakout Strength of Anchors in Tension 202
      10.4.7 Using Anchor Reinforcement for Tension 207
   10.5 Design of Anchors for Shear per ACI 318-08 Appendix D 214
      10.5.1 Introduction 214
      10.5.2 Steel Strength of Anchors in Shear 215
      10.5.3 Concrete Breakout Strength in Shear: General 216
      10.5.4 Basic Concrete Breakout Strength in Shear V_b 220
      10.5.5 Concrete Breakout Strength in Shear for Anchors Close to Edge on Three or More Sides 221

		10.5.6	Concrete Breakout Strength in Shear: Modification Factors	222
		10.5.7	Using Anchor Reinforcement for Concrete Breakout Strength in Shear	224
		10.5.8	Using a Combination of Edge Reinforcement and Anchor Reinforcement for Concrete Breakout Strength in Shear	227
		10.5.9	Concrete Pryout Strength in Shear	227
		10.5.10	Combined Tension and Shear	228
		10.5.11	Minimum Edge Distances and Spacing of Anchors	228
		10.5.12	Concluding Remarks	233
	References			234
11	**Concrete Embedments in Metal Building Systems**			**235**
	11.1	The Role of Concrete Embedments		235
		11.1.1	Prior Practices vs. Today's Code Requirements	235
		11.1.2	Two Options for Resisting High Horizontal Column Reactions	235
		11.1.3	Transfer of Uplift Forces to Foundations: No Alternative to Anchor Bolts?	236
	11.2	Using Anchor Bolts to Transfer Horizontal Column Reactions to Foundations		237
		11.2.1	Some Problems with Shear Resistance of Anchor Bolts	237
		11.2.2	Possible Solutions to Enable Resistance of Anchor Bolts to Horizontal Forces	238
		11.2.3	Design of Anchor Bolts for Bending	240
	11.3	Concrete Embedments for the Transfer of Horizontal Column Reactions to Foundations: An Overview		242
	11.4	Shear Lugs and the Newman Lug		243
		11.4.1	Construction of Shear Lugs	243
		11.4.2	Minimum Anchor Bolt Spacing and Column Sizes Used with Shear Lugs	245
		11.4.3	Design of Shear Lugs: General Procedure	247
		11.4.4	Determination of Bearing Strength	249
		11.4.5	Determination of Concrete Shear Strength	249
		11.4.6	The Newman Lug	250
	11.5	Recessed Column Base		254
		11.5.1	Construction	254
		11.5.2	Design	255
	11.6	Other Embedments		257
		11.6.1	Cap Plate	257
		11.6.2	Embedded Plate with Welded-On Studs	260
	References			261
A	**Frame Reaction Tables**			**263**
	Index			**293**

Preface

The design of foundations for metal building systems, also known as pre-engineered metal buildings, is a subject of much controversy and misunderstanding. The primary reason: Reliable foundations for these structures might be more difficult to design and more expensive to construct than the foundations for conventional buildings.

And yet the building owners and contractors, who tend to perceive metal building systems as inexpensive structures, are rarely aware of the foundation challenges. Their main efforts are usually directed to getting the best price on the steel superstructure; the foundations are an afterthought. As a result, these owners and contractors are sometimes alarmed by the cost of their metal building foundations and might even question the competency of the foundation designers. After a few such experiences, the engineers who design these foundations might start to question their own design methods and settle for the cheapest solutions.

What makes the foundations for metal building systems difficult to design? The design complexities stem from two main issues. First, the most popular types of the primary frames used in metal building systems tend to exert significant horizontal column reactions on the foundations. While some conventional building foundations also resist lateral forces generated by wind and seismic loading, those forces are generally smaller than the typical frame reactions in metal buildings. Second, metal building systems are extremely lightweight, which means that their supports are routinely subjected to a net uplift loading caused by wind. It follows that metal building foundations are usually designed to resist a combination of lateral and vertical (either downward or uplift) forces. When the primary frame columns are fixed at the bottom, rather than pinned, still more complications arise for the foundation design.

These unique challenges and complexities might not be familiar to the engineers accustomed to designing foundations for conventional buildings. Unfortunately, little authoritative design guidance exists on the subject, which undoubtedly adds to the controversy. At present, only one chapter in the author's book *Metal Building Systems: Design and Specifications,* 2d ed. (McGraw-Hill, 2004), provides some guidance about this area of design. In the past, some metal building manufacturers used to publish their own brochures dealing with the subject, but no longer do so, perhaps because the manufacturer's scope of work normally excludes the building foundations.

The lack of clear design procedures naturally results in uneven design solutions. Some foundation designs for metal buildings have been overly complicated. But more commonly in the author's experience, they have been barely adequate for the imposed loads (or not adequate at all).

This book is intended to fill the knowledge void by exploring the complexities of designing foundations for metal building systems and providing clear design procedures for various foundation types. The book covers the most common foundation systems used for metal buildings and illustrates their design through step-by-step examples. It describes a variety of possible solutions, ranging from the most reliable (and often the most expensive) to the less reliable (and typically less expensive) approaches.

The book is intended to serve as a desk reference source for structural engineers, both inside and outside the manufacturers' offices, and for other design and construction professionals, such as architects, facility managers, building officials, and technically inclined contractors. It will help these professionals in the selection of the most appropriate and cost-effective system for their projects and in developing realistic cost expectations.

Alexander Newman, P.E., F.ASCE

Foundation and Anchor Design Guide for Metal Building Systems

CHAPTER 1
Introduction to Metal Building Systems

1.1 Two Main Classes of Metal Building Systems

The purpose of this chapter is to introduce the reader to metal building systems (MBS) and to explain how they work. These structures, also known as pre-engineered metal buildings and sometimes abbreviated as PEMBs, are both designed and manufactured by their suppliers. In this book the terms metal building systems, metal buildings, pre-engineered buildings, and pre-engineered metal buildings are used interchangeably.

Metal buildings are extremely popular in the United States. Many, if not most, low-rise nonresidential buildings there contain pre-engineered structures. Reasons for the popularity of these systems include their cost-effectiveness, speed of construction, single-source responsibility for the superstructure, ease of maintenance, and the ability to span long distances without intermediate supports.

Metal building systems can be found in all types of buildings. Approximately one-third of them are used in commercial applications, such as office buildings, retail stores, and garages. Another one-third include manufacturing uses, such as factories, production plants, and material recycling facilities (see Fig. 1.1). The rest are used in community buildings (schools, churches, municipal), agricultural, and storage occupancies (Newman, 2004).

The two main classes of metal building systems are frame-and-purlin and Quonset hut–type buildings. Both are described in this chapter, but our main focus is on the former. From the foundation designers' point of view, there is a fundamental difference between the two classes of metal buildings.

1.2 Frame-and-Purlin Buildings: Primary and Secondary Framing

As the name suggests, this class of metal building systems includes separate primary frames and secondary members. The primary frames (also called *main frames*) can be of various types. The foundation design is greatly influenced by the type of primary frame selected for the building. For example, some frame types tend to exert significant lateral reactions on the supports and others do not.

Figure 1.1 Metal building system used in an industrial facility.

The primary frames carry secondary roof members called *purlins* and secondary wall members called *girts*. At the intersection of the roof and wall panels typically exists the third type of secondary member, known by the names of *eave girt, eave purlin,* or *eave strut*. The various names reflect the versatile nature of this important element that can wear three hats: a girt, a purlin, and an axially loaded structural member. Axial loading comes into play when the strut resists lateral forces as part of the cross-bracing assembly in the wall.

The secondary members support metal panels or other roof and wall cladding. In pre-engineered buildings, *endwalls* are at the ends of the building parallel to the primary frames, while *sidewalls* are perpendicular to the frames. The typical components of the frame-and-purlin system are shown in Fig. 1.2.

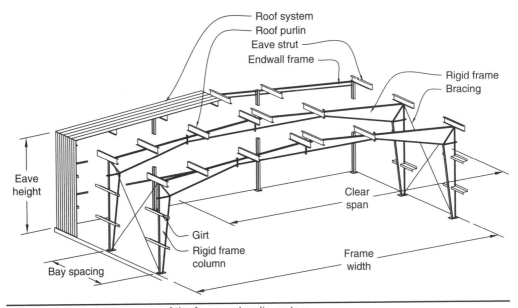

Figure 1.2 Typical components of the frame-and-purlin system.

The Metal Building Manufacturers Association (MBMA) is the trade group that includes most major manufacturers of pre-engineered metal buildings. *Metal Building System Manual* (MBMA, 2006) provides a wealth of information related to the design, fabrication, and procurement of frame-and-purlin pre-engineered metal buildings.

1.2.1 Primary Frames: Usage and Terminology

Primary frames are the backbone of the frame-and-purlin metal building system. The available primary frame systems vary significantly in their design assumptions, structural behavior, column configuration, and, perhaps most important for the reader, in the magnitude of the horizontal column reactions exerted on the foundations. Some of the considerations involved in the selection of the primary frames include:

- Width, length, and height of the building
- Acceptability of tapered columns
- Slope of the roof
- Clear-span requirements
- Wall and roof materials

Each frame consists of a rafter and at least two columns. The joint between an exterior column and the rafter is called the *knee*. A *clear span* is the distance between the inside faces of the frame columns. Each of the primary frame types has an optimum range of clear spans. The clear span can be measured between the frame knees, as shown in Fig. 1.2, but it can also be measured between the inside faces of the columns at the base.

A similar term, *frame width*, refers to the distance between the outside faces of the wall girts, a plane that is called the *sidewall structural line*. Looking at Fig. 1.2, please also note that for a single-story building the *clear height* is the distance between the floor and the lowest point of the frame at the roof level. Also, the *eave height* is the distance between the bottom of the column base plate and the top of the eave strut.

The rafter is usually spliced, with the dimensions of the individual segments fine-tuned for the applied bending moments and shears at those locations. The variability of the frame cross-section tends to complicate the analysis but saves metal by improving the framing efficiency. The splices are typically made at the peak, at the knees, and at the intermediate locations determined by the manufacturer's preferences.

Certain types of frames, such as a single-span rigid frame, can resist lateral (wind and seismic) forces acting in the direction parallel to the frames. Others require separate vertical and horizontal lateral-force-resisting elements in addition to the frames, as discussed in Sec. 1.3.

1.2.2 Single-Span Rigid Frames

The *single-span rigid frame* system is perhaps the most familiar and versatile type of primary framing used in pre-engineered buildings. These frames can be found in a wide variety of applications, from warehouses to churches. The frame profile is tapered to approximate the shape of the frame's moment diagram. A single-span rigid frame can be of either two- or three-hinge design. An example of two-hinge rigid frame can be found in Fig. 1.2; a three-hinge frame design is shown in Fig. 1.3a.

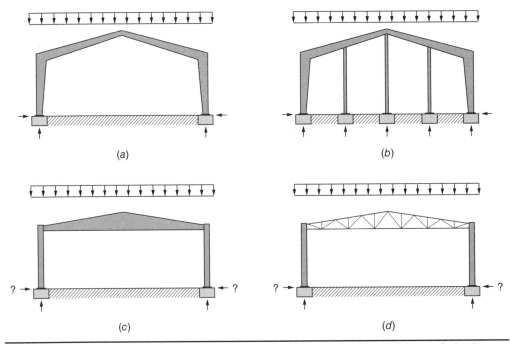

Figure 1.3 Common types of primary frames used in metal building systems and the gravity-load reactions they generate: (a) Single-span rigid frame; (b) multiple-span rigid frame; (c) tapered beam; (d) truss.

The knee is typically the deepest point of the tapered frame profile, making this system appropriate for the buildings where the clear height at the exterior walls is not critical. In this type of construction, the tallest equipment can be placed near the middle of the span or between the frames.

The single-span rigid frame provides a significant column-free area and the maximum flexibility for the user. The most economical frame width for this system lies between 60 and 120 ft, although frames over 200 ft wide have been produced. The most common range of the eave heights is from 10 to 24 ft.

The degree of rafter slope depends on the type of the roof used in the building. The low-slope frames, with a pitch from ¼:12 to 1:12, are appropriate for *waterproof* (hydrostatic) metal and nonmetal roofs. The medium-slope frame profile has a pitch of about 2:12, which works well with "waterproof" metal roofs in the regions where heavy snow accumulation is common. And finally, the high-profile frames, with a pitch of 3:12 or 4:12, are used with *steep-slope* (hydrokinetic) roofing systems. The different types of metal roofing are discussed in Newman (2004) and in *Roofing Systems Design Manual* (MBMA, 2000).

A single-span rigid frame typically exerts significant horizontal reactions on the foundations under gravity loading. The smaller the eave height, the larger these horizontal reactions tend to become.

1.2.3 Multiple-Span Rigid Frames

When the frame width of the building exceeds the limitations of the single-span rigid frame system, intermediate columns can be added. The resulting system is called the *multiple-span rigid frame*, also known as continuous-beam, post-and-beam, and modular frame (Fig. 1.3b). Multiple-span rigid frames can be more economical than their single-span siblings, even within the theoretical range of the optimum efficiency of the latter. However, the added columns obviously diminish the building's flexibility of use, a factor that might outweigh any potential cost savings. Accordingly, multiple-span rigid frames are best used in buildings with relatively static layout, such as factories with mature processes, large warehouses, and resource-recovery facilities.

Even with the multiple-span rigid frame system, the overall building width is constrained by the limitations imposed by thermal movement considerations. A very long frame without external restraints to movement will tend to expand and contract so much with the seasonal changes in temperature that the building's dimensional stability might become compromised. Alternatively, if the combination of the secondary framing, metal cladding, and bracing restrains the temperature-induced movements of the frame, thermal stresses might build up significantly. As a practical matter, expansion joints are usually needed when the building width exceeds about 300 ft.

Exterior columns in this system are usually tapered, and interior columns are straight, made of either pipe or W shapes. Interior columns are often assumed to have pinned connections to the rafters, and their horizontal reactions on the foundations are typically negligible. By contrast, the horizontal reactions at the bases of the exterior columns under gravity loading could be significant.

Unlike single-span rigid frames, multiple-span rigid frames can be adversely affected by the differential settlements of their supports. As we discuss in Chap. 2, some amount of foundation settlement is often unavoidable. When the interior frame columns settle at a rate different from that of the exterior columns, the continuous frame might become overstressed. Vulnerability to differential settlement is one of the few disadvantages of this otherwise very economical structural system.

1.2.4 Tapered Beam

The tapered beam, sometimes called a *wedge beam*, is similar to a conventional built-up plate girder, with two important differences. One, as the name suggests, is the tapered profile of the rafter. The second difference has to do with the manner in which the tapered beam is connected to the column, as discussed next.

In the most common version, the top surface of the beam is tapered, and the bottom surface is horizontal. This design solution allows the top of the roof to slope for drainage. A less common design has both flanges tapered, analogous to a scissors truss. A splice is typically made at the midspan of the beam. The columns in this system are usually of the straight profile. The tapered-beam system works best with a low-slope roof profile, as the rafter becomes too deep in steep-slope applications.

The tapered-beam system can be economical for relatively small frame widths—between 30 and 60 ft. Beyond 60 ft, single-span rigid frames typically become more cost-effective. The eave height in this system usually does not exceed 20 ft. Tapered beams are well suited for small office buildings and other commercial applications, particularly where ceilings are desired.

The tapered beam is typically connected to the column by bolted end plates. Some manufacturers consider this a semirigid (partially restrained) connection—rigid enough to resist lateral loads but flexible enough to allow for a simple-span beam behavior under gravity loading. The idea of the "wind" connections of this kind has been popular in the past, but today such simplistic assumptions are out of favor. Contemporary *AISC Specification* (2005) recognizes this type of attachment as "partially restrained (PR) moment connection." These are treated as moment connections, but with an understanding that the rotation between the connected members is not negligible as in fully restrained moment connections. According to the Specification, the design of a PR connection must consider its force-deformation response characteristics, properly documented by prior research, analysis, or testing.

From a practical standpoint, the end-plate connection behavior depends on the design details. For example, the *AISC Manual* (2005) includes the design provisions for extended end-plate fully restrained moment connections and provides further references to the AISC Design Guides. But the manual also recognizes, and provides the design tables for, *shear* (simple-span) end-plate connections, where the moment resistance is neglected. To qualify for this assumption, the end-plate length must be less than the depth of the beam to which it is attached.

It follows that a rafter and two columns joined by the full-depth or extended end-plate connections should probably be treated as a PR rigid frame, similar to the single-span rigid frame examined previously. Under this assumption the columns are designed for the fixity moments at the knees.

What is the importance of this discussion for the foundation designer? The actual behavior of the tapered-beam system affects the magnitude of the column reactions. The moment-frame behavior produces horizontal reactions at the column bases from gravity loading, while a simple-span beam behavior does not.

As the question marks shown in Fig. 1.3c imply, depending on which details are used, the horizontal reactions may or may not be present in the tapered-beam system. Despite the common manufacturer's assumptions to the contrary, only a tapered beam with true "shear" end-plate connections will *not* generate horizontal reactions. As just explained, that would be a tapered beam with the partial-depth end plates. A tapered beam with the full-depth end-plate connections probably *will* generate horizontal reactions. The foundation engineers should be aware of these design nuances.

1.2.5 Trusses

All the primary framing systems previously described—single- and multiple-span rigid frames and tapered beams—have solid webs. Similar primary-frame systems can be produced with trusslike webs. The open webs allow for a passage of pipes and conduits that otherwise would have to run underneath the frames. As a result, the clear height and volume of the building could both decrease, which in turn could reduce its heating and cooling costs. Among the disadvantages of the open-web solution is the increased labor required to assemble the framing. To use their advantages, trusses are best specified in the occupancies that require a lot of piping and conduits, such as in manufacturing and distribution facilities.

In conventional (non-pre-engineered) buildings, open-web steel joists are often used in combination with joist girders. In this approach the whole floor and roof structure can be framed with trusslike elements. Some pre-engineered metal building manufacturers offer proprietary primary and secondary truss framing as a complete system as well.

The foundation reactions exerted by the truss-type frames are similar to those of the corresponding frames with solid webs, and the previous discussion applies here as well. For the trusses that function akin to the tapered-beam frames, the rafter-to-column connections should be evaluated as described previously for the tapered-beam system. Depending on the connection design, horizontal reactions on the foundations under gravity loading may or may not exist, as shown by the question marks in Fig. 1.3*d*. When the truss system is not designed for the moment-frame action (as in simply supported joist girders), gravity loading generates only the vertical reactions. To ensure stability under lateral loads, buildings with truss-type framing typically require separate vertical lateral-force-resisting elements, such as braced frames or shear walls.

1.2.6 Other Primary Framing Systems

The previous discussion describes the most common primary framing systems, but there are a few others:

- *Single-slope rigid frames.* These primary frames are sometimes selected for shopping malls and similar applications that require roof drainage to be diverted to the back of the building. In terms of structural function and foundation reactions, they behave similarly to single- and multiple-span double-slope gable frames discussed previously.

- *Lean-to frames.* These frames typically consist of one single-slope rafter and one column, with the other column being that of another building frame that supports the lean-to structure. Because lean-to frames require another primary frame system for stability, their foundations should be designed in concert with those of the "parent" building. Lean-to framing is occasionally used in the additions to existing pre-engineered buildings—often an unwise course of action, for several reasons. One of them is the fact that an extra load is being placed on the existing column and its foundation, both of which might become overstressed as a result. It is certainly possible to strengthen the foundations for the increased loading, but a much more economical and straightforward solution is generally to frame the addition structure as an independent system with its own foundations.

- *Proprietary systems.* There are a few truly unique primary framing systems, where the framing consists of proprietary members (see Newman, 2004). The foundations for these systems are best designed only after the system's behavior and the loads applied to the supports are thoroughly understood by the foundation designer.

1.2.7 Endwall and Sidewall Framing

The typical endwall and sidewall framing consists of vertical and horizontal secondary members. The horizontal members are girts, which laterally brace the metal siding or other wall materials. The vertical members are girt supports. In single-story buildings they span from the foundation to the roof. Girts are usually made of cold-formed C or Z sections.

In sidewalls, girts generally span horizontally between the primary frame columns, and no vertical secondary members are required. There are exceptions, of course, such as when open-end steel joists are used in the roof. In that case the so-called *wind columns* can reduce the horizontal girt span. Wind columns carry little vertical loading: Their chief role is to reduce the span of the girts they support.

In endwalls, vertical secondary members are generally needed, unless the building is so narrow (less than 25 to 30 ft), that the single-span girts can cover its whole width. The vertical secondary members used for girt support are made of either cold-formed

FIGURE 1.4 Expandable endwall framing.

or hot-rolled channels or I sections. In addition to providing lateral support for the wall girts, they might carry some vertical loads as well. The typical spacing of the endwall posts is about 20 ft, controlled by the flexural capacities of the girts. Endwalls can include wall bracing.

There are two types of the endwall framing: expandable and nonexpandable. *Expandable endwall framing* includes a regular interior frame supported by the foundations designed for the interior-frame loading. Horizontal and vertical secondary members are placed within or outside of the frame to support the wall siding or other wall materials that enclose the building (Fig. 1.4). These secondary members primarily resist their tributary lateral forces applied perpendicular to the wall. When needed, the secondary members can be removed and the building expanded by placing additional primary frames parallel to the existing. The former endwall primary frame then becomes an interior frame in the expanded building.

Nonexpandable endwall framing contains no primary frames and consists solely of horizontal and vertical secondary members supporting single-span roof beams spanning between them. The roof beams support purlins at the top of the wall. As with expandable endwalls, these horizontal and vertical secondary members primarily resist lateral loads applied perpendicular to the plane of the endwall. The vertical members also support the tributary gravity loads carried to them by the roof beams. The foundation reactions from the endwall posts thus depend on whether the endwall is expandable or not.

For both endwalls and sidewalls, girts can be positioned relative to the columns or other vertical supports in three possible ways (insets): bypass, flush, and semiflush. As the name suggests, a bypass girt is placed wholly outside of the column's outer flange (Fig. 1.5). A flush girt is framed so that its outside flange is approximately even with the column's

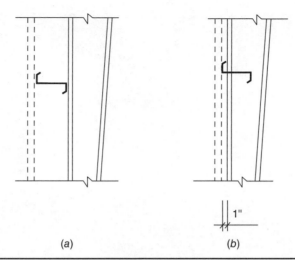

FIGURE 1.5 Girt inset: (*a*) Bypass girt; (*b*) flush girt.

(sometimes it protrudes outward by about 1 in., to allow for field tolerances). A semi-flush girt falls between the two extremes.

1.3 Frame-and-Purlin Buildings: Lateral-Force-Resisting Systems

As stated previously, many primary framing systems found in metal buildings can resist not only gravity, but also lateral forces applied in the plane of the frames. This advantage makes single-span and multiple-span rigid frames particularly appealing. Other primary framing systems rely on separate vertical lateral-force-resisting elements in the direction parallel to the frames. These elements are typically placed in the endwalls, and a roof diaphragm is needed to complement the vertical elements and provide a complete lateral-force-resisting system. In any case, the separate vertical elements are usually needed in the direction *perpendicular* to the frames.

The vertical lateral-force-resisting elements that are separate from the primary frames can be of various types. The most common is the wall cross bracing (Fig. 1.2). A typical cross-bracing assembly ("bent") consists of two columns, diagonal bracing, and the eave strut. This system can be classified as concentrically braced frame with tension-only bracing. The diagonal bracing can be made of steel cables, rods, or angles (listed in the order of increasing reliability).

Rods or cables are often attached directly to the column webs via the so-called hillside washers (Fig. 1.6). Some of the pitfalls of this arrangement and the suggested improvements are discussed in Newman (2004). In a more reliable detail, the rod is attached to a tab welded to the column base plate via a bolted connection between the tab and a small plate welded to the end of the rod (see Fig. 1.7). The angle brace can be attached in a similar fashion.

Another attachment alternative is to connect the wall bracing directly to the concrete floor structure by means of foundation clips. A typical clip looks like an inverted T, with four or more anchor rods securing it into concrete (Fig. 1.8). For this solution to work, an ample area of the foundation concrete should be available for the clip to be

FIGURE 1.6 Connection of the wall rod bracing to the column web via a hillside washer.

FIGURE 1.7 Connection of the wall rod bracing to the column base plate via a bolted tab.

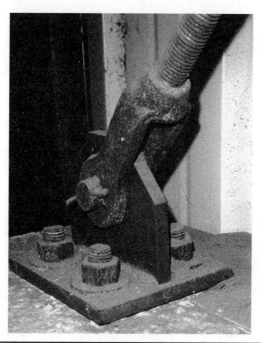

Figure 1.8 Foundation clip is directly attached to concrete.

embedded in. This means that the foundation pier, if one is present, should be enlarged to allow for the anchorage, as discussed in the chapters that follow.

Some metal buildings, such as car dealerships, storerooms, and offices, have large windows that fill the space between the sidewall columns along the whole length of the building. Since the windows leave no place for the wall bracing, other solutions must be employed. One of them uses *wind posts* (not to be confused with the wind column described earlier). The wind post is a fixed-base column cantilevered from the foundation. The base fixity of the wind post places significant demands on the foundation, which must be designed for the shear force and the fixity moment at the base. As discussed in Chap. 7, moment-resisting foundations are quite costly, and this solution should be used sparingly, with the full understanding of the cost consequences.

Portal frames—rigid frames placed in between and perpendicular to the primary framing—represent another, more common alternative solution (Fig. 1.9). The attachments between the portal frame and the primary frames require some thought, so as not to introduce torsion into the portal frame and to provide a proper lateral bracing for it. Some relevant details are included in Newman (2004).

The connections at the base of the portal frames are quite important for the foundation designers. Portal frames can extend all the way down to the foundation (Fig. 1.9*a*), or be stopped short and be attached to the primary frame columns (Fig. 1.9*b*). As with foundation clips, additional concrete area must be provided wherever full-height portal frames are specified. If the partial-height portal frames are used, the supporting primary frame columns should be designed for a weak-axis bending imposed by the portal frame reactions.

12 Chapter One

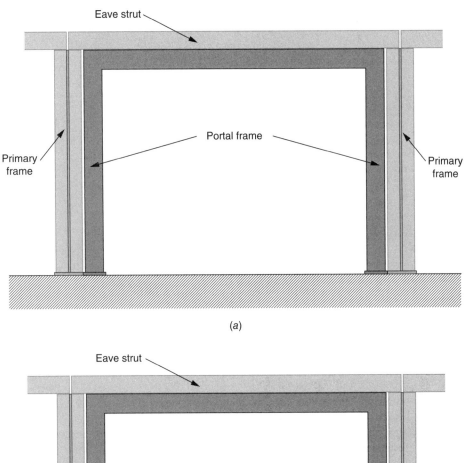

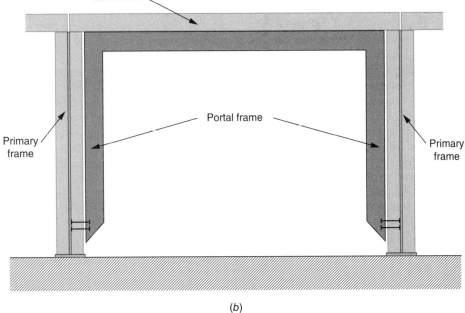

FIGURE 1.9 Portal frames: (a) Full-height, bearing on foundation; (b) stopped short of the foundation, attached to the primary column.

1.4 Quonset Hut–Type Buildings

The original Quonset hut was a popular World War II structure that provided inexpensive shelter to U.S. soldiers all around the world. Hundreds of thousands of these unassuming prefabricated buildings were produced during the war. They were made of corrugated steel sheets formed into arches. The arches could be easily assembled and disassembled using only ordinary hand tools, making them practical and portable shelters.

Today, Quonset hut–type buildings are still produced—and still need foundations. The main foundation-related difference between these structures and the frame-and-purlin systems previously described is the nature of their reactions at the base.

In the original Quonset hut and in many of today's versions, the corrugated steel sheets serve as roofing, siding, and primary and secondary framing all in one. They *are* the structure. Since there are no columns, the corrugated arches exert distributed reactions on the supporting foundations. The bearing connections at the base in today's versions vary. In the most basic approach, the ends of the corrugated arches are simply inserted into the slots formed within the foundations. Alternatively, the ends of the sheets are attached to continuous steel angles anchored to the concrete. In a more sophisticated detail, the corrugated sheets are connected to a special base section, a combination of flashing and bearing strips matching the profile of the sheets.

Another version of the contemporary Quonset hut–type buildings relies on discrete steel arches that support corrugated or ribbed sheets running horizontally (Fig. 1.10).

Figure 1.10 A version of Quonset hut–type building that uses discrete steel arches and horizontally spanning steel sheets.

The structure consists of the arches bearing on the sidewall concrete walls and the vertical posts at the end walls, with the corrugated sheets acting as cladding. The foundation loading for this version is similar to that of the frame-and-purlin system, but the loads are generally smaller owing to the modest building sizes.

While the foundation design for the various versions of the Quonset hut–type buildings is not included in this book, the basic design approach would be similar. The typical foundation system for these structures is either the slab with haunch (Chap. 8) or continuous wall footing (Chap. 5). A separate rectangular grade beam, while commonly encountered, may not provide the required strength and stability.

A word of caution: Some manufacturers of Quonset hut–type buildings rely on fixity moments at the base to reduce the steel weight. As explained in the chapters that follow, a fixity moment at the base greatly increases the demand on the foundation and consequently, its size and cost. The foundation designers for Quonset hut–type buildings should undertake a careful investigation of the reactions these structures impose.

References

Metal Building System Manual, Metal Building Manufacturers Association (MBMA), Cleveland, OH, 2006.

Newman, Alexander, *Metal Building Systems: Design and Specifications*, 2d ed., McGraw-Hill, New York, 2004.

Roofing Systems Design Manual, Metal Building Manufacturers Association (MBMA), Cleveland, OH, 2000.

Specification for Structural Steel Buildings, American Institute of Steel Construction (AISC), Chicago, IL, 2005.

Steel Construction Manual, 13th ed., American Institute of Steel Construction (AISC), Chicago, IL, 2005.

CHAPTER 2
Foundation Design Basics

2.1 Soil Types and Properties

2.1.1 Introduction

A building foundation is only as good as the soil that supports it. The origins of many foundation failures lie in a basic lack of understanding of the relevant soil properties by the designers and builders alike. Anyone wishing to design a foundation should first become familiar with the soils at the site and should possess at least a rudimentary comprehension of the geotechnical issues involved. Those who attempt to design foundations for a metal building system without such information are jeopardizing both the building and their professional reputations.

The field of soils engineering is quite complicated, and some of it is art rather than science (Merritt, 1983). Throughout this chapter we strongly recommend that a qualified geotechnical engineer be engaged to help determine the soil properties including the allowable bearing values. Our goal is to help the foundation designer work with the soils professional and understand some of the most relevant topics presented in the geotechnical reports. We try to avoid getting overly technical or attempting to squeeze the content of many geotechnical books into a few pages. Instead, we explain how various soils influence the strength and settlement of foundations and why these soils behave differently.

2.1.2 Some Relevant Soil Properties

What is the soil made of? Its most basic components are the ground-down particles of rock, water, and air. The presence of voids, filled with air or water, explains why the soil weighs much less than a solid rock (typically 90–130 lb/ft^3 vs. about 165 lb/ft^3). When the soil is loaded by the foundation, the voids are compressed, and settlement takes place. The size and shape of the soil particles influence the soil properties. Some of the properties that are of utmost importance to the foundation designers are listed as follows (with common, not necessarily academically rigorous, definitions):

- Allowable bearing capacity: The amount of unit pressure the soil can support.
- Compressibility: How much the soil can be compressed under load.
- Density: The weight of soil per unit of volume (can be measured as wet or dry density, i.e., with or without the weight of water in the soil).

- Expansiveness: How much the soil expands or shrinks with the changes of the moisture content.
- Moisture content: The weight of water in the soil sample divided by the weight of dry soil in it.
- Porosity: The volume of air and water in the sample divided by the total volume of the sample. A high porosity usually suggests a low bearing capacity.
- Shear strength: The soil shearing resistance at failure, established by laboratory testing.
- Void ratio: The volume of the voids in the soil sample divided by the volume of the solids.

Some of these properties are further explained, and a few other properties are introduced, in the sections that follow.

2.1.3 Soil Classification

There are several ways to classify soils, perhaps the most common being the Unified Soil Classification System. Under this system, soils are divided into three groups, based on the particle size and other characteristics. The sizes of soil particles are determined by passing soil samples through a series of sieves with standardized opening sizes. The smallest sieve size, No. 200, has 200 openings per inch, each opening being 0.074-mm wide. A particle of this size cannot be seen by the naked eye (NAVFAC DM-7.1). The two main soil groups in the Unified Soil Classification System are:

A. Coarse-grained soils (defined as those where more than 50 percent of the sample by weight is retained on the No. 200 sieve).
B. Fine-grained soils (where more than 50 percent of the sample passes through the No. 200 sieve). A subset of this group includes highly organic soils, such as peat, discussed separately later.

Table 2.1 further classifies soil materials and their fractions (fine, medium, coarse) as a function of the sieve size.

Material	Fraction	Sieve Size
Boulders		12″ +
Cobbles		3″–12″
Gravel	Coarse	¾″– 3″
	Fine	No. 4 to ¾″
Sand	Coarse	No. 10 to No. 4
	Medium	No. 40 to No. 10
	Fine	No. 200 to No. 40
Fines (silt and clay)		Passing No. 200

Source: NAVFAC DM-7.1.

TABLE 2.1 Definitions of Soil Components and Fractions by Grain Size

2.1.4 Characteristics of Coarse-Grained Soils

Coarse-grained soils include some the best soils on which to build—sands and gravels—as well as boulders and cobbles. The particles of these soils have a spherical shape, and there is no adhesion ("stickiness") between them. Accordingly, these soils are often called *cohesionless*. The border separating sands and gravels is No. 4 sieve size. As Table 2.1 shows, the border between coarse and medium sand is No. 10 sieve size, and between fine and medium sand, No. 40.

Cohesionless soils derive their shear strength from friction between the particles. Their bearing strength depends on the angle of internal friction φ (phi), which for clean sand is approximately 30°. A pile of sand will typically stabilize at an angle of a similar magnitude.

The distribution of the various particle sizes within the sample is determined by a sieve analysis. The analysis shows how much soil is retained on each of the various sieves as a percentage of the total sample weight. A plot of the size distribution is then constructed and the soil gradation established. *Well-graded* sands and gravels will show a wide distribution of grain sizes, while *poorly graded* soils might have one or two sizes that predominate. *Gap-graded* coarse-grained soils typically have lots of fine and coarse particles, with little in between.

Well-graded sands and gravels are preferred for foundation support and as backfill. Why? Poorly and gap-graded soils have many voids between the larger particles that remain unfilled because of absence or scarcity of fine and intermediate factions. Since the bearing strength of cohesionless soils depends on friction of closely packed particles, any gaps and voids tend to reduce it.

Regardless of the gradation, sands and gravels are still considered better soils for foundation support than cohesive soils. Sands and gravels are highly permeable and thus drain well, are less susceptible to frost, and can be readily compacted.

The Unified Soil Classification System allows for a ready identification of soil gradation by assigning the appropriate letters corresponding to various gradation conditions (*W* for well-graded, *P* for poorly graded, and *G* for gap-graded soil). These letters follow the symbols for the type of soil (e.g., *G* for gravel, *S* for sand). For example, the soils within the group symbol *GW* represent well-graded gravels, while those with the symbol *SP* define poorly graded sand. A summary of the soil properties classified according to the Unified Soil Classification System is shown in Table 2.2.

Coarse-grained soils can have different degrees of compactness, such as loose, medium, or dense. Naturally, dense sands and gravels tend to have the highest load-bearing capacities, and foundation failures in those soils have been infrequent. Compactness of sands can be estimated from the number of blow counts (see Sec. 2.3). Many geotechnical texts, such as NAVFAC DM-7.1, contain charts that help establish the estimated degree of compactness in sands and gravels.

2.1.5 Characteristics of Fine-Grained Soils

Fine-grained soils include silts and clays. Silts are composed of the larger-size particles than clays and tend to be more stable. According to the U.S. Bureau of Soils, a typical particle of clay is 0.005 mm in size, a particle of silt 0.05 mm, and a grain of medium sand 0.5 mm. Put differently, the clay particles are the finest, those of silt are 10 times larger, and those of sand are 10 times larger than silt. The fine-grained soil category also includes loess—wind-driven silt.

Soil Group	Unified Soil Classification System Symbol	Soil Description	Drainage Characteristics[a]	Frost Heave Potential	Volume Change Potential Expansion[b]
Group I	GW	Well-graded gravels, gravel sand mixtures, little or no fines.	Good	Low	Low
	GP	Poorly graded gravels or gravel sand mixtures, little or no fines.	Good	Low	Low
	SW	Well-graded sands, gravelly sands, little or no fines.	Good	Low	Low
	SP	Poorly graded sands or gravelly sands, little or no fines.	Good	Low	Low
	GM	Silty gravels, gravel-sand-silt mixtures.	Good	Medium	Low
	SM	Silty sand, sand-silt mixtures.	Good	Medium	Low
Group II	GC	Clayey gravels, gravel-sand-clay mixtures.	Medium	Medium	Low
	SC	Clayey sands, sand-clay mixture.	Medium	Medium	Low
	ML	Inorganic silts and very fine sands, rock flour, silty or clayey fine sands or clayey silts with slight plasticity.	Medium	High	Low
	CL	Inorganic clays of low to medium plasticity, gravelly clays, sandy clays, silty clays, lean clays.	Medium	Medium	Medium to low
Group M	CH	Inorganic clays of high plasticity, fat clays.	Poor	Medium	High
		Inorganic silts, micaceous or diatomaceous fine sandy or silty soils, elastic silts.	Poor	High	High
Group IV	OL	Organic silts and organic silty clays of low plasticity.	Poor	Medium	Medium
	OH	Organic clays of medium to high plasticity, organic silts.	Unsatisfactory	Medium	High
	Pt	Peat and other highly organic soils.	Unsatisfactory	Medium	High

Source: Massachusetts State Building Code, 7th ed., 2008.
For SI: 1 in. = 25.4 mm.
[a] The percolation rate for good drainage is over 4 in. per hour, medium drainage is 2 in. to 4 in. per hour, and poor is less than 2 in. per hour.
[b] Soils with a low potential expansion typically have a plasticity index (PI) of 0 to 15, soils with a medium potential expansion have a PI of 10 to 35 and soils with a high potential expansion have a PI greater than 20.

TABLE 2.2 Summary of Soil Properties Classified according to the Unified Soil Classification System

In addition to the particle size, particle *shape* of fine-grained soils is also different from that of sand. Unlike the globules of sand, the particles of silt and clay are flat-shaped. The strength of silts and clays is derived from cohesion between these flat particles, not from friction between them. Accordingly, silts and clays are called *cohesive soils*.

Silts and clays differ in the degree of their *plasticity*. Plasticity characteristics of cohesive soils are determined by the *Atterberg limits* tests. The Atterberg limits, which are also used to determine the degree of soil expansiveness, are described in the next section. Some clays are called "fat" or "lean," although these terms are considered less than technical, as discussed in Sec. 2.2.2; they shrink and swell significantly with changes in soil moisture.

Fine-grained soils also differ by the degree of *consistency*, which ranges from very soft (when squeezed, the soil can be extruded between fingers) to hard (it can be indented with difficulty by a thumbnail). The clay with medium consistency can be molded by a strong finger pressure.

The higher the soil consistency, the higher the strength, as measured by the blow count. A medium-consistency clay typically has an estimated range of unconfined compressive strength of ½ to 1 tons/ft^2 and a blow count of 4 to 8. A table in NAVFAC DM-7.1 (based on Terzaghi and Peck, 1967) provides a guide for estimating consistency of fine-grained soils.

Certain clays lose much of their strength after being disturbed by vibration. Vibration could be caused by common construction activities, such as pile driving or even heavy truck traffic. These *sensitive clays* typically recover most of their loss of strength after some time has passed.

In general, clay soils are nearly impermeable and some are expansive, which makes them poor candidates for foundation support. The higher their plasticity, the less desirable they are. Clays are also generally avoided as backfill, because they develop high lateral pressures that can damage foundation walls. Excavations in clay soils drain very slowly and are difficult to dewater.

2.1.6 The Atterberg Limits

The Atterberg limits are specific moisture contents at which a cohesive soil changes its state. Soils typically exist in a solid state, of course, but they can also be liquid if enough moisture is added. They can also be in a plastic state, which is defined as the state where soil can be rolled in a thread with the diameter equal to 1/8 in. The minimum moisture content at which this becomes possible is called the *plastic limit* (PL).

Adding more water to a sample in a plastic state will eventually liquefy it. The moisture content at which the soil starts to flow is called the *liquid limit* (LL). The soil containing more water than the liquid limit would be expected to be in a liquid state. A liquid limit of 50 is typical for cohesive soils. At that moisture content the soil is two-thirds solid particles and one-third water (Butler, undated).

Conversely, starting with a soil sample in the semisolid state (where the soil can be easily molded) and *removing* moisture at first causes the soil to shrink. Eventually, the volume stops decreasing, signifying that the soil has transitioned into a solid state. The moisture content at which this occurs is called *shrinkage limit* (SL).

In summary, as shown in Fig. 2.1, shrinkage limit separates soil in solid state from semisolid state; plastic limit is a border between semisolid and plastic states; and liquid limit lies between plastic and liquid states.

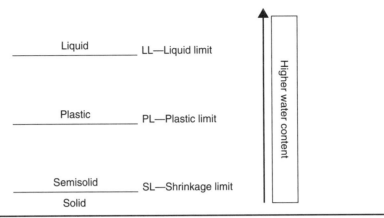

FIGURE 2.1 Atterberg limits (after *Brown*).

The Atterberg limits can be used to arrive at two important indices: plasticity index and liquidity index. Plasticity index (PI) is equal to the liquid limit (LL) minus the plastic limit (PL):

$$PI = LL - PL$$

Plasticity index defines the high and low boundaries of the plastic state of the soil. A high PI is typical of expansive clays.

Liquidity index (LI) is established from the following equation, using the abbreviations defined previously and the soil's natural water content W:

$$LI = (W - PL)/PI$$

2.1.7 Soil Mixtures

Many real-life soils contain mixtures of various materials. Quite often, coarse-grained soils contain fines. A mixture that is predominantly gravel but also includes silt, and a mixture that contains gravel, sand, and silt, are assigned group symbol *GM*. Silty sands and sand-silt mixtures are called *SM*. Similarly, clayey gravels and gravel-sand-clay compounds are designated as *GC*, and clayey sands and sand-clay mixtures are given the symbol *SC*.

Naturally, the soil component that predominates governs the designation of the mixed soil. Depending on the percentage of the secondary soil components, the following descriptive adjectives are used:

Trace	1 to 10%
Little	10 to 20%
Some	20 to 35%
And	35 to 50%

For example, a sand mixture containing 15 percent silt is called "sand, little silt."

Some common terms referring to stratified soils, as defined in NAVFAC DM-7.1, include *varved clay* (alternating seams or layers of sand, silt, and clay) and soil *pocket* (a small erratic deposit, usually less than 1 ft deep). A soil *stratum* refers to material thicker than 12 in.; a *layer* is ½ to 12 in. thick; and a *seam* is 1/16 to ½ in. thick.

2.1.8 Structural Fill

Structural fill can be used whenever in situ soil is unsuitable for foundation support. The fill might also be required when the building is located on a sloped site. The fill material can be simply a soil brought from another location, or it can be reconstructed on site by adding the appropriate amount of water. Whatever the origin, structural fill requires proper compaction to achieve its desired load-bearing capacity. Proper compaction packs the soil particles closer together and helps them develop the desired degree of friction. Countless foundations have undergone significant settlements because the fill under them was not placed or compacted properly. Structural fill typically consists of well-graded sand and gravel.

A typical project specification might call for structural fill to be placed in 6-, 8-, or 12-in.-thick maximum layers and compacted until the specified dry-density value is met. A common criterion specifies the dry density of each layer over the entire area being at least equal to 95 percent of the dry density achieved by ASTM D1557 Method D (Modified Proctor Density).

The goal of compaction and adding water is to produce the fill with the maximum practically attainable unit weight (density), which generally signifies the strongest possible material. For each fill composition there is an optimal water content that produces the maximum density. Too little water does not allow the soil particles to be properly "lubricated"; too much water results in a soupy fill material, which is also undesirable. In many cases the optimal water content is established by laboratory testing.

To what depth should the fill be carried? Obviously, it must stop at some point. The degree of in situ soil removal depends on the particulars of the site and is generally specified in the geotechnical reports.

2.1.9 Rock

Foundation designers tend to welcome the discovery of solid ledge (top of rock) at the site as great news. While rock has indeed the highest allowable bearing capacities, foundations bearing on ledge have their own challenges. Here are a few issues to consider.

First, which rock is it? There are various types of rock, with different characteristics, hardness and bearing capacities. The latter can vary by two orders of magnitude. According to NAVFAC DM-7.1, a rock with "very soft" hardness (it can be peeled with a knife) has a range of uniaxial compressive strength ranging from 10 to 250 tons/ft^2. By contrast, "extremely hard" rock (it requires many blows of geological hammer to break an intact specimen) has a strength exceeding 2000 tons/ft^2.

Second, the rock on site might be weathered or partly decomposed. A decision needs to be made whether some of the rock should be removed to get a better allowable bearing capacity, or whether a reduced capacity can be used instead. A competent geotechnical investigation should help answer this question.

Third, the rock surface is rarely perfectly level, and sometimes it slopes significantly. A common problem arises when some column footings are placed on rock and others on soil. A similar challenge is present when one part of a wall footing bears on rock and another on soil. In both cases differential settlement of the structure is probable, and the solutions are not easy. Quite often, the highest ridges of the rock are removed, so that at least *some* soil cushion exists between the bottom of the footing and the ledge. The bigger the cushion, the less likely that the foundation will settle unevenly. However, rock removal is costly and could damage the adjacent structures, and the pros and cons of this approach should be considered very carefully.

2.2 Problem Soils

2.2.1 Expansive Soils: The Main Issues

The flat shape of clay particles helps attract and hold free moisture. Clay soils are known to shrink and swell with moisture changes, but some clays move much more than others. *Expansive clays* present a particular challenge for the foundation designers.

When wet, these soils swell and exert upward pressures on slabs on ground and lateral pressures on basement walls, a phenomenon that has caused numerous foundation failures. When moisture disappears, the soils shrink and the foundations settle, losing support and causing more failures. Much of this movement—heaving and settlement—is caused by temporary or seasonal moisture changes. For example, short downpours in an otherwise arid climate can trigger it, as can leaking pipes, irrigation systems, vegetation near the building, and so on.

Expansive soils can be found in many areas of the United States, but they are most common in the Midwest and the South, in states such as Texas, Mississippi, Louisiana, Oklahoma, and Colorado. Many areas in California are also affected.

What differentiates these highly unstable soils from the other clays is the inclusion of some swelling materials that have even smaller particle sizes than clay. In decreasing order of expansiveness, these are montmorillonite, illite, attapulgite, chlorite, and kaolinite (Brown, 1997). The worst offender, montmorillonite, is reportedly able to expand up to 20 times the original volume! The more of these materials are in the clay, the more expansiveness can be expected.

Some highly expansive soils can swell more than 30 percent and exert pressures measured in tens of kips/ft^2. The expansive potential of soil is sometimes expressed as *potential vertical rise* (PVR). In many cases PVR is relatively modest—a few inches at most, but in some locales it is measured in feet.

The soils with high expansion potential are identified in Table 2.2. The topics of soil investigation, testing, and the design of foundations in expansive clays are covered in depth in UFC 3-220-07 (Formerly TM 5-818-7), *Foundations in Expansive Soils,* and other authoritative publications.

2.2.2 Measuring Expansive Potential of Soil

Soil expansiveness can be measured by several parameters. The soil *plasticity index* is one of the Atterberg limits described earlier. Another useful soil property is *expansion index* (EI). Expansion index is determined by the standardized procedure in ASTM D 4829, *Standard Test Method for Expansion Index of Soils.*

In this method, a sample of soil is placed into a standard metal ring, saturated to 50 percent and placed in a consolidometer. The degree of saturation is measured as the volume of water divided by the volume of voids, so at 50 percent saturation one-half of the voids is filled with water. A 1-psi (lb/in^2) pressure is applied, and then the sample is fully saturated. The sample's deformations are recorded for 24 h, or until the rate of swell becomes less than 0.005 mm/h. The expansion index is computed as the percentage of swell multiplied by the percentage of the soil fraction passing No. 4 sieve. (The openings in sieve No. 4 are equal to 4.76 mm or 3/16 in.) The following correlation between the expansion index numbers and the degree of potential expansion is provided:

   0–20   Very low
   21–50   Low
   51–90   Medium
   91–130   High
   > 131   Very high

The formal parameters that define expansive soils are established by building codes. Paragraph 1803.5.3 of the 2009 International Building Code® (IBC-09) defines expansive soils as those meeting all four of the following criteria:

1. Plasticity index determined per ASTM D 4318 is 15 or greater.
2. More than 10 percent of particles pass No. 200 sieve (determined per ASTM D 422).
3. More than 10 percent of particles are less than 5 micrometers [0.005 mm] in size (determined per ASTM D 422).
4. Expansion index determined per ASTM D 4829 exceeds 20.

Item 2 refers to the "fines" fraction of the soil (recall the definition in Table 2.1). Item 3 considers the amount of even smaller particles—those associated with expansive components. When the test for item 4 is conducted, IBC allows the other three tests to be waived.

The clays with a liquid limit of greater than 50 are sometimes called "fat clays," while "lean clays" are those with low plasticity indices. Many geotechnical engineers frown on the use of the terms "fat clay" and "lean clay" and prefer calling those materials clays with high or low plasticity indices.

2.2.3 Organics

Few words are less welcome on the construction site than *organics*. A more technical name is *highly organic soils*, a soil class that includes peat and organic silt. "Organic" means that these materials contain vegetable matter and typically have low specific gravities. Peat is readily identified by organic color, odor, fibrous texture, and spongy feel.

In a submerged state, organics are stable and can even carry significant loading. Their blow counts could be high. But when the water level drops, organic soils dry out—and sometimes decompose—and lose their structural capacity, producing settlements sometimes measured in *feet*. Decomposition can also occur when peat is subjected to drying and wetting cycles (Butler, undated).

Organic soils are generally considered unsuitable for building support, although minor amounts might be acceptable when justified by a careful analysis. When significant amounts of organics are found at the site, special steps must be taken (soil removal, preloading, improvement, using deep foundations, and so on). Figure 2.2 shows organics and other unsuitable material excavated and awaiting removal. Dealing with deep organic deposits can increase the construction costs so dramatically that it might be more economical to find another site to build on.

Figure 2.2 Organics and other unsuitable material.

2.2.4 Collapsing Soils and Karst

Some soils exhibit significant reduction in volume and in void ratio with increased moisture content, even under stable loading. Such decrease in volume leads to major settlement and loss of bearing. Collapsing soils include loess, some weakly cemented sands, and silts with soluble cementing agents (such as gypsum or halite), as well as some granite residual soils. In these soils loose bulky grains are held together primarily by capillary stresses (NAVFAC DM-7.1).

Collapsible soils are found mostly in the arid regions. When encountered on site, these soils should be carefully investigated and appropriate steps taken to mitigate the situation.

Another problem soil is *karst*, made of soluble bedrock—limestone or dolomite. Some areas of the United States (e.g., parts of Kentucky) that contain karst are known for sinkholes, or depressions caused by the soil collapses. Some of the most popular underground caves exist in those areas.

2.3 Soil Investigation

2.3.1 Types of Investigation

Geotechnical investigation establishes the soil conditions at the site and ultimately, the allowable bearing capacity of the soil. The extent of the investigative efforts depends on the available information about the site, the size of the building, its use, location, seismic design category, and a host of other factors. A small structure planned in the area of known excellent soils with little variability often does not require the investigative efforts needed for a large pre-engineered building intended for industrial use in an area of questionable soils.

Chapter 18, Soils and Foundations of IBC-09, specifies the conditions when a soil investigation is required. It also establishes the basic parameters for investigation, sampling, and reporting. The code gives the local building official the power to make some critical decisions related to soil investigation. For example, the important issue of whether classification and investigation of the soil requires involvement of a registered design professional is left to the building official's discretion (IBC-09 Section 1803).

A geotechnical investigative program often consists of at least two phases: preliminary and detailed. A preliminary investigation is (or should be) one of the first steps in determination of the site's suitability. Credible signs of unsuitable materials, such as highly organic soils or loose fills, detected at this stage might stop the proposed project in its tracks and prompt the prospective purchaser to look elsewhere. At the very least the proposed construction budget will need to be reexamined. A detailed investigation might include making soil borings or test pits, collecting soil samples, and performing laboratory and office analyses. Both these phases are further described later.

2.3.2 Preliminary Exploration

In a preliminary exploration, the existing sources of information are reviewed. These might include soil borings from the adjacent sites, topographic maps, aerial photos, and other sources. In many urban areas, the results from previous investigations are readily available, and these could give a basic idea of what to expect. A Web site maintained by the U.S. Department of Agriculture can also be useful in identifying the probable soils present at the site.

A site visit can supply plenty of information related to the site's suitability. An experienced engineer or contractor, who observes that the site is located on a hill or contains a virgin forest, will probably conclude that a layer of organics is unlikely to be encountered underneath. The same conclusion could be reached when large stones and glacial till appear on the surface, as is common in many areas of New England.

The site visit could include interviews with the neighbors, who might know about the soil under their buildings, and an observation of their buildings. Any signs of serious settlement, such as corner cracks in masonry buildings, deep ruts under vehicle tires, and puddles of standing water in dry weather, should be carefully noted. These signs could indicate the presence of loose fill or fine-grained soils. Again, the neighbors could shed some light on this issue.

A cost-effective tactic might be to have a few test pits excavated at the site. A test pit (basically, a large hole in the ground) allows a direct observation of the soil. It can be dug manually or by a backhoe. A backhoe can reach about 10 to 12 ft down and can usually make five to eight pits per day. A test pit that contains full-depth unsuitable material strongly suggests continuing the exploration program with soil borings. The pits should be filled after their purpose is achieved, to avoid introducing unnecessary site hazards.

Unlike a soil boring, which has been likened to a bullet-hole view, a test pit offers a broad perspective of the soil conditions near the surface. If the groundwater level is high, this becomes quite clear. Test pits can be invaluable in the situations where the competent soil at some depth is overlain by a highly variable or stratified material near the surface. The pits can also be useful for retrieving relatively undisturbed samples from their sides or bottom, which can be used for testing. Test pits also expose the soil that can be tested on site by some rudimentary techniques. A couple of such quick tests and observations are:

- Try to roll a soil sample into a narrow tube (about 1/8-in. diameter). If the tube can be made, the soil is cohesive. Try to shake off the remains from the hands. If the soil sticks to the hands, it is probably silt.
- Place a sample of soil into a container with water and stir. How long does it take for the soil to settle? Sand and gravel settles very quickly, within 30 s. Silt usually settles in 15 to 60 min, while clay can stay suspended for hours.

The soil color observed at the walls of the pits can yield some useful information. Gray or bluish color indicates presence of at least some clay (this is the color of Boston blue clay, for example). Soils with black or dark brown color often contain organic materials. Red color typically indicates a presence of iron.

Preliminary soil exploration often makes use of special equipment such as a pocket penetrometer. This portable device, somewhat similar to a tire pressure gauge, can be pushed by hand into the sides or bottom of a test pit, and the unconfined compression strength and the allowable bearing pressures read directly. Obviously, this basic device provides only a very rough idea about the soil's capacity. It is particularly helpful in evaluating consistency of cohesive soils. More sophisticated equipment is generally used in the next phase of investigation.

2.3.3 Detailed Exploration: Soil Borings and Other Methods

Detailed soil exploration builds on the results obtained from the preliminary investigation. At this stage, soil borings are ordinarily used, sometimes supplemented by test pits to answer some field questions. For example, a test pit can clarify whether the "refusal" of the soil boring is the result of encountering a rock ledge or a boulder. A series of test pits alone might occasionally suffice when the confidence in the soil composition is very high, and it is only the soil conditions near the surface that are unclear.

A typical soil-boring rig could progress perhaps 40 ft per day in average soil. The soil-boring process follows the procedure established by ASTM D 1586 for Standard Penetration Test (SPT). In this procedure, a 140-lb hammer drops a distance of 30 in. in a free fall. The part that penetrates the soil can be fitted with a standard split-barrel sampler with an outside diameter of 2 in.

The operator measures the amount of hammer blows called *the blow count* (N) required to drive the split-spoon sampler 12 in. into the soil. The measurements start after an initial 6-in. penetration has been made. The blow count is recorded in a boring log as a number of blows for every 6 in. of penetration. (Since the log states the blow count per 6 in. of penetration, while the analysis considers the number of blows per **12 in.**, a potential for confusion exists among the uninitiated.) A refusal is defined **as a condition** where 100 blows produce less than 6 in. of movement. Figure 2.3 shows **a typical soil-boring log.**

The blow count can be used in the formulas that correlate it with the allowable bearing pressures. Such correlation is high for sands and gravels but low for silts and clays. For those cohesive soils, other methods of soil exploration are used, such as Shelby tube. Shelby tube is a thin-wall tube that is hydraulically pushed into a cohesive soil to obtain an undisturbed sample. The sample is sealed in the field to prevent moisture loss and is sent to a laboratory for an unconfined compression test.

In addition to yielding the blow count, the Standard Penetration Test can be useful for retrieving soil samples—disturbed or undisturbed. To obtain an undisturbed sample,

Foundation Design Basics 27

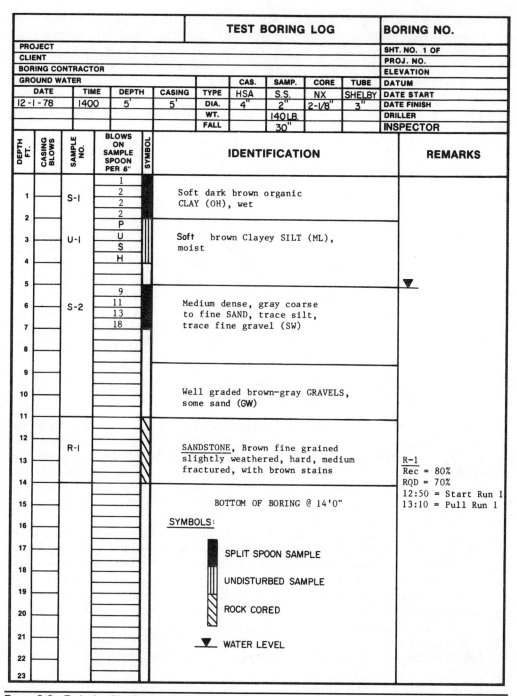

FIGURE 2.3 Typical soil-boring log (*NAVFAC DM-7.1*).

a brass liner is placed within the split-spoon sampler. Undisturbed samples can be sent to a lab for strength and compressibility tests or used for field tests, such as those involving pocket penetrometers.

Disturbed samples are useful for soil classification. They are typically taken every 5 ft and at every change in soil strata. When rock is encountered, the split-spoon sampler is replaced with a rock core. To make certain that the observed "refusal" is indeed solid rock and to compile some data about the material, one or two cores are often drilled about 5 ft deep into the rock.

How deep should the soil borings be? The depth depends on the soil conditions, and it should be preferably established as part of the geotechnical report's recommendations. In general, it is useful to carry the borings to a depth where the vertical soil stresses caused by the proposed foundation loading become relatively low, such as less than 10 percent of the effective overburden stress. For low-rise buildings, which make up the majority of metal building systems, some engineers specify a depth of 30 ft below the bottom of the foundation, unless rock is found. To help safeguard against an unanticipated poor soil located at a greater depth, at least one boring could be drilled to a depth equal to the lesser of 100 ft, the least building dimension, or to refusal.

In any event, a soil boring should never terminate in an unsuitable material. If the unsuitable soil is still present at the bottom of the specified boring depth, the depth must be extended. A decision like that—and many other practical challenges encountered during the investigation—requires engineering expertise at the site. For this reason, a technically competent person should be present on site when soil borings are taken. As IBC-09 states in Section 1803.4:

> The *registered design professional* shall have a fully qualified representative on the site during all boring and soil sampling operations.

How many soil borings should be used? The practices vary, and the number depends on the soil variability at the site, as determined by the preliminary exploration. At the very least, one boring should be placed at each corner of the building and one in the middle. For larger buildings, additional borings should be placed at regular intervals. The practice of many engineers in this respect is to use a 50 ft by 50 ft grid, so that one boring is placed for each 2500 ft^2 of the building footprint. This spacing used to be mandated by some older model codes.

Other methods of soil exploration include the Cone Penetrometer Test (CPT), whereby a standard conelike device is forced into the ground and the rate of pressure is measured for each increment of penetration. The resistance to the cone's penetration can come either from the soil pressure on the point or the frictional resistance on the sides. The ratio between the two shows which resistance mechanism predominates. This *friction ratio* is equal to the frictional resistance divided by the point resistance.

The higher the friction ratio, the more cohesive the soil is. Accordingly, it is low in sands and gravels and high in clays. Among the advantages of CPT is speed and low cost, but the test obviously does not yield any samples, and it is difficult to use in hard soils.

2.3.4 Laboratory Testing

The samples retrieved by the soil borings or other methods described earlier can be analyzed in the laboratory to scientifically establish the soil properties. These properties help establish load-bearing capacity of the soil and estimate the amount of foundation settlement in it.

The available tests range from basic (unit weight, particle size, water content) to the tests that determine the Atterberg limits, to those that find the unconfined compression strength of the soil. For cohesive soils, the latter involves taking a soil specimen obtained by use of a Shelby tube and loading it until it shortens by 20 percent. The geotechnical engineer interprets the results to determine the allowable bearing pressure. As demonstrated in Butler, with certain assumptions about the foundation size and depth and the factor of safety of 3.0, the allowable bearing pressure can be equal to the unconfined compression strength determined by testing.

Other tests include direct shear test, which helps compute the φ (phi) angle for cohesionless soils, and other specific tests that might be needed, as called for by the project geotechnical engineer.

2.4 Settlement and Heave Issues

2.4.1 What Causes Settlement?

All materials deform elastically under pressure, and soil is no exception. The foundation load or the weight of any added layers of fill causes stress in the soil, which undergoes elastic strain. But soil compressibility greatly exceeds the amount of its elastic shortening alone, because soil contains voids filled with air and water. Soil consolidates under pressure, as the air and water are forced from the voids and the particles are packed closer together. The speed of consolidation depends on soil permeability: The faster the water can migrate away from the compressed area, the faster the soil consolidates. Soil compressibility can be expressed as the amount of consolidation caused by a unit of pressure.

Another settlement mechanism comes into play when the pore water exits the soil, even when the foundation pressure stays constant. The dropping water level could be caused by natural climatic fluctuations or by the construction activities nearby. A classic problem occurs when dewatering at a nearby site temporarily lowers the water level in the vicinity and leads to settlement of nearby structures.

A similar situation involves vibrations at a construction site (a pile-driving operation, perhaps). Vibrations can cause settlement in the adjacent structures underlain by certain soils, such as loose sand.

2.4.2 Settlement in Sands and Gravels

In sands and gravels—highly permeable materials—consolidation and settlement occur very quickly, almost instantaneously. The amount of settlement can be reliably predicted by simple formulas and in general is relatively small. One such formula (from Brown, 1997) is:

$$\Delta H = \frac{4qB^2}{K_v(B+1)^2}$$

where B = foundation width (ft)
q = actual soil pressure (kips/ft²)
K_v = modulus of vertical subgrade reaction (kips/ft³)

Compactness	Relative Density (kips/ft³)	K_v
Loose	<35	100
Medium/dense	35–65	150–300
Dense	65–85	350–550
Very dense	>85	600–700

Source: Adapted from Brown, 1997.
Notes:
1. 1 kip = 1000 lb and 1 kip/ft³ = 157 kN/m³.
2. Use one-half of the listed values if the water table is at the base of footing.

TABLE 2.3 Modulus of Vertical Subgrade Reaction K_v for Sand

The equation assumes that the depth of the foundation below the ground level is less than its width, and that the width does not exceed 20 ft. The values of K_v are established by geotechnical investigation. These values are correlated with the relative density and compactness of sand or gravel: The higher those are, the higher the value of K_v is. Table 2.3 suggests the following values of K_v for sandy soils, based on compactness and density of the material.

Design Example 2.1 illustrates the process of computing settlements in sandy soil.

Design Example 2.1: Settlement of Column Footing

Problem Compute the settlement of a square column footing in dense sand with allowable bearing pressure of 4 kips/ft² under a combined dead and live load of 100 kips, including the weight of the foundation. (1 kip = 1000 lb.)

Solution Find the required footing area for $q = 4$ kips/ft².
$A_{req'd} = 100/4 = 25$ (ft²)

Use 5 × 5 ft footing ($B = 5$ ft).

The value of K_v in Table 2.3 for dense sand ranges from 350 to 550 kips/ft³.

Use the average value of 450 kips/ft³.

The settlement is computed as:

$$\Delta H = \frac{4qB^2}{K_v(B+1)^2} = \frac{4 \times 4 \times 5^2}{450(5+1)^2} = 0.0247 \text{ (ft)} = 0.30 \text{ (in)} \quad \blacktriangle$$

2.4.3 Settlement in Silts and Clays

Soil deformation in silts and clays occurs differently than in sands and gravels. Unlike quick-settling cohesionless soils, cohesive soils tend to gradually compress, creep, and deform under sustained load. Loading silts and clays above the pressures in their natural state leads to overconsolidation and settlement. When the load is removed—during excavation, for example—the soil may rebound, leading to swelling and heaving. Why does it happen?

Clay has little permeability, and it is difficult for its pore water to diffuse away from the loaded area. The process is lengthy, as the water must find a path through small soil fissures and tiny spaces between the clay particles. Recall that those particles are only about 0.005 mm in size. Settlement in cohesive soils can take a very long time, often measured in many years.

A classic example of gradual settlement in clay is the world-famous Tower of Pisa, which started to lean during construction and has continued to do so ever since. As was later discovered, a pocket of saturated clay existed about 30 ft below the surface, sandwiched between two good soils—sand and granite. The pocket extended only under a part of the tower's footprint, which explained the unequal settlement. (This should serve as a cautionary tale to those who wish to skimp on the soil exploration efforts today.)

In order to compute the amount of foundation settlement in clay caused by added pressure, one determines the thickness of the clay stratum, the exact soil properties, and the soil pressures in the original state and after the loading. Changes in moisture content, another source of clay consolidation discussed earlier, should also be considered. Also, some clayey soils in the northern states are already preconsolidated by the pressures exerted by the glaciers over thousands of years during the Ice Age. Preconsolidated clays can be expected to settle only a fraction of regular clays. Almost all soils in Wisconsin, Minnesota, and Michigan are preconsolidated (Butler, undated). All of these issues mean that the settlements analysis in clay is quite complex and is best left to the qualified geotechnical engineers.

2.4.4 Differential Settlement

As demonstrated in Design Example 2.1, some settlement can be expected even in a very competent soil material. A *uniform* settlement of the whole building might not present a major problem. Even if a pre-engineered building uniformly "sinks" 2 to 3 in., the steel structure would be affected very little. However, some damage to buried pipes, utilities, and driveways could be expected.

Differential settlement is another matter. When one column of a multiple-span rigid frame settles differently from the others, it produces significant stresses in the frame—and also in the continuous purlins bearing on that frame.

In many if not most practical situations involving metal building systems *some* differential settlement is likely. Why? Here are a few contributing factors:

- The actual loads on the various foundations of the same size might differ. Foundation designers often try to minimize the number of the footing sizes, specifying identical foundations under the frame columns with slightly different tributary areas. Also, heavy suspended pipes and equipment might be present in one area of the building but not in another, snow accumulates (or roof live load occurs) nonuniformly throughout the roof, and so on.

- The soil composition and the degree of compaction could vary slightly at different locations.

- The loads other than the downward-acting dead and live loads could govern the foundation design. Sometimes, wind uplift in combination with 60 percent of the dead load controls the sizes of interior footings in a building, while some other load combinations control the sizes of the exterior foundations. Accordingly, under the downward loading alone, the soil stresses would differ at the interior and exterior foundations.

- For lightly loaded buildings, the minimum sizes of the exterior foundations might be prescribed by the governing code or be given in the geotechnical report, while the sizes of the interior foundations are derived from analysis. In this common scenario the soil pressure under the interior footings often exceeds that of the exterior foundations.

2.4.5 Some Criteria for Tolerable Differential Settlement

If differential settlement is nearly impossible to avoid, then what amount could be considered harmless? The building codes and ASCE 7 (ASCE, 2006) are silent on the issue, and rightly so, as the amount of differential settlement a structure can tolerate depends on its composition, the presence of brittle finishes, the owner's attitude toward the appearance of the inevitable cracks, and so on.

Some guidance is given in UFC 3-220-07 (formerly TM 5-818-7), *Foundations in Expansive Soils*, since angular deflection is one of the problems that need to be controlled in those soils. The document uses the term "the tolerable angular deflection (Δ/L)," where Δ is the deflection and L is the length between the building columns. We will use this ratio for an evaluation of differential settlements in any soil. Logically, brittle structures can tolerate smaller differential settlements than flexible ones.

According to UFC 3-220-07, the following maximum Δ/L ratios are suggested as a rough guide and as a function of *superstructure* rigidity (as interpreted and narrowed by the author of this book for metal buildings):

- Rigid (unreinforced masonry, plaster, slabs on grade not isolated from walls): 1/600 to 1/1000.
- Semirigid (reinforced masonry and concrete, slabs on grade isolated from walls): 1/360 to 1/600.
- Flexible (steel framing; brick veneer with joints; metal panels, gypsum board on metal or wood studs, with all water pipes and drains having flexible joints, suspended floor or slabs on grade isolated from walls): 1/150 to 1/360.

These suggestions, while seemingly simple, contain plenty of caveats. While most metal building systems can be classified as flexible, they could contain brittle masonry walls and finishes and thus fall into the other categories. Also note that if the slab on grade is integral with the foundation, the building probably belongs in the "rigid" category. Plus, in the "flexible" category it is *presumed* that cracking in slabs on grade is probable and accounted for in the design—not a common approach, in the author's experience. In any event, UFC 3-220-07 cautions that "a Δ/L value exceeding 1/250 is not recommended for normal practice, and a Δ/L value exceeding 1/150 often leads to structural damage."

The document also offers guidance for tolerable angular deflection of various foundation systems. For shallow foundations or continuous footings, a suggested Δ/L value is between 1/600 and 1/1000, not to exceed ½ in. When these foundation systems are used, fractures can appear in walls that are not designed for differential movement after Δ/L ratios exceed 1/600 or about ½ in. For sites where larger angular deflection ratios are probable, stiffened mats or deep foundations could be useful.

Alternatively, UFC 3-220-07 mentions a simple commonly used differential-settlement limit of 1/500. However, it also mentions that cracking in unreinforced masonry can still be expected at Δ/L ratios between 1/600 and 1/1000.

Other suggested limits on differential movements can be found elsewhere. For example, Merritt, quoting *Bjerrum* (1963), suggests an angular distortion of 1/500 as a limit for wall cracking and 1/600 as a limit for diagonally braced frames. *Gaylord* (1997), quoting *Sowers* (1979) offers the following limits (as adapted by the author of this book for metal buildings):

- Simple steel frame: 1/200
- Continuous steel frame: 1/500
- Gypsum plaster cracking: 1/1000
- High continuous brick walls: 1/1000 to 1/2000

Metal building systems should presumably fall into the "continuous steel frame" category, since the purlins, if not the frames, are typically designed as continuous structural members. Accordingly, a Δ/L limit of 1/500 appears appropriate for metal buildings without brittle finishes, such as drywall and brick veneer. Where these finishes are present, the Δ/L limit of 1/1000 might be more suitable. Again, these numbers are offered only as a rough guide, and other sources might suggest different criteria, but we recommend these limits as a useful starting point.

To illustrate the differential-settlement ratios, the 1/500 limit in a metal building without hard walls and gypsum-board partitions and with 25-ft bay spacing would require that the foundations under the adjacent frame columns do not exhibit differential settlement in excess of $(25 \times 12/500) = 0.6$ in. The 1/1000 limit would halve that, to 0.3 in.

2.5 Determination of Allowable Bearing Value

2.5.1 Why Not Simply Use the Code Tables?

The main goal of a typical soil investigation program is to establish the allowable bearing pressure on the soil. In some construction projects with limited budgets, attempts are made to dispense with formal geotechnical explorations. If the soil in the area is well known, the reasoning goes, why spend money investigating it again and again? And if we know from experience that the soil is, say, clean sand for miles around, why not simply use the allowable bearing values listed in the building code?

True, most model building codes such as IBC-09 contain tables of the allowable bearing pressures for various soils. For example, IBC-09 has Table 1806.2, Presumptive Load-Bearing Values (in IBC-06, Table 1804.2, Allowable Foundation and Lateral Pressure). However, the values listed in that table, and in other similar code tables, are very conservative, as they attempt to encompass many disparate conditions and to allow for some misidentification of the soil. The allowable pressures for non-rock soils listed in the table range from a low of 1500 lb/ft^2 (1.5 ksf) for clay and silt materials to 3000 lb/ft^2 (3 ksf) for sandy gravel and gravel. Even for sedimentary and foliated rock the number is only 4000 lb/ft^2, with a higher capacity assigned to crystalline bedrock.

In reality, all these soils might support higher pressures—perhaps much higher—but justifying the higher values requires a geotechnical involvement. Is this worth doing? For a small structure the allowable pressures might not matter much, but for a large pre-engineered building the difference between the low code values and the higher values established by a geotechnical exploration program may prove significant.

According to one study (Ruddy, 1985), "An increase in an allowable bearing pressure from 3 to 6 ksf can result in a savings of $0.08/sf for a shallow spread footing foundation system in a one-story building." (We shall add, that's stated in 1985 dollars.) As demonstrated in the chapters that follow, this statement is even more relevant to metal building systems, where the foundations often apply nonuniform pressures on the soil.

As further discussed in Sec. 3.1.3 of this book, IBC-09 allows a one-third stress increase for alternative basic load combinations using allowable stress design (ASD) that include wind or seismic loads. The stress increase applies both to vertical foundation pressures and lateral bearing pressures listed in IBC-09 Table 1806.2, Presumptive Load-Bearing Values.

How about poor and organic soils? The IBC-09 Section 1806.2 states that no presumptive load-bearing capacity shall be given to "[m]ud, organic silt, organic clays, peat or unprepared fill," except for lightweight and temporary structures and when approved by the building official.

2.5.2 Special Provisions for Seismic Areas

The International Building Code® requires performing a soil investigation in the areas of high and moderate seismicity; the exact provisions depend on the Seismic Design Category (SDC) of the building. The SDC is established by following the IBC Section 1613, as a function of the building's Occupancy Category and the spectral response acceleration parameters of the site. Seismic Design Categories range from A (the lowest seismic hazard) to F (the highest). The following indicates a rough equivalency between an SDC and the seismic risk:

$$\text{SDC A and B} = \text{low risk}$$
$$\text{SDC C} = \text{moderate risk}$$
$$\text{SDC D, E, F} = \text{high risk}$$

For structures in Seismic Design Categories C through F, the geotechnical investigation is required by IBC-09. The scope of the investigation includes an evaluation of four potential geologic and seismic hazards: slope instability, liquefaction, differential settlement, and surface rupture. In SDCs D, E and F some additional requirements are prescribed.

2.5.3 What Constitutes a Foundation Failure?

What does the allowable bearing capacity of the soil represent? Will the foundation loaded 1 pound per square foot (psf) above the capacity fail? And what does "foundation failure" mean?

To answer these questions, we must keep in mind that soils are natural materials. Even the soils of the same classification are not identical, and their properties vary, at least slightly. Accordingly, the factors of safety for soils are higher than those for man-made materials with predictable and tightly controlled properties, such as structural steel. *AISC Specification for Structural Steel Buildings* (AISC, 2005) lists the factors of safety for various conditions. For example, a compact structural steel beam loaded in flexure and designed by the allowable stress method is given the factor of safety Ω_b of 1.67.

By contrast, the factor of safety in soil is typically much higher—*at least* 3.0. The relatively high value accounts for soil variability, imprecise knowledge of the subsurface conditions, and high cost of foundation remedial work (Butler, undated). The building-code tables, which are used all around the continental United States and beyond, probably incorporate even higher safety factors, but they are not explicitly stated. The absence of precise factors of safety for soil pressures makes it difficult to state with certainty whether the foundation will actually fail when the actual soil stresses exceed the given value or not.

The failure of a foundation typically manifests itself as excessive settlement, accompanied by a shear failure of the bearing strata and heaving of the adjoining soil at the surface. The configuration of the possible shear plane depends on the foundation size. For example, the wider a spread column footing is, the larger the shear failure plane becomes. But the shear failure plane would get even larger were the footing to be placed deeper. In general, the bearing capacity of the foundation depends on shear strength of the soil, footing width and depth, soil density, and the water table (Butler, undated).

2.5.4 Summary

To summarize the foregoing discussion, the appropriate method of arriving at the allowable bearing value of soils depends on many variables. These variables include the degree of soil uniformity in the area, the availability of qualified geotechnical personnel and soil-boring contractors, and the practices followed by the local building officials. The size of the proposed building matters as well.

For a large pre-engineered building planned in the area of variable or difficult soil conditions, a formal soil exploration program is usually warranted. The program might include soil borings, laboratory testing, and geotechnical analysis.

For a small structure proposed in the area of well-known nonexpansive competent soils of uniform composition, an investigation program relying on penetrometers, test pits, or even previous experience, might suffice, provided the governing building code allows it. In that case, and in other possible scenarios, guidance from the local building official should be sought. For areas of high and moderate seismicity, regardless of the building size, special code provisions may apply, resulting in a more extensive soil exploration program.

2.6 Shallow vs. Deep Foundations

Depending on the foundation loads and the soil composition, either a shallow or deep foundation system might be appropriate. Shallow foundations include column footings, continuous wall footings, grade beams, and mats. Column footings in pre-engineered buildings are typically single "spread" foundations, but in some situations other solutions could be desirable. For example, a solution for the tight site where the lot line is at the exterior face of the building could be a series of combined or cantilevered footings, both of which incorporate two or more columns bearing on the same foundation.

In some situations foundation mats might be appropriate. Mats can be of the common design similar to a thick slab on grade reinforced in both directions, or they could be ribbed "rafts."

Deep foundations include piers (caissons) and piles. Piers are typically placed one at a time, while piles are commonly placed in groups. Piers can be classified as either deep or shallow foundations. One way to differentiate between the two (perhaps somewhat artificially) is to define shallow foundations as those with depth-to-width ratios not more than five. Deep foundations would be those with depth-to-width ratios exceeding five. The following chapters address the design of all these systems.

References

2009 International Building Code®, International Code Council, Country Club Hills, IL, 2009.

AISC 360-05, *Specification for Structural Steel Buildings*, American Institute of Steel Construction, Chicago, IL, 2005.

ASCE/SEI 7-05, *Minimum Design Loads for Buildings and Other Structures*, American Society of Civil Engineers, 2006.

Bjerrum, L., *European Conference on Soil Mechanics and Foundation Engineering*, vol. 2, Wiesbaden, Germany, 1963.

Brown, Robert W., *Foundation Behavior and Repair*, 3d ed., McGraw-Hill, New York, 1997.

Butler Manufacturing Company and Computerized Structural Design, Inc., *Foundation Design and Construction Manual*, 2d ed., Kansas City, MO, and Milwaukee, WI, undated.

Gaylord, Edwin H., Jr. et al. (eds.), *Structural Engineering Handbook*, 4th ed., McGraw-Hill, New York, 1997.

Merritt, Frederick S. (ed.), *Standard Handbook for Civil Engineers*, 3d ed., McGraw-Hill, New York, 1983.

NAVFAC DM-7.1, *Soil Mechanics*, Department of the Navy, Naval Facilities Engineering Command, Alexandria, VA, 1982.

Ruddy, John L., "Evaluation of Structural Concepts for Buildings: Low-Rise Buildings," BSCE/ASCE Structural Group Lecture Series at MIT, Cambridge, MA, 1985.

Sowers, G. B., *Introductory Soil Mechanics and Foundations*, 4th ed., Macmillan, New York, 1979.

Terzaghi, Karl and Peck, Ralph B., *Soil Mechanics in Engineering Practice*, Wiley, New York, 1967.

Unified Facilities Criteria UFC 3-220-07 (Formerly TM 5-818-7), *Foundations in Expansive Soils*, U.S. Department of Defense, 2004.

CHAPTER 3

Foundations for Metal Building Systems: The Main Issues

3.1 The Differences between Foundations for Conventional Buildings and Metal Building Systems

Some issues in the design of conventional building foundations are addressed in Chap. 2 and will be revisited in the following chapters. The discussion here focuses on the unique aspects of the foundation design for metal building systems (MBS). The engineers who have not designed the foundations for metal buildings might not be fully aware of these unique challenges.

3.1.1 Light Weight Means Large Net Uplift

One of the advantages of metal building systems is their cost efficiency. As discussed in Chap. 1, the systems typically include cold-formed secondary members and lightweight metal roofing and siding. The total weight (dead load) of the building shell—the primary and secondary framing, roofing, and siding—often adds up to only 3 to 5 lb/ft^2 [pounds per square foot (psf)]. Obviously, the weight of the framing increases with larger the frame spans.

Strong winds tend to generate upward loading (uplift) on gable roofs, in the same phenomenon that allows an airplane to fly. Typical distribution of wind pressures on a single-story building with a gable roof is shown in Fig. 3.1. The dead load of the system is generally insufficient to counteract the effects of wind-generated uplift, which could easily exceed the weight of the structure many times over. In essence, a large pre-engineered building acts as a huge sail tied to its foundations. The foundations must be heavy enough, or be anchored into the soil, to resist the uplift load.

To make matters more complicated, only a portion of the dead load likely to be in place at the time when the wind is acting may be used to resist the uplift. The *2009 International Building Code®* (IBC-09) includes the following "basic" load combination for the allowable stress design (ASD) method:

$$0.6D + W + H$$

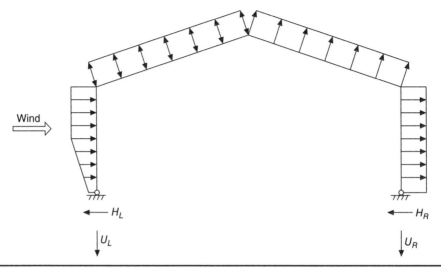

Figure 3.1 Typical distribution of wind pressures on a gable-roof building.

A related IBC "alternative basic" load combination (as simplified by the author for metal building foundations where roof live load or snow load is not present and wind loads are *not* computed in accordance with ASCE 7 Chapter 6*) is:

$$2/3D + W + H$$

In these formulas D is dead, W is wind, L is live, and H is lateral earth pressure loading. In both these load combinations the dead and wind loads counteract one another, despite the "plus" sign. (Chapter 4 of this book examines the relevant IBC load combinations in more detail.)

The weight of the building comprises the dead load likely to be in place during the wind event; it should *not* include the so-called *collateral load*. The latter is an allowance for pipes, conduits, lights, suspended ceilings, and similar superimposed dead loading that may or may not be in place during the design hurricane. It is prudent to neglect any beneficial contribution of such items toward the total weight of the dead-load "ballast."

Seismic loads, while not addressed in detail in this book, could govern the design of some pre-engineered buildings located in the areas of high seismicity. A very light dead load of metal building systems, which makes wind-generated uplift loading on these buildings so dangerous, also results in a rather small seismic loading. When seismic loads are of concern, the reader is encouraged to consult *Seismic Design Guide for Metal Building Systems* (Bachman, 2008).

To be sure, foundations for conventional buildings can be subjected to net wind-generated uplift, too. However, its magnitude is typically smaller, because the structure usually weighs more and the spans are smaller than in metal building systems.

---

*IBC-09 Paragraph 1605.3.2 requires the wind load in alternate load combinations to be multiplied by 1.3 if the wind loads are computed in accordance with ASCE 7 Chapter 6. If they are computed by other methods (such as IBC-09 Section 1609), the coefficient is equal to 1.0.

In buildings taller than a single story, the weight of the supported floors is often sufficient to overcome the wind uplift loading, although even there a net uplift could be present in the foundations under the lateral-force resisting elements.

A classic telltale sign of the foundation designer's familiarity (or lack thereof) with these issues is the size of isolated spread footings under the interior columns in multispan rigid-frame systems. Whenever the foundation drawings show tiny interior footings in the areas of high winds, it is likely that a load combination involving wind uplift has not been properly considered. The footings might be adequate for the downward acting loading but not for the wind uplift, a problem that plagues the designs of some inexperienced foundation engineers. This problem is demonstrated in Design Example 4.1.

Figure 3.2 illustrates the components of the foundation dead load "ballast" that helps resist the wind uplift force. They include the weights of the footing, the column pier and the soil on top of the footing ledges. Some engineers also include the effects of the shearing resistance of the soil, shown as the soil wedges on the perimeter of the footing. The angle of the wedges may be taken as 30° for cohesive soils and 20° for cohesionless soils (NAVFAC DM-7.2). It is conservative to ignore the shearing resistance, as most engineers do (Building Systems Institute, 1990). We recommend neglecting the weight of the soil wedges as well, because it might be difficult to properly compact the soil around the foundations and to fully develop the shearing resistance.

In the areas subjected to flooding, the weights of soil and concrete should be taken as "submerged" (that is, reduced by the buoyancy forces). Since soil is normally less expensive than concrete, even including the excavation and backfilling costs, the most cost-effective way of increasing the uplift resistance of the footing might be to make it deeper. Making the footing wider is often more expensive. The other methods of resisting uplift forces, such as using deep foundations and soil anchors, are generally even more costly.

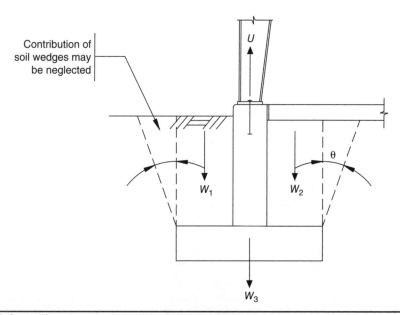

Figure 3.2 Uplift resistance at an isolated column footing.

3.1.2 Large Lateral Reactions

As noted in Chap. 1, gable rigid frames tend to generate significant lateral reactions on the foundations. These horizontal frame reactions will exist under both gravity and lateral (wind and seismic) loading, although the magnitude and the direction of the reactions will differ (Fig. 3.3). In some circumstances, the horizontal frame reactions exceed the vertical ones. Under gravity loads, the horizontal reactions tend to decrease with increasing eave height. In other words, the higher the building the less the gravity-generated horizontal reactions, all other variables being the same.

But don't horizontal column reactions exist in conventional buildings as well? The answer is yes, but not necessarily of the same magnitude and direction. Consider the

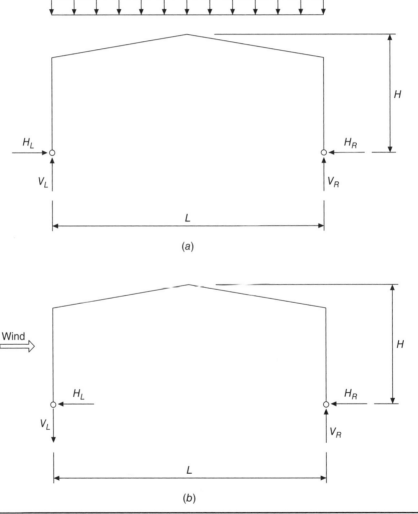

FIGURE 3.3 Column reactions in a single-span gable rigid frame: (a) From gravity loading; (b) from lateral loading.

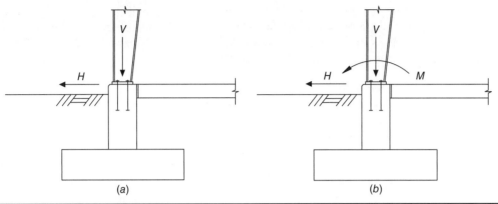

FIGURE 3.4 Reactions from rigid-frame columns: (a) At pin-base columns; (b) at fixed-base columns.

differences in the value and direction of the horizontal column reactions in two identically looking buildings with gable roofs, with the same overall dimensions and roof slope. One building is framed with conventional single-span trusses supplemented by wall bracing, and another is framed with a metal building system using single-span rigid frames. The horizontal reactions at the truss supports will be caused only by lateral but not gravity loads, and these reactions will likely be acting in the same direction. By contrast, the horizontal reactions on the rigid frame supports will exist under both lateral and gravity loads. In a single-span rigid frame, the lateral frame reactions from gravity loads act in the opposing directions.

Where exactly are the horizontal frame reactions applied? Typically, at the bottom of the column base plates, which are located well above the bottoms of the foundations (Fig. 3.4). For this reason, the horizontal frame reactions tend to overturn the foundations and push them sideways. Thus the metal building foundations are designed to resist not only downward pressures, but also overturning and sliding, as explained later.

3.1.3 Factors of Safety and One-Third Stress Increase

The load combinations of the 2009 International Building Code® for the ASD method that include wind and dead loads acting together have certain implied safety factors. These are 1.67 for the "basic" load combination of $0.6D + W$, and 1.5 for the "alternative basic" load combination of $2/3D + W$. Therefore, the foundation design using the "alternative basic" load combination that includes wind uplift has a slightly lower factor of safety, and it has an inherent cost advantage over the corresponding design that uses the "basic" load combination.

Should the foundations for metal building systems be designed for any *additional* factors of safety on top of those suggested by the IBC load combinations? Most engineers consider the implied factors of safety to be sufficient for the design of foundations against overturning, sliding, and uplift.

There are some special cases. At least one popular MBS foundation—the moment-resisting foundation described in Chap. 7—behaves similarly to a cantilever retaining wall. An argument could be made that it should be designed in accordance with the code provisions for retaining walls, since no other code section seems to explicitly apply to such a foundation. (This is yet another complication awaiting the designers of the metal building foundations.) According to some authorities, such as *CRSI Design Handbook* (CRSI, 2002), a minimum factor of safety at service loads against overturning of retaining walls is 2.0 and against sliding, 1.5. However, Paragraph 1807.2.3 of IBC-09 requires a factor of safety of 1.5 for both overturning and sliding of retaining walls, without using the load combinations of IBC Section 1605. Instead, 100 percent of the nominal loads should be used (except 70 percent of seismic load).

The question of whether the allowable soil pressures can be increased by one-third in load combinations that include wind or seismic loading causes some confusion. In the past, building codes allowed a one-third stress increase for these load combinations using the ASD method, but then the situation has become less clear. The IBC-06 and IBC-09 Paragraph 1605.3.1.1 does not allow such a stress increase for "basic" load combinations, except for wood structures. For "alternative basic" load combinations, IBC-06 and IBC-09 Paragraph 1605.3.2 allows a stress increase only where permitted by the material chapter of the IBC or by the referenced standard applicable to the material in question.

In IBC-06, Chapter 18, which deals with soils and foundations, is silent on the matter. Since there is no referenced standard for soil design, a reasonable conclusion could be made that IBC-06 does not allow a one-third stress increase for soil stresses under foundations.

However, IBC-09 has revisited the issue. IBC-09 Section 1806.1 specifically *permits* a one-third stress increase for the alternative basic load combinations using the ASD method that include wind or seismic loads. The stress increase applies to both vertical foundation pressures and lateral bearing pressures listed in IBC-09 Table 1806.2, Presumptive Load-Bearing Values. It is not clear whether a one-third stress increase may be used for the allowable soil pressures determined by means of geotechnical investigation. As always, the final determination of the questions related to the code provisions belongs to the local building official.

3.1.4 In Some Circumstances, Uncertainty of Reactions

The *Metal Building System Manual* published by the Metal Building Manufacturers Association (MBMA, 2006) is an authoritative design guide for the manufacturers of pre-engineered buildings. According to Section 3.2.2 of its Common Industry Practices, the typical scope of work of the metal building manufacturer excludes any foundation design. The manufacturer is responsible for providing the frame reactions to the Builder (the entity that signs the contract with the owner). The foundation designer uses these column reactions to design the metal building foundations.

Ideally, the manufacturer first designs the metal building system, determines the column reactions, and then provides them to the foundation designer. However, in some circumstances the order is reversed: The foundations must be designed first, before the building manufacturer is even selected.

This is typical in public construction, where the law mandates competitive bidding. In this situation, the contract documents must be completed to give all the bidders access to the same design information. Both the prospective foundation contractors and the metal building manufacturers receive the design information at the same time. It means that the foundation design must be performed well *before* the column reactions are available.

A similar situation exists even in some private fast-track construction projects. A developer who wants to save construction time will often insist that the building designers produce a so-called "early foundation design package" set of construction documents before the rest of the building is fully designed. For the developer it makes sense, because the foundations are the first items being built. But the foundation designer in this case faces the same problem as a designer of metal building foundations in public construction: a lack of information about the column reactions for which to design the foundations.

What to do in both these unfortunate but common situations? When the foundations have to be designed before the column reactions become available, the reactions must be estimated, as discussed in the next section. The need to estimate frame reactions is one of the biggest challenges of designing foundations for metal building systems.

3.2 Estimating Column Reactions

3.2.1 Methods of Estimating Reactions

As just discussed, there are circumstances where foundations for metal building systems have to be designed before the column reactions are available from the manufacturer. There are several methods of estimating reactions:

- *Using the manufacturers' tables*. Some metal building manufacturers publish design manuals that include tables of column reactions. These tables typically cover single- and multiple-span rigid frames and sometimes other frame types. The tables are limited to the most common spans, eave heights, roof slopes and loading conditions. Naturally, the tables are developed using the building codes adopted at the time of their compilation. Between the time the tables are developed and published to the time when the designer uses them, the code editions on which the tables are based often become obsolete. This time lag primarily limits the usability of the tables for wind-load reactions, because the code provisions for wind loads change often. Fortunately, the tables are still valuable for estimating horizontal and vertical reactions from gravity loads. The appendix includes a set of column reaction tables from one large manufacturer.

- *Using general analysis software*. Some foundation designers prefer computing the column reactions using general frame-analysis software. This approach is best suited for multiple-span rigid frames with unequal spans, a condition not generally covered in the manufacturers' tables. Since only the overall frame dimensions are known at this stage, rather than the actual sizes of the tapered frame members, the designers typically use rectangular cross-sections for modeling. This simplifying assumption leads to some inaccuracy in the analysis results. Another obvious disadvantage of this approach is the need to determine the design loading on the frame. While the manufacturers' design software can perform this task in seconds, it might take much longer in the foundation designer's office, given today's code complexity.

- *Using specialized design computer software for metal building systems*. The engineers who specialize in metal building foundation design might find it advantageous to invest in the software similar to that used by the manufacturers. The software will provide a lot of information unnecessary for the task of designing foundations, because only the frame reactions are needed. The software is also quite expensive. Still, if available and properly set up, such computer software is a useful tool for estimating column reactions. (But see Sec. 3.2.2 on the accuracy of estimating the reactions.)

- *Using standard frame formulas.* A common single-span rigid frame can be readily analyzed by the basic frame formulas to determine horizontal reactions under a variety of loads. (The vertical reactions can be found without any software at all; they are the same as for a simply supported beam!) A classic source of frame formulas is Kleinlogel (1964). Using standard formulas allows analyzing the frames that are not covered by the manufacturers' tables. However, the foundation designer needs to determine the loading on the frames, which takes time.

- *Asking a manufacturer for help.* A simple alternative is to simply ask a metal building manufacturer to determine the reactions using its design software. The manufacturer could do it very quickly, sometimes within minutes. Obviously, a working relationship first needs to be established between the foundation designer's and the manufacturer's offices for this to happen, and the manufacturer's goodwill should not be abused by continually asking for such favors.

3.2.2   How Accurate Are the Estimates?

Some of the methods of estimating column reactions discussed previously are more accurate than others, but in all probability none will yield the exact values that the selected metal building manufacturer eventually supplies. Each of the methods assumes a certain frame cross-section that will be probably different from the actual frame, the loading assumptions might slightly differ, and so on. For instance, a single-span rigid frame with straight columns, or columns with a small amount of taper, will typically produce smaller horizontal reactions than a similar-size frame with a more pronounced column taper.

Despite the best efforts of the foundation designers, their reaction estimates are just that—the estimates. They are virtually guaranteed to differ from the final numbers. This is true even if another manufacturer provides them (see the last option of Sec. 3.2.1).

There are two basic methods of dealing with this challenge. Some engineers preface their foundation designs in the contract drawings by a note that the foundation sizes are given for bid purposes only and that the actual sizes are to be determined by the foundation contractor for the final column reactions. As an example, a note in an Appendix to USACE TI 809-30 (1998) states:

> The metal building foundations shown on this sheet are estimates for the building and are to be used as a guide for the building supplied. The contractor will redesign the metal building foundations for the actual building loads provided by the Metal Building Manufacturer. Foundations for items other than the metal building will be constructed as shown.

Obviously, the assumption in this note is that the foundation contractor will engage another engineer to redesign the foundations for the actual loading conditions. But this common approach is fraught with potential problems.

Since the foundation design practices vary among the engineers, the possibility of a disagreement between the original foundation designer and the one retained by the contractor is quite real. Who is responsible for the foundation design in that case? And what happens if the contractor's designer comes up with a much more costly system than the one shown in the contract drawings? A claim for extra compensation might follow. A real possibility of getting involved in these thorny issues argues for using extra care with this approach.

Another way to meet the challenge of working with the estimated numbers relies on good old-fashioned design conservatism: The foundation designer simply uses the values of the column reactions that exceed the estimates by some comfortable margin. This approach might not be the most economical. But it is prudent when any change to the foundation design is likely to be met with a request for cost "extras" in a change order from the concrete contractor. In the worst-case scenario, the foundations are already built under the "early foundation design package" mentioned previously, and they need to be physically removed or enlarged. The cost impact of the contractor's change order for this work can only be imagined.

3.3 Effects of Column Fixity on Foundations

3.3.1 Is There a Cost Advantage?

The metal building industry is a highly competitive business. The building manufacturers pride themselves on optimizing the design of steel superstructure and providing the most economical solution to their customers. But in an effort to gain a cost advantage over the competition, some might choose a system that generally should be avoided. In this system the column bases are assumed to be fixed into the foundation.

The difference between a pin-base and fixed-base column is this: A pin-base column can resist only horizontal and vertical reactions, while a fixed-base column can also resist the bending moments resulting from the base fixity (Fig. 3.4). A rigid frame fixed at the base will typically result in some (probably rather modest) cost savings for the steel superstructure, but will *substantially* increase the foundation costs. Typically, the foundations would become much larger. In the author's opinion, the overall project cost will generally increase when the base fixity is assumed.

3.3.2 Feasibility of Fixed-Base Columns in MBS

In addition to the cost issues, an argument against using fixed-base columns can be made from a technical standpoint as well. A true base fixity requires that the angle between the top of the foundation and the column centerline at the base stay exactly at 90°. Even a minor rotation—only a few degrees—transforms the connection from a fully to partially restrained, as defined in AISC 360. Yet a minor rotation of this magnitude can easily take place because of the flexibility of the base plate, absence of grout under the column base plate (which might allow for some column "play"), and unequal soil deformation under the foundation. These issues are further debated and illustrated in Newman (2004).

To be sure, there are circumstances when the assumption of column fixity might be appropriate for some columns. One example is a building with heavy overhead cranes, where the columns supporting the cranes could be designed with fixed bases to reduce the lateral drift and building movements under the crane loading. However, even in this situation the assumption of the column base fixity should be made at the outset and the foundations designed with this assumption in mind.

3.3.3 Communication Breakdown

The biggest problem, in the author's experience, is not with using column fixity per se, but with a lack of communication on this topic between the metal building manufacturer and the foundation designer. A truly unfortunate situation exists when the building foundations are already designed—or worse, built—using the assumption that the columns had pinned bases, but the manufacturer designs the frames with column fixity.

This can easily occur if a structural engineer of record, who would generally specify the design criteria for the metal building design, simply does not exist on a project. In an all too common scenario, the owner simply procures the pre-engineered building from the contractor and hires a local engineer to design the foundations afterward. The foundation designer typically assumes the pin-base conditions, to keep the foundation sizes reasonable, but the manufacturer might choose a fixed-base design, also to save some money on *its* structure. In the absence of any contractual requirements to the contrary, the manufacturer is certainly free to do so. This sort of miscommunication tends to result in a construction claim and even litigation.

All this heartache can easily be avoided by hiring a design team at the outset, including a structural engineer of record, as further discussed in Newman (2004). The team would establish the design requirements for the project and spell out which assumptions are made about the column bases.

3.4 General Procedure for Foundation Design

The actual process of designing foundations for metal building systems depends on a number of factors, including the local building code provisions, the frost depth, the building's location relative to the other structures, whether the foundation is designed before or after the building reactions become available, and so on. What follows is an outline of the general design process.

3.4.1 Assign Responsibilities

Whether the building owner realizes it or not, the first step in designing metal building foundations involves an allocation of the basic design responsibilities among the various parties involved. The "outside" (outside the manufacturer's office, that is) structural engineer can be engaged in one of two ways:

1. Only to design the foundations—an unfortunate situation, as discussed previously
2. To serve as a structural engineer of record—the preferred way of retaining the services of an outside design team

When the owner contracts directly with a builder and retains a local engineer only to design the foundations, the first approach is chosen. In the second approach the outside structural engineer's responsibilities include specifying the design criteria for the metal building superstructure. The second approach allows establishing the important building design criteria, such as whether fixed-base columns are allowed, in advance, and any controversy on these issues is thus avoided.

3.4.2 Collect Design Information

The next step is to collect the design data about the building and the proposed site. This information includes the size and use of the building, its proximity of the property line, and of course the information about the soil. At this stage any available soil-related studies and reports are examined and a program of geotechnical investigation is established.

As discussed in Chap. 2, the extent of this program depends on such factors as the type of soils at the proposed construction site, variability of soil conditions, the building code requirements related to soil investigations, and the designer's experience with the soils in the area. In any event, *some* program of soil investigation is needed to determine and classify the soil at the site. If anything, it permits using the building-code tables of the allowable soil capacities, should this conservative method of arriving at the design bearing value of the soil be chosen.

3.4.3 Research Relevant Code Provisions and Determine Reactions

In the next step the foundation designer determines which code and which edition applies to the project. In most cases this task is very straightforward, but occasionally the project takes place during a transition period between the two code editions. In this situation some research into the relevant differences between the two codes is in order, as is coordination with the other team members. It might be quite embarrassing for the foundation designer to use the just-adopted edition of the code, while the rest of the design team and the metal building manufacturer elect to use the outgoing edition. A similar situation arises in some rural areas, where no building code is adopted at all, and the coordination within the team becomes even more critical.

As already discussed, the natural progression of the events is to have the foundations designed *after* the building reactions become available from the metal building manufacturer. In this case the foundation engineer simply uses the manufacturer's reactions (perhaps after a quick check for reasonableness). The main challenge is to combine the reactions in accordance with the code-mandated load combinations, as discussed next. If the foundation designer must develop the loading on the frames (perhaps because the frame configuration precludes using the column reaction tables), the effort becomes much more involved.

3.4.4 Determine Controlling Load Combinations

A study of the local building code provisions includes the structural loads and the load combinations that must be considered. If the reactions are already available from the metal building manufacturer, they are commonly listed load by load, rather than arranged in specific load combinations. The foundation designer is then expected to combine the loads per the governing building code requirements.

When using the International Building Code, the designer must choose between the "basic" and "alternative basic" load combinations. Some of the differences between the two are described previously in Sec. 3.1.3. Recall our conclusion that an "alternative basic" load combination involving wind uplift has a slightly lower factor of safety and thus might be more economical than a corresponding "basic" load combination.

The foundation designer should consider all the code-mandated load combinations, but two of them typically control the foundation design in single-story metal building systems with single-span rigid frames:

1. Dead + collateral + snow (or roof live load)
2. Dead + wind + lateral earth pressure

The first load combination typically produces the maximum downward vertical reaction on the foundation combined with the maximum outward-acting horizontal reaction (Fig. 3.5a). The second load combination becomes either an IBC "basic" load combination of $0.6D + W + H$ or an "alternative basic" load combination (as simplified by the author in Sec. 3.1.1) of $2/3D + W + H$. (Here D is dead, W is wind, and H is lateral earth pressure loading.)

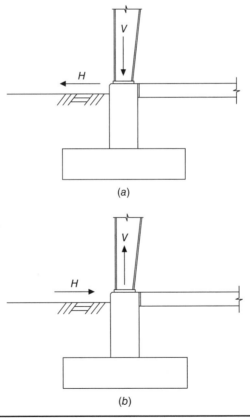

Figure 3.5 Two load combinations that typically control the foundation design in single-story MBS with single-span rigid frames: (a) Dead + collateral + snow (or roof live load); (b) dead + wind + lateral earth pressure.

The second load combination typically produces the maximum upward vertical reaction (uplift) combined with the maximum inward-acting horizontal reaction on the foundation (Fig. 3.5b). As discussed in Sec. 3.1.1, the dead load in this load combination should not include any collateral loading.

3.4.5 Choose Shallow or Deep Foundations

The next step involves selection of the basic foundation type. Depending on the soil at the site and the loads the foundations must resist, either a shallow or a deep foundation is chosen. The differences between shallow and deep foundations are explained in Chap. 2. Briefly, shallow foundations include column and wall footings, grade beams, and mats; their design is examined in Chaps. 4 through 8. Deep foundations include piers (caissons) and piles; these are discussed in Chap. 9.

Naturally, the designer attempts to select the most economical foundation system among the many available types of shallow and deep foundations. The foundation cost has to be balanced against reliability and other factors, as discussed in Sec. 3.5.

3.4.6 Establish Minimum Foundation Depth

The criteria for determination of the minimum depth of shallow and deep foundations are somewhat different. For deep foundations the depth is largely dictated by the location of the appropriate bearing stratum. There are exceptions, of course. For example, the depth of friction piles depends on the overall length of their embedment in the soil.

For shallow foundations the factors are more diverse. In general, the bottom of shallow foundations should be below:

- The minimum frost depth, as established by a local building code or other authoritative document
- At least 12 in. below the undisturbed ground surface (IBC-09 Section 1809.4)
- The depth of any unsuitable soil materials, such as organics, expansive clays, uncontrolled fills, and collapse-prone soils

The elevations of the adjacent building foundations and utilities also play a role in establishing the minimum depth of shallow foundations. Ideally, the new foundations should neither undermine the existing footings by being below them, nor impose an additional loading on them from above. For granular soils, IBC-09 Section 1809.6 requires the slope of the line drawn between the lower edges of the adjoining footings not to exceed 30° with the horizontal, unless lateral bracing or support is provided for the higher footing. The shallow foundations of a new building located next to the existing may have to be sloped or stepped to meet these requirements.

In practice, the depth of isolated column footings is often controlled by the minimum amount of "ballast" needed to counteract wind uplift. Thus even in the areas without deep frost lines the foundations may have to be placed several feet deep, as demonstrated in Design Example 4.1.

3.4.7 Design the Foundation

At this point enough information is gathered to select the foundation system and to design the foundations. A comprehensive design considers not only the stresses in the soil and the strength of the foundation itself, but also such issues as foundation settlement under load (see a discussion in Chap. 2) and the transfer of the frame reactions from the building columns to the foundations (see Chap. 10). The actual design procedures and the design examples are included in the chapters that follow.

3.5 Reliability, Versatility, and Cost

3.5.1 Definitions

There are many foundation systems, both shallow and deep, that can support pre-engineered buildings. The cost and reliability of these systems varies. Not surprisingly, the foundation systems that are less expensive also tend to be less reliable.

What do we mean by *reliability* in a foundation system? For our purposes, structural reliability of a foundation can be defined as the probability that the foundation will perform as intended for a specified period of time under various field conditions. The most reliable systems are able to tolerate predictable irregularities in construction, loading, and maintenance. The overall reliability of metal building foundations can be distilled into three interrelated factors pertaining to the system's ability to function as specified in adverse circumstances:

1. *Simplicity of installation*. Foundations that are difficult to install, or those that require perfection in construction, tend to be less reliable. Some placement errors are quite common.

2. *Redundancy*. The foundation's ability to transfer the applied forces to the soil via more than one load path. If one load path becomes invalid, another mechanism of load transfer takes over.

3. *Survivability*. The foundation's ability to maintain its load-carrying capacity after some of its elements have become corroded or damaged. Survivability also means that the foundation system will function even when some part of the building beyond the actual foundation becomes damaged.

These three factors are further described in the following section.

In addition to reliability, the second foundation property of interest to us is *versatility*. A versatile foundation system can be used with various building floor and soil conditions. The third relevant foundation property is construction *cost*, self-explanatory.

3.5.2 Some Examples

Typically, transfer of downward gravity loads presents less of a problem with reliability than transfer of lateral column reactions. There are exceptions, of course, such as the single-pile system without grade beams. Here, a pile is placed directly under each column and a small pile cap. In theory, it is a very economical system for lightly loaded buildings, where even a single pile might have enough capacity to support the vertical column reactions. In practice, pile installation is subjected to tolerances of several inches.

It means that there will be inevitable eccentricities between the column and the pile, leading to a potential pile overstress and even failure. This system is an example of the foundation that requires essentially a perfect installation.

A much more reliable system has at least *three* piles under each column, even if a single pile would have sufficient capacity. With this pile tripod the placement tolerances in two directions can be readily accommodated, but of course three times as many piles are now required. Another possible solution has a rectangular pile cap with two piles under each column and with rigid grade beams placed perpendicular to the long direction of the pile caps. This system can also accommodate the column eccentricities in two directions. The two- and three-pile systems cost much more than a single-pile foundation but offer greater reliability. (See further discussion on this topic in Chap. 9.)

An example of a redundant system is the moment-resisting foundation with perimeter grade beams described in Chap. 7. This foundation system can transfer lateral loading to the soil by both the friction between the foundation and the soil and by the passive pressure of the grade beam against the soil. By contrast, an isolated exterior footing without grade beams has less redundancy.

An example of lack of survivability? How about a hairpin system (Chap. 6)? Here, the slab on grade is relied on to transfer the horizontal column reactions from one side of the building to another. Were the slab to be cut at any time (for pipe maintenance, and so on), the system would cease to function, without any alternate load path.

To illustrate the concept of versatility, consider a tie-rod system (Chap. 6). Its versatility is limited, because tie rods cannot be used in buildings with trenches, deep floor depressions, and pits. Another less-versatile system is the mat foundation (Chap. 8), which might be difficult to use in the areas with deep frost lines.

Table 3.1 provides a snapshot of the comparative reliabilities, degree of versatility, and cost for several foundation systems commonly used to support pre-engineered buildings.

Foundation System	Cost	Reliability	Versatility
Tie rod	Low to high	Low to high	Low to medium
Hairpins and slab ties	Low	Low	Low
Moment-resisting foundation	High	High	High
Slab with haunch	Medium	Low to high	Low to medium
Trench footing	Medium to high	High	High
Mat	High	High	Low
Deep foundations	High	High	High

Source: Author.
Note: The ratings reflect the specific designs discussed in Chaps. 6 to 9.

TABLE 3.1 Comparative Cost, Reliability, and Degree of Versatility of Common Foundation Systems for Metal Building Systems

3.6 Column Pedestals (Piers)

3.6.1 The Area Inviting Controversy

The area immediately below the column base plate is where the design responsibility of the metal building manufacturer typically ends (MBMA, 2006). The foundation designer is generally responsible for everything that lies below, the concrete on which the column bears and the anchorage securing the column to the concrete.

There may or may not be a grout pad under the column base plate; in most cases, there is no grout. Why not? Grouting is excluded from the scope of work of the metal building manufacturer per MBMA Common Industry Practices (MBMA, 2006). The typical base plate details of many metal building manufacturers reflect this approach and do not show any grout underneath. The owners and general contractors usually do not see a need to add something that is not required by those details.

The result? Without a grout pad, the column simply rests on top of the foundation concrete, which in all likelihood is not perfectly level. In many cases the concrete surface is uneven, and the base plate bears on it only partially. This can lead to some rotation ("play") of the column base under load (see a related discussion in Sec. 3.3.2). In extreme cases the gaps under some parts of the base plate are quite large, as further discussed and illustrated in Newman (2004).

To be sure, grouted column bases under metal building frame columns are used in some circumstances, such as at the columns supporting overhead cranes. However, we believe that grouted bases should be used much more often and recommend specifying them for most applications.

Anchor bolts typically attach the metal building column to the foundation. The design of anchors is complex (see Chap. 10). Column piers are yet another area inviting controversy and misunderstanding, as discussed next.

3.6.2 Two Methods of Supporting Steel Columns in Shallow Foundations

There are two basic methods of supporting a metal building column on a shallow foundation. One is to place the column directly on top of the foundation (Fig. 3.6) and another, on a concrete pedestal (Fig. 3.7). As just discussed, in both cases there may or may not be a grout pad under the base plates.

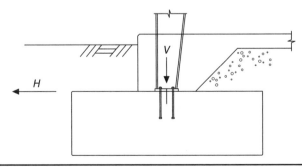

FIGURE 3.6 Column placed directly on top of the footing.

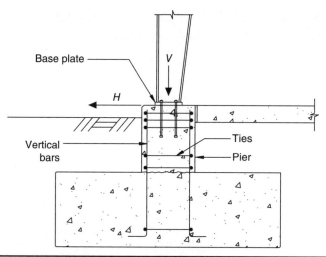

Figure 3.7 Column placed on a pedestal on top of the footing.

These pedestals are short concrete column-like structures extending from the top of the footing or other shallow foundation to the bottom of the column base plate. These piers, as they are frequently called, should not be confused with the deep foundation piers (caissons) described in Chap. 9.

Both methods of column support have their advantages and disadvantages. One advantage of placing the column directly on top of the footing is that it minimizes the distance between the point of application of the horizontal column reaction and the bottom of the footing. This reduces the overturning moment on the foundation. One disadvantage: Some degree of column fixity could result if the slab on grade envelops the column base, as in Fig. 3.6. This fixity might overstress the footing not designed with this assumption in mind, as discussed in Sec. 3.3.

Also, unless an isolation joint is provided around the column, any column movement under wind load or temperature fluctuations could cause cracking in the floor slab that envelops the column. Yet another disadvantage of this design is the fact that the column, base plate and anchors are all located below the ground level and thus exposed to corrosion. Corrosion could be caused by groundwater or floor-washing water seeping through the cracks in the slab, between the slab and the footing, or around the column.

Placing the column atop a concrete pedestal reduces corrosion concerns. In industrial buildings the pedestals often protrude above the slab, so that any water on the floor does not enter the area around the base. Another advantage: The column can be placed on the pedestal after all concrete work is done, while in the direct-on-footing installation some of the slab must be placed after the steel is installed. Concrete work requiring multiple workers' trips to the site tends to be more expensive. Also, an elevated column pedestal provides some protection against column damage by moving equipment.

The obvious disadvantage of using the piers is the increased distance between the point of application of the horizontal column reaction and the bottom of the footing. This increases the overturning moment on the foundation.

The situation is simpler in deep foundations. Here, the columns are generally placed on top of pile or deep-pier caps.

3.6.3 Establishing Sizes of Column Pedestals (Piers)

The first step in designing a column pedestal is to establish its dimensions. This effort involves some challenges.

It is relatively easy to establish the height, once the designer decides where the bottom of the column is. In many metal building systems the bottom of the column base plate is at the top of the floor (again, with or without a grout pad in between). Alternatively, as just discussed, the top of the pier could be elevated above the slab if the floor might become wet in service.

Establishing the cross-section of the pier is more complicated. It involves two separate issues:

1. The pedestal must be large enough for a comfortable placement of the steel column on it. This is easily done when the column size is known in advance. But what about a scenario when the foundation designer must estimate the column reaction and design the foundations before the metal building manufacturer is selected (see Secs. 3.1 and 3.2)? In that case not only the frame reactions but also the column sizes are unknown at the time the pedestal is designed. As with the column reactions, the sizes of the column base plates need to be estimated. While most major manufacturers keep the base plate depth (measured in the direction parallel to the frame) modest, such as 9 to 12 in., others might use much larger sizes.

2. The pedestal must be long and wide enough to allow the development of the anchor bolts or other embedments that transfer the column reactions to the foundation. The edge distances of anchor bolts are particularly important in this respect. Therefore, the pier's footprint could be controlled by the design of the embedments, as discussed in Chap. 10.

Many foundation designers feel that if they specify a pedestal with a large cross-sectional area, they would be wasting concrete. (Remember the old joke about an engineer's take on whether the glass was half-full or half-empty? The engineer said the glass was twice as big as it needed to be.) However, nobody is happy when the building columns end up being wider than their supports (Fig. 3.8).

This has happened even to the author of this guide! From then on the author started adding a certain note on his foundation drawings, whenever the metal building manufacturer was not yet selected at the time of the foundation design. The note required the building column base plates not to exceed the specified dimensions and to fit on the piers. As a practical matter, a 2 ft by 2 ft pier should suffice for many applications, but the design of anchor bolts or other embedments might still require a larger size.

3.6.4 Minimum Reinforcement of Piers

Once the cross-section of the pedestal is established, its reinforcement can be designed. In most cases vertical pier reinforcement is not required for resisting downward loading, but it might be needed for resisting uplift, as explained later. Providing some minimal vertical pier reinforcement and ties makes sense in most circumstances. The *Building Code Requirements for Structural Concrete and Commentary*

Figure 3.8 This unfortunate situation could have been avoided if the concrete support were larger.

(ACI 318-08), Paragraph 10.9.1, prescribes the limits for reinforcement of compression members as follows:

- The minimum area of longitudinal reinforcement = $0.01A_g$
- The maximum area = $0.08A_g$

where A_g = the gross area of the member

The minimum reinforcement ratio of $0.01A_g$ has been in every edition of ACI 318 since 1936, but it might be excessive for lightly loaded column piers. In this case the code (Paragraph 10.8.4) allows the minimum reinforcement percentage to be based on a reduced cross-section of the compression member. The reduced section should not be less than one-half of the gross area. Therefore, the minimum reinforcement ratio of lightly loaded column piers is $0.005A_g$. This provision is echoed by ACI 318-08 Paragraph 15.8.2.1 that requires the area of reinforcement in a pedestal across the interface with the footing (i.e., dowels or vertical bars extending from the footing into the pedestal) to be at least $0.005A_g$.

ACI 318 Paragraph 10.8.4 contains provisions for designing compression members built monolithically with the wall, as might occur when the pedestal forms a part of an exterior foundation wall. According to these provisions, the outer limits of such members shall not exceed the distance of 1.5 in. outside the lateral pier reinforcement dimensions.

What about the size and spacing of this lateral pier reinforcement? According to ACI 318 Paragraph 7.10.5, ties in the compression members shall be at least No. 3 bars

Pedestal Size (Cross Section), in.	Gross Area of Vertical Bars, sq. in.	Vertical Bar Options*	No. 3 Tie Spacing, in.†‡
24 × 24	2.88	4 #8	16
		6 #7	14
22 × 22	2.42	4 #7	14
		6 #6	12
20 × 20	2.00	4 #7	14
		6 #6	12
18 × 18	1.62	4 #6	12

Source: Author.
Notes:
*Min. area of vertical bars to satisfy $0.005A_g$.
†No. 4 ties can also be used, same spacing.
‡In all cases, 16 longitudinal bar diameters controls spacing.

TABLE 3.2 Minimum Reinforcement of Some Common Column Pedestal Sizes Used in Foundations for Metal Building Systems

for longitudinal bars No. 10 and smaller, and No. 4 for larger bars. The vertical spacing of the ties shall not exceed:

- 16 longitudinal bar diameters
- 48 tie bar diameters
- Least dimension of the pier

The first tie above the footing must be placed not more than one-half of this maximum vertical spacing. At the top of the pedestal, additional ties are needed, if anchor bolts are placed there. The additional ties shall consist of at least two No. 4 bars or three No. 3 bars, distributed within 5 in. of the top of the pedestal. Table 3.2 provides minimum reinforcement of some common column pedestal sizes used in foundations for metal building systems.

Concrete cover, measured to the ties, is typically specified as 3 in., although a 2-in. and even a 1.5-in. cover might be acceptable for the piers placed in forms, rather than against the soil (see ACI 318, Paragraph 7.7.1). The minimum number of vertical bars is four for compression members with rectangular ties, as is typical in column pedestals. A basic cross-section of the column pedestal with four vertical bars is shown in Fig. 3.9.

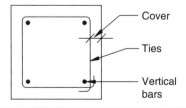

FIGURE 3.9 Typical cross-section of a column pedestal with four vertical bars.

References

2009 International Building Code® (IBC), International Code Council, Country Club Hills, IL, 2009.

ACI 318-08, *Building Code Requirements for Structural Concrete and Commentary*, American Concrete Institute, Farmington Hills, MI.

AISC 360-05, *Specification for Structural Steel Buildings*, American Institute of Steel Construction, Chicago, IL, 2005.

Bachman, Robert E., et al., *Seismic Design Guide for Metal Building Systems*, International Code Council, Country Club Hills, IL, 2008.

CRSI Design Handbook, Concrete Reinforcing Steel Institute, Schaumburg, IL, 2002.

Kleinlogel, Adolf, *Rigid Frame Formulas*, Frederick Ungar Publishing Co., New York, 1964.

Metal Building System Manual, Metal Building Manufacturers Association (MBMA), Cleveland, OH, 2006.

Metal Building Systems, 2d ed., Building Systems Institute, Inc., Cleveland, OH, 1990.

NAVFAC DM-7.2, *Foundations and Earth Structures*, Department of the Navy, Naval Facilities Engineering Command, Alexandria, VA, 1982.

Newman, Alexander, *Metal Building Systems: Design and Specifications*, 2d ed., McGraw-Hill, New York, 2004.

TI 809-30, *Metal Building Systems*, U.S. Army Corps of Engineers, 1998.

CHAPTER 4
Design of Isolated Column Footings

4.1 The Basics of Footing Design and Construction

4.1.1 Basic Design Requirements

The isolated column footing is a staple of conventionally framed buildings. These foundations are also used in metal building systems (MBS), most commonly at the interior columns of multiple-span primary rigid frames. Their design is rather straightforward, which makes for a good starting point in our exploration of various foundation systems used in pre-engineered buildings. The design basics described here will be helpful in our later discussions dealing with more complicated foundations.

Isolated column footings, also known as spread or individual footings, are typically square in plan, but could also be rectangular if needed. These foundations are generally made of reinforced concrete, but plain-concrete footings may be used in minor structures when allowed by the building codes. The general and specialized concrete codes contain the basic provisions for footing design and construction. The authoritative concrete code is ACI 318, *Building Code Requirements for Structural Concrete and Commentary* (ACI 318-08) published by the American Concrete Institute.

Chapter 3 discusses some of the design requirements for isolated footings, such as those related to their minimum depth. A few additional provisions are contained in Section 1809 of the *2009 International Building Code®* (IBC-09), as follows:

- The top surface of a footing must be level, but the bottom may slope; the maximum slope is 10 percent. Where a larger slope is required, the footing should be stepped.
- The footings adjacent to slopes greater than 33.3 percent require special consideration regarding their setbacks, elevations, and clearances (see IBC-09 Section 1808.7).
- The minimum width of footings is 12 in.
- Concrete used in foundations of structures assigned to Seismic Design Categories (SDCs) A, B, or C must have a minimum specified 28-day compressive strength (f'_c) of 2500 psi. For nonresidential structures assigned to SDCs D, E, or F the minimum f'_c is 3000 psi.
- Certain spread footings in zones of high seismicity must be interconnected by ties (see discussion in Sec. 4.1.3).

4.1.2 Construction Requirements

In addition to the construction-related publications of the American Concrete Institute, much information on the topic of foundation construction can be found in IBC-09. We will mention just three of the relevant IBC provisions:

- Placing of concrete through water is prohibited, unless special approved construction methods such as a tremie are used, and steps are taken to minimize concrete segregation and turbulence of water. Water should not be allowed to flow through fresh concrete.
- Concrete footings require protection from freezing during placement and for at least five days afterward.
- When approved by the building official, concrete footings can be placed directly against the soil, without formwork. When required, concrete forming should conform to Chapter 6 of ACI 318.

4.1.3 Seismic Ties

Spread footings in buildings assigned Seismic Design Categories D, E, or F and founded on soils with Site Classes E or F must be interconnected by ties. The ties must be capable of resisting a tension or compression force equal to the lesser of:

1. Vertical load carried by the larger of the footings multiplied by the seismic coefficient S_{DS} and divided by 10. (S_{DS} is the design spectral response acceleration at short periods and 5-percent damping, as described in IBC.)
2. Twenty-five percent of the design gravity load of the smaller footing.

Alternatively, it must be shown that an equivalent restraint is provided by reinforced concrete slabs on grade or by reinforced concrete beams located within reinforced slabs on grade.

4.1.4 Reinforced-Concrete Footings

Chapter 15, Footings, of ACI 318-08 contains specific design requirements for isolated column footings. Many popular reference books, such as *CRSI Design Handbook*, contain tables of reinforced square footings far various allowable soil-bearing capacities. The general design procedures are described in Sec. 4.2.

According to the long-running provisions of ACI 318, the minimum thickness of a reinforced footing on soil is 6 in. above the bottom reinforcement. Since Section 7.7 of ACI 318 specifies a minimum concrete cover of 3 in. for bottom reinforcing bars in concrete cast against and permanently exposed to earth, the minimum footing thickness becomes 9 in. plus the diameter of the bottom reinforcement, or about 10 in. total.

4.1.5 Plain-Concrete and Other Footings

Chapter 22, Structural Plain Concrete, of ACI 318-08 describes three specific conditions in which plain concrete may be used. They are:

1. Members continuously supported by soil or other structural members
2. Members in compression resulting from arching action under all loading conditions
3. Walls and pedestals (but not columns)

Accordingly, footings and pedestals used in metal building systems and meeting these requirements may be made of plain concrete. The minimum thickness of plain-concrete footings is 10 in. ACI 318 requires that the design thickness of these footings be taken 2 in. less than the actual thickness when they are cast against the soil (the typical case, of course). This sometimes overlooked provision means that a 12-in.-thick footing made of plain concrete must be analyzed as if it were only 10 in. thick.

For plain-concrete pedestals the maximum height-to-thickness ratio is three (ACI 318-08 Paragraph 22.8.2). Foundations made of plain concrete are not permitted in Seismic Design Categories D, E, and F, with an exception granted to the residential foundations meeting certain conditions and to wall footings.

Section 1809 of IBC-09 describes design and construction requirements for some other types of column footings. These include footings of masonry units, steel grillages, and wood. A present time, none of these column footings is commonly used in the United States for the support of metal building systems.

4.1.6 Nominal vs. Factored Loading

The building codes differentiate between nominal and factored structural loads. *Nominal* (sometimes called "service") loads are those specified in the codes (such as IBC Chapter 16 or ASCE 7). These are the maximum loads likely to be experienced by the structure within a certain degree of probability. Nominal loads are used in proportioning foundations—establishing their sizes, checking stability, and calculating settlements.

Nominal loads can be used directly in allowable stress design (ASD) method. When various nominal loads are combined as prescribed by the governing code, the computed stresses should be less than the maximum allowable stresses stipulated in the code. The load combinations using allowable stress design can be found in Section 1605.3 of IBC. Two sets of load combinations are included: *basic* and *alternative basic*, each with slightly different methods of combining loads.

Factored loads are those where nominal loads are multiplied by load factors. Load factors account for uncertainties of the actual load levels, a simplified nature of analysis, and a possibility that more than one extreme load occurs at the same time (IBC-09). The load factors associated with the loads of which we are relatively certain are lower than those assigned to the more variable loads. For example, the dead load is multiplied by a load factor of 1.2, while snow has a load factor of 1.6. Factored loads are used in strength design methods, such as contemporary concrete design codes.

The basic strength design equation stipulates that the "ultimate" (factored) loading effect U should be less than the design strength of the structural member. The design strength is computed as the nominal strength of the member multiplied by the strength reduction factor ϕ (phi) assigned to the limit state being checked:

$$\phi(\text{Nominal strength}) \geq U$$

At present time, the model codes and specialized codes share the same load factors for the same loads. This was not always the case. For example, the 1999 and earlier editions of ACI 318 had the load factors of 1.4 for dead and 1.7 for live load. In 2002 edition of ACI 318, these factors were changed to 1.2 and 1.6, respectively, to comply with the load combinations of ASCE 7-02. The corresponding strength reduction factors were changed as well. The previously used load factors have been moved to ACI 318 Appendix C.

The subsequent editions of ACI 318 introduced some minor revisions to the Appendix C load factors and the corresponding strength reduction factors. The designers may still use Appendix C, but there is rarely a need to do so.

4.2 The Design Process

4.2.1 General Design Procedure

The basic procedure for the design of isolated column footings is as follows:

1. Compute the design load on the footing, in terms of both nominal (unfactored) and factored loading.
2. Determine the required footing area by dividing the unfactored load by the allowable bearing pressure of the soil.
3. Decide on presence and size of the column pedestal (pier), following the discussion in Chap. 3.
4. Determine the critical sections for moment and shear (see the discussion in Secs. 4.2.5 and 4.2.6) and the design length of the cantilevered footing ledges.
5. Establish a preliminary thickness of the footing from previous experience, design tables found in the engineering reference sources, or trial and error.
6. Check the trial section for punching shear and beam-type shear, refine the footing thickness if needed, then design the footing for flexure.

4.2.2 Using ASD Load Combinations

The required footing area for downward loading is found by dividing the maximum unfactored load by the allowable bearing pressure of the soil. The maximum load is the largest effect of all applicable combinations using the allowable stress design method.

The load combinations of IBC Section 1605.3 are listed as follows, separately for basic and alternative basic combinations. These are used for stability calculations and for finding the soil pressure. The *basic* ASD load combinations are:

$$D + F$$

$$D + H + F + L + T$$

$$D + H + F + (L_r \text{ or } S \text{ or } R)$$

$$D + H + F + 0.75(L + T) + 0.75(L_r \text{ or } S \text{ or } R)$$

$$D + H + F + (W \text{ or } 0.7E)$$

$$D + H + F + 0.75(W \text{ or } 0.7E) + 0.75L + 0.75(L_r \text{ or } S \text{ or } R)$$

$$0.6D + W + H$$

$$0.6D + 0.7E + H$$

The *alternative basic* ASD load combinations are:

$$D + L + (L_r \text{ or } S \text{ or } R)$$
$$D + L + \omega W$$
$$D + L + \omega W + S/2$$
$$D + L + S + \omega W/2$$
$$D + L + S + E/1.4$$
$$0.9D + E/1.4$$

where D = dead load
  F = load caused by fluids with well-defined pressures and heights
  H = load from horizontal soil pressure, or pressure from ground water or bulk materials
  L = live load (except roof live load), including any allowed reductions
  L_r = roof live load, including any allowed reductions
  S = snow load
  R = rain load
  T = self-straining force from expansion or contraction due to temperature or moisture changes, shrinkage, creep, differential settlement, or a combination of these
  W = wind load
  E = earthquake load effects (horizontal and vertical)
  ω = wind load coefficient taken as 1.3 or 1.0, depending on the procedure used to compute the load

4.2.3 Using Load Combinations for Strength Design

Strength design load combinations are contained in Section 1605.2 of IBC-09, as listed here. These are used for concrete design. The notations are the same as those in Sec. 4.2.2.

$$1.4(D + F)$$
$$1.2(D + F + T) + 1.6(L + H) + 0.5(L_r \text{ or } S \text{ or } R)$$
$$1.2D + 1.6\,(L_r \text{ or } S \text{ or } R) + (f_1 L \text{ or } 0.8W)$$
$$1.2D + 1.6W + f_1 L + 0.5(L_r \text{ or } S \text{ or } R)$$
$$1.2D + 1.0E + f_1 L + f_2 S$$
$$0.9D + 1.6W + 1.6H$$
$$0.9D + 1.0E + 1.6H$$

The coefficients f_1 and f_2 are explained in the code. Note that ACI 318-08 has similar load combinations, but they are written in a slightly different format.

4.2.4 What Is Included in the Dead Load?

As just stated, the total nominal (unfactored) load on the footing consists of various service loads (dead, live, and so on), combined in various code-prescribed load combinations. When appropriate, live-load reductions may be taken.

But what about the weight of the footing itself: Should it be included in the dead load? What of the soil on top of the footing ledges? Surprisingly, the answers are somewhat subjective.

Since the footing displaces the same volume of soil, some engineers feel that the weight of concrete need not be included. Or perhaps only the difference between the unit weights of concrete and the soil should be counted toward the increased pressure on the bottom of the footing. However, this logic applies only to the situations where a footing is placed in virgin soil.

When structural fill is used, the soil at the bottom of the footing has not experienced any precompression from the weight of the soil above, and the weight of the footing should now be included. At the time the foundation is designed, it may not be clear whether or not fill will be used at the site. Thus it might be wise to conservatively assume it *will* be used and to include the weight of the footing and the soil above in the dead load.

Perhaps reflecting the uncertainty, the related provisions of the International Building Code have undergone slight changes in recent years. Paragraph 1805.4.1.1 of IBC-03 directs the designer to include the weight of foundations, footings, and the overlying fill in the design loading. Paragraph 1805.4.1.1 of IBC-06 and Section 1808.3 of IBC-09 both simply *permit* the inclusion of these weights as part of the dead load.

Some designers expand the argument and ask whether the live load acting on the slab on grade should also be included in the foundation loading. The code is silent on the matter. Since the live load on the slab does not cause flexure in the footing, this load (as well as the weight of the footing and of the overlying fill) certainly need not be included in the column load used in concrete design. Instead, if considered at all, these loads will be directly subtracted from the allowable soil pressure under the footing, reducing the allowable pressure accordingly.

4.2.5 Designing for Moment

Isolated footings are assumed to be rigid for the purpose of their analysis, and for concentrically loaded isolated footings the soil pressure is uniform (*PCA Notes*). According to ACI 318-08 Section 15.4, the maximum bending moment at the column footing is found by passing a vertical plane through the footing at the critical section for moment. The location of the critical section for moment depends on whether a concrete pedestal is used under the column or the column is bearing directly on the footing (Fig. 4.1):

1. If the steel column bears on top of a concrete pedestal, the critical section for moment is located at the face of the pedestal (Fig. 4.1*a*).

2. If the steel column with a base plate bears directly on top of the footing, the critical section for moment is located halfway between the face of the column and the edge of the base plate (Fig. 4.1*b*).

These provisions assume that the steel column and base plate are square or rectangular in plan. According to ACI 318-08 Section 15.3, for the purpose of locating critical sections for moment and shear, circular or polygon-shaped concrete pedestals can be treated as square members of the same area. No explicit guidance is given about the treatment of round *steel* columns with square base plates bearing directly on footings, but presumably a similar approach should work.

Design of Isolated Column Footings

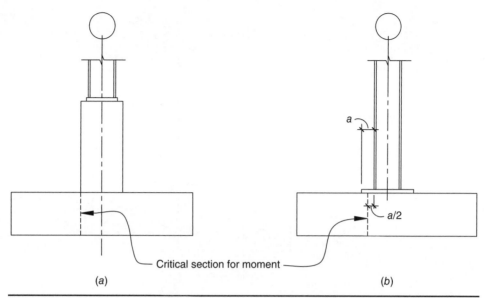

FIGURE 4.1 Critical sections for flexure in isolated footings: (a) Steel column bearing on a concrete pedestal supported by a footing; (b) steel column bearing directly on the footing.

Once the critical section for moment is located, the length L of the footing cantilever subjected to bending can be determined (Fig. 4.2). The maximum bending moment M_u per foot of footing width is then computed as:

$$M_u = q_u L^2/2$$

Here, q_u is the *factored* soil pressure under the footing, found from the strength design load combinations. In the basic equation, the nominal flexural strength of concrete M_n multiplied by the strength reduction factor ϕ must be equal or exceed the factored moment:

$$\phi M_n \geq M_u$$

The strength reduction factor ϕ is 0.9 for flexure in reinforced concrete and 0.6 for flexure in plain concrete. The design of concrete for flexure can become quite involved. In the absence of appropriate computer software the reader is referred to *ACI Design Handbook*, *PCA Notes*, and *CRSI Design Handbook* for helpful charts and tables that can significantly reduce the design time.

4.2.6 Designing for Shear

Design provisions for shear in footings are addressed in ACI 318-08 Section 11.11. Two types of shear strength are considered in the design of isolated footings: wide-beam action and two-way action. Wide-beam action shear strength is computed assuming the footing behaves as a wide beam with cantilevered ledges. It is similar to shear in a reinforced-concrete cantilevered beam, except that the shear stress in the footing is

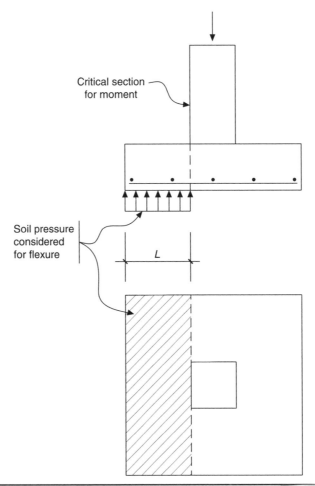

FIGURE 4.2 The length L of the footing cantilever subjected to bending.

typically resisted by concrete alone. In two-way action shear, the column or pedestal attempts to "punch" through the footing, hence another name for this limit state, punching shear.

The critical section for beam-action shear is located the distance d (the distance from the extreme compression fiber to the centroid of the tension reinforcement) from the critical section for moment. It can be found by drawing a line that extends at 45° from the top of the footing at the critical section for moment until it intersects the bottom reinforcement (Fig. 4.3a). The critical section for wide-beam action shear extends through the whole width of the footing. The equation for beam-type shear strength is:

$$\phi V_n \geq V_u$$

Design of Isolated Column Footings

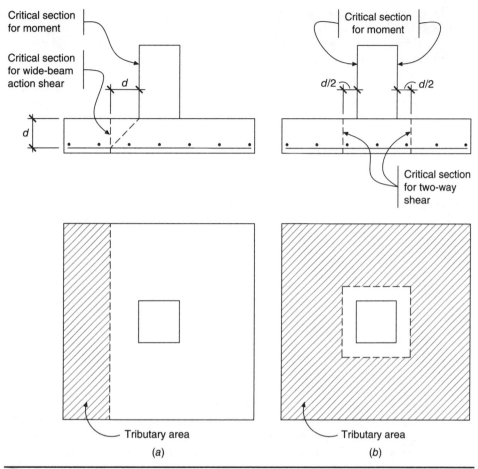

FIGURE 4.3 Critical sections for shear: (a) Wide-beam action shear; (b) two-way action (punching) shear.

When the footing is subjected to flexure and shear only, the factored shear loading is carried solely by concrete ($V_n = V_c$). Assuming that concrete is of normal weight, the beam-type shear strength V_n (which is equal to V_c) can be computed as:

$$V_c = 2\lambda\sqrt{f'_c}b_w d$$

where f'_c = 28-day compressive strength of concrete
 b_w = the width of the footing
 d = the distance from the extreme compression fiber to the centroid of the tension reinforcement
 λ = a coefficient accounting for reduced strength of lightweight concrete

λ is equal to 0.85 for sand-lightweight concrete and 0.75 for all-lightweight concrete. For normal weight concrete typically used in footings $\lambda = 1$.

In some circumstances, a more complex formula given by ACI 318-08 Eq. 11-5 can be useful.

The maximum two-way action shear stresses occur in the vicinity of the column or pedestal. The critical section for two-way action shear is located at a distance $d/2$ from the critical section for moment on all four sides (Fig. 4.3b). The concrete area that resists two-way action shear is the product of the perimeter of the critical section b_o and the depth d. According to ACI 318-08, the two-way action concrete shear strength for footings V_c is the smallest of the following three values:

$$V_c = \left(2 + \frac{4}{\beta}\right)\lambda\sqrt{f'_c}b_o d$$

$$V_c = \left(\frac{\alpha_s d}{b_o} + 2\right)\lambda\sqrt{f'_c}b_o d$$

$$V_c = 4\lambda\sqrt{f'_c}b_o d$$

where β = the ratio of long side to short side of the column or rectangular pedestal
α_s = 40 for interior columns, 30 for edge columns and 20 for corner columns

Other symbols are explained earlier. It can be seen that using rectangular columns might decrease punching shear capacity of concrete. The strength reduction factor ϕ is 0.75 for shear in reinforced concrete and 0.6 for shear in plain concrete.

4.2.7 Minimum Footing Reinforcement

According to ACI 318-08 Section 15.10.4, reinforcement in each principal direction needs to be sufficient to meet the same minimum reinforcement ratio as required for slabs on grade (0.0018 of gross concrete area). This reinforcement may be placed near the top or bottom of the footing, or at both locations (see Commentary to ACI 318-08 Section 15.10.4). The maximum bar spacing is 18 in.

For square footings, it is typical to provide the same reinforcement in both directions at the bottom. A question commonly asked is: Where exactly is the effective depth to reinforcement d measured to? Is it from the top of the footing to the centerline of the bars in the lowest layer? Upper layer? Midway between the two?

The practices vary, but measuring midway between the two layers seems like a reasonable compromise. This "average effective depth" approach is used in *CRSI Design Handbook*. Under this method the effective depth is equal to the footing thickness minus one bar diameter. Measuring to the centerline of the bars in the upper layer is of course always conservative.

4.2.8 Distribution of Reinforcement in Rectangular Footings

In reinforced-concrete footings, the distribution of reinforcement is prescribed by ACI 318-08 Section 15.4. For square footings reinforcement is distributed uniformly in both directions.

For rectangular footings reinforcement is distributed uniformly in the long direction across the width of the footing, but nonuniformly in the short direction. In short direction, a certain fraction of the total reinforcement A_s is placed in the band located at the center of the footing. The fraction that must be placed in the center band is equal to $\gamma_s A_{s'}$ where

$$\gamma_s = \frac{2}{(\beta+1)}$$

Here, β is the ratio of the lengths of the long to short side of the rectangular footing. The remaining reinforcement placed in the short direction is distributed uniformly over the two areas on each side of the center band.

In an attempt to simplify reinforcement placement in the field, some engineers determine the size and spacing of reinforcement in the central band as required by code, and then continue that spacing throughout the footing. This uses more reinforcing steel but simplifies rebar placement and helps avoid any associated field problems.

4.2.9 Designing for Uplift

Some load combinations involving lateral forces may result in the net uplift loading on the footing. Performing a stability check for uplift is discussed in Chap. 3, and Fig. 3.2 shows the components of the "ballast" provided by the weights of the footing, the column pedestal (if any is present), and the soil on the footing ledges. We recommend neglecting all other contributing factors such as the soil "wedges" shown on Fig. 3.2 and any flexural restraint that might be provided by the slab on grade. Indeed, we suggest *reducing* the dead load by the amount of the probable buoyancy force in the areas with high water levels or subject to flooding.

Design of column footings for this condition is not well covered in the building codes. Conceptually, the design approach is the same as for downward loading. Figure 4.4 illustrates an isolated footing with pedestal under uplift loading. The critical

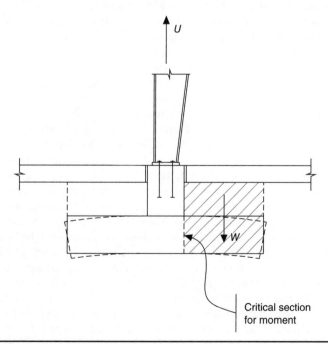

Figure 4.4 Isolated footing resisting uplift loading.

section for moment is at the face of the pedestal, and it is conservative to take critical section for wide-beam action shear at that location as well. The critical section for two-way action (punching) shear can be taken at the same location as for downward loading. A punching-shear failure caused by uplift loading might occur in some thin and large footings, where the upward design load exceeds the downward one.

What about the columns bearing directly on spread footings? When the design wind uplift force is acting, the isolated footing is literally hanging from the column by the anchor bolts. For this situation it is prudent to take the critical sections for moment and shear at the center of the footing.

At the bottom of the pedestal, the load path for uplift continues through the vertical pier reinforcement, which should be extended into the footing and developed there with properly embedded hooks. The uplift loading is generally low enough to allow the minimum pier reinforcement to accomplish this task, but this check should still be made.

Concrete is not commonly designed for tension, but the building codes cover some conditions where tension is considered, such as in brackets and corbels. According to ACI 318-08 Section 11.8, the design strength of tension reinforcement must exceed the applied factored tensile force N_{uc}:

$$\phi A_n f_y \geq N_{uc}$$

where $\phi = 0.75$
A_n = area of steel reinforcement

The factored tensile force N_{uc} is computed from the appropriate load combination, such as $0.9D + 1.6W$.

4.2.10 Reinforcement at Top of Footings

Can an isolated footing reinforced only at the bottom resist the bending moment caused by wind uplift? For moderate uplift loads it is often possible to make the plain-concrete section thick enough to resist the bending. But is the plain-concrete behavior acceptable in this situation? In other words, must the top surface of the footing be reinforced even if plain-concrete flexural capacity is adequate?

The isolated footing is among the structures where ACI 318-08 Section 22.2 specifically allows using structural plain concrete for members "continuously supported by soil." The main function of the isolated footing is to transmit the downward-acting loads to the soil during almost the entire useful life of the building. An argument can be made that a rare occasion when a net uplift loading exists could be considered an incidental use of the foundation that ordinarily is continuously supported by soil. According to this argument, assuming that the plain-concrete section is strong enough to resist the bending caused by uplift forces, no top reinforcement is needed.

Of course, the opposite argument can also be made that during the uplift the footing is not continuously supported by soil. Therefore, the footing must be reinforced at the top.

In the absence of clear code provisions both arguments have merit, and it is certainly conservative to reinforce the top of any isolated column footing subjected to net uplift. Some designers specify top reinforcement as standard practice.

We should note that placing reinforcing bars at the top of the footing requires special high chairs or similar bar support elements that are not needed for bottom reinforcement, where inexpensive short chairs of even concrete bricks can be used. Large and relatively thin column footings subjected to high net uplift forces would benefit the most from reinforcing their top surfaces.

Design Example 4.1: Column Footing

Problem Design the isolated column footing to support an interior column of the single-story multiple-span rigid frame. The spacing of the interior columns within the frame is 60 ft, and the spacing between the frames is 25 ft. The following loads act on the roof: 3 pounds per square foot (psf) dead load, 30 psf design roof snow load, and 14 psf wind uplift. The depth of the footing must be at a minimum 3 ft below the floor. The column is supported by a 20 in. by 20 in. concrete pedestal, the top of which is flush with the floor slab. Use f'_c = 3000 psi and allowable soil bearing capacity of 4000 psf. Assume the average weight of the soil, slab on grade and foundation as 130 lb/ft³. The building is not located in the flood zone. Use IBC basic load combinations.

Solution Find the design loads on the column for a tributary area of (60 × 25) = 1500 ft².

The design dead load D = 3 × 1500 = 4500 lb or 4.5 kips (1 kip = 1000 lb).

The design snow load S = 30 × 1500 = 45,000 lb or 45 kips.

The design wind uplift load W = –14 × 1500 = –21,000 lb or –21 kips. (The negative sign denotes an upward-acting force.)

Total downward load $D + S$ = 4.5 + 45 = <u>49.5</u> kips.

Total uplift load on foundation $(0.6D + W)$ = 0.6 × 4.5 – 21 = <u>–18.3</u> kips.

Weight of the soil, slab on grade and foundation is 130 lb/ft³ × 3 ft = 390 lb/ft² or psf.

The net available soil pressure is 4000 – 390 = 3610 psf or 3.61 ksf (1 ksf = 1000 lb/ft²).

Required area of the footing for downward load is 49.5/3.61 = 13.71 ft².

∴ For downward load only, could use a 3.7 by 3.7 ft footing.

Check stability against wind uplift. Minimum required weight of the foundation, soil on its ledges and tributary slab on grade can be is found from:

$$0.6 D_{min,\, found} + W = 0$$

$$D_{min,\, found} = 18.3/0.6 = 30.5 \text{ kips or } 30{,}500 \text{ lb}$$

This corresponds to 30,500/130 = 234.62 ft³ of the average weight of "ballast."

With the depth of footing 3 ft below the floor, this requires the minimum square footing size of $(234.62/3)^{1/2}$ = 8.84 ft.

To reduce the footing size, try lowering the bottom of the footing by 1 ft. Then the minimum required square footing is $(234.62/4)^{1/2}$ = 7.66 ft.

To arrive at a round number, 8.0 by 8.0 ft footing, its depth must be:

$$234.62/(8)^2 = 3.66 \text{ ft. Use the footing depth of 3 ft 8 in., or 3.67 ft}$$

Recompute the weight of the soil, slab on grade and foundation as:

$$130 \text{ lb/ft}^3 \times 3.67 \text{ ft} = 477.1 \text{ psf}$$

The net available soil pressure is 4000 − 477.1 = 3522.9 psf = 3.52 ksf.

∴ Proceed to concrete design with 8.0 × 8.0 ft footing 3 ft 8 in. deep.

Note that if the designer were to consider only the downward load, the footing dimensions would have been less than half as large!

Load Combination 1: 1.2D + 1.6S (Downward-Acting Factored Loads)

$$P_u = 1.2(4.5) + 1.6(45) = 77.4 \text{ kips}$$

$$q_s = \frac{P_u}{A_f} = \frac{77.4}{64} = 1.21 \text{ ksf} < 3.52 \text{ ksf} \quad \text{OK}$$

The footing thickness is typically controlled by shear, so start concrete design with a shear check for a preliminary thickness established by trial and error or a reference look-up method. Try a footing with depth $d = 14$ in, which corresponds to approximately 18 in. overall footing thickness.

1. Two-Way Action Shear The critical section is located the distance $d/2$, or 7 in. from the faces of the 20 × 20 in. pedestal, so each side of the critical section is (20 + 7 + 7) = 34 in.

The critical perimeter $b_o = 4(34) = 136$ in.

Tributary area = $8 \times 8 - \frac{34^2}{144} = 55.97$ ft².

$V_u = q_s \times$ tributary area = $1.21 \times 55.97 = 67.72$ kips.

Find the smallest of three concrete punching-shear capacities:

(a) $V_c = \left(2 + \frac{4}{\beta}\right)\lambda\sqrt{f'_c}b_o d$

For a square column $\beta = 1$ and this equation does not control over third equation

(b) $V_c = \left(\frac{\alpha_s d}{b_o} + 2\right)\lambda\sqrt{f'_c}b_o d$

Using $\alpha_s = 40$ for interior column

$$V_c = \left[\frac{40(14)}{136} + 2\right](1)\sqrt{3000}(136)(14) = 637{,}987 \text{ lb} = 637.99 \text{ kips}$$

(c) $V_c = 4\lambda\sqrt{f'_c}b_o d$

$= 4(1)\sqrt{3000}(136)(14) = 417{,}146$ lb $= 417.15$ kips   Controls

$\phi V_c = 0.75 \times 417.15 = 312.86$ kips $> V_u = 67.72$ kips   OK

2. Wide-Beam Action Shear

Tributary area = $8(4 - 10/12 - 14/12) = 16.0$ ft²

$V_u = q_s \times$ tributary area $= 1.21 \times 16.0 = 19.36$ kips

$V_c = 2\lambda\sqrt{f'_c}b_w d$

$\phi V_c = 0.75 \times 2(1)\sqrt{3000}(96)(14) = 110{,}421$ lb $= 110.42$ kips

$\phi V_c = 110.42$ kips $> V_u = 19.36$ kips   OK

3. Flexure

For critical section at face of pier, $L = \tfrac{1}{2}(8 \text{ ft}) - (10 \text{ in.})/12 = 3.17$ ft.

$M_u = 1.21 \times 8 \times 3.17^2/2 = 48.64$ kip-ft

This moment is very small, so try minimum reinforcement first.

$A_{min} = 0.0018 \times 8 \text{ ft} \times 12 \text{ in.} \times 18 \text{ in.} = 3.11$ in.²

Try eight No. 6 bars each way ($A_s = 8 \times 0.44 = 3.52$ in²).
Center-to-center spacing of eight No. 6 bars with 3 in. clearance at the footing sides is:

$[8 \times 12 - 2 \times 3 - 6/8]/7 = 12.75$ in. < 18 in.   OK

Using Table 4.1, adapted from the information in *ACI Design Handbook* (the author finds the older versions of the *Handbook* to be most user-friendly for fast calculations), compute coefficient a_u as a function of ρ:

For $\rho = \dfrac{3.52 \text{ in}^2}{8(12)(14)} = 0.0026$ $a_u = 4.36$

$\phi M_n = A_s \times d \times a_u = 3.52 \times 14 \times 4.36 = 214.86$ kip-ft $> M_u = 48.64$ kip-ft   OK

Check development length of reinforcement.
The development length of reinforcement can be found by the ACI 318-08 formula:

$$l_d = \left[\dfrac{3 f_y}{40\,\lambda\sqrt{f'_c}}\dfrac{\Psi_t \Psi_e \Psi_s}{\left(\dfrac{c_b + K_{tr}}{d_b}\right)}\right]d_b$$

where $\Psi_t = 1.0$ (< 12 in. of concrete below bars)
 $\Psi_e = 1.0$ (uncoated bars)
 $\Psi_s = 0.8$ (bars less than No. 7)
 $\lambda = 1$ (normal weight concrete)
 $c_b = 3 + 6/16 = 3.375$ in. [distance from bottom of concrete to center of bar, controls over ½ bar spacing of ½(12.75) = 6.375 in.]
 $K_{tr} = 0$ (no transverse reinforcement)

74 Chapter Four

	ρ	
f'_c = 3000 psi	f'_c = 4000 psi	a_u
0.0010	0.0013	4.45
0.0015	0.0020	4.42
0.0020	0.0027	4.39
0.0025	0.0033	4.37
0.0030	0.0040	4.34
0.0035	0.0047	4.31
0.0040	0.0053	4.29
0.0045	0.0060	4.26
0.0050	0.0067	4.23
0.0055	0.0073	4.21
0.0060	0.0080	4.18
0.0065	0.0087	4.15
0.0070	0.0093	4.13
0.0075	0.0100	4.10
0.0080	0.0107	4.08
0.0085	0.0113	4.05

Source: Adapted from *ACI Design Handbook*, American Concrete Institute, Farmington Hills, MI 48331, 1971.
Notes:
(1) $\phi M_n = A_s \times d \times a_u$.
(2) Interpolation between listed values is acceptable.

TABLE 4.1 Coefficients for Flexural Design of Sections without Compression Reinforcement for f_y = 60,000 psi

$$\text{Check } \frac{c_b + K_{tr}}{d_b} = \frac{3.375 + 0}{0.75} = 4.5 > 2.5 \quad \therefore \text{use 2.5 in.}$$

$$l_d = \left[\frac{3}{40} \frac{60,000}{1\sqrt{3000}} \frac{1 \times 1 \times 0.8}{2.5} \right] 0.75 = 19.7 \text{ in.} > 12.0 \text{ in.}$$

l_d = 19.7 in. < available embedment length of (48 –10 – 3) = 35 in.   OK

The transfer of downward force from pier to footing is OK by observation, owing to trivial force level.

Load Combination 2: Dead Load and Wind Uplift   Since stability has already been checked, the design for this load combination focuses on checking the footing for moment and wide-beam shear in negative bending. While the load combination 0.9D + 1.6W might seem appropriate (H = 0 here), the check will consider the condition when the footing is suspended in the air and must resist its own weight plus the weight of the soil on its ledges. To maximize the load effects, the following load combination should be considered instead:

$$1.2D + 1.6W + f_1L + 0.5(L_r \text{ or } S \text{ or } R)$$

where the second, third, and fourth components are not acting. This approach follows the statement in ACI 318-08 Paragraph 9.2.1 that the effect of some loads in a load combination not acting simultaneously must be investigated. The factored dead load is then:

$$1.2 \times 477.1 \text{ psf} = 572.52 \text{ psf} = 0.572 \text{ ksf}$$

1. Two-Way Action Shear The punching-shear load, using the same critical section as the preceding, is

$$V_u = 0.572 \text{ ksf} \times \text{tributary area} = 0.572 \times 55.97 = 32.01 \text{ kips}$$

Noting that 32.01 < 67.72 kips from Load Combination 1, which has been checked already, we conclude that two-way action shear is OK.

2. Wide-Beam Action Shear As discussed previously in Sec. 4.2.9, in case of uplift it is conservative to take the critical section for wide-beam action shear at the face of the pedestal.

$$\text{Tributary area} = 8(4 - 10/12) = 25.33 \text{ ft}^2$$
$$V_u = q_s \times \text{tributary area} = 0.572 \times 25.33 = 14.49 \text{ kips}$$

Noting that 14.49 < 19.36 kips from Load Combination 1, which has been checked already, we conclude that wide-beam action shear is also OK.

3. Flexure For critical section at the face of pier, $L = \frac{1}{2}(8 \text{ ft}) - (10 \text{ in.})/12 = 3.17 \text{ ft}$.

$$M_u = 0.572 \times 8 \times 3.17^2/2 = 22.99 \text{ kip-ft}$$

This moment is very small, so try plain-concrete behavior first.

As discussed previously, the design depth of the 18-in. thick footing placed directly on soil is reduced by 2 to 16 in.

Section properties for 8-ft-wide 16-in.-deep footing are:

$S_m = bd^2/6 = 8(12)(16^2)/6 = 4096 \text{ in.}^3$

$A_g = 8(12)(16) = 1536 \text{ in.}^2$

$\phi M_n \geq M_u$

$\phi M_n = 0.6 \times 5\ \lambda \sqrt{f'_c} S_m = 0.6 \times 5 \times 1.0\ \sqrt{3000}(4{,}096) = 673{,}041 \text{ lb-in} = 56.09 \text{ kip-ft}$

$\phi M_n = 56.09 > M_u = 22.99 \text{ kip-ft}$ OK

Plain-concrete behavior would be sufficient, but to make the section act as reinforced concrete, we could provide top reinforcement.

Since minimum reinforcement (eight No. 6 bars each way) was provided at the bottom for downward loading, the same bars could also be placed at the top. The moment capacity of this reinforcement (ϕM_n = 214.86 kip-ft) is more than sufficient to resist M_u = 22.99 kip-ft from Load Combination 2.

A 2-in. minimum cover required for concrete permanently exposed to earth, but not cast against it.

Those seeking to reduce reinforcement requirements might recall that the minimum footing reinforcement (0.0018 times gross area of footing) may be divided between the top and bottom of the footing. Therefore, it might be possible to provide reinforcement of smaller size than No. 6 at top and bottom.

Try eight No. 4 bars each way top and bottom of the footing:

$$A_s = 8 \times 0.2 = 1.6 \text{ in}^2 \text{ top and bottom, for a total of } 1.6 \times 2 = 3.2 \text{ in}^2$$
$$A_s > A_{min} = 3.11 \text{ in}^2 \quad \text{OK}$$

Since bottom reinforcement controls, recheck eight No. 4 bars for that condition only, conservatively using the same d.

$$\text{For } \rho = \frac{1.6 \text{ in}^2}{8(12)(14)} = 0.0012 \quad a_u = 4.44$$
$$\phi M_n = A_s \times d \times a_u = 1.6 \times 14 \times 4.44 = 99.46 \text{ kip-ft}$$
$$\phi M_n > M_u = 48.64 \text{ kip-ft} \quad \text{OK}$$

∴ Can use eight No. 4 bars each way top and bottom of the footing.

The development length of smaller No. 4 bars is OK, since it was already checked for larger No. 6 bars.

The last design item is the pedestal (pier). Since the size is given, no horizontal load is present, and the gravity load is small, for Load Combination 1 the pedestal is OK by observation. For Load Combination 2 the pedestal must resist uplift, which is transferred through its vertical reinforcement. From Table 3.2, select one of the options for the minimum reinforcement of a 20 × 20 pedestal, which is four No. 7 bars with No. 3 ties at 14 in. o.c. (on center). Check these bars for adequacy to resist a factored tension load of $0.9D + 1.6W$:

$$N_{uc} = 0.9 \times 4.5 - 1.6 \times 21 = -29.55 \text{ kips}$$
$$\phi A_n f_y = 0.75 \times 4 \times 0.6 \times 60 = 108 \text{ kips} > 29.55 \text{ kips} \quad \text{OK}$$

As a final note, the footing size and depth are controlled by overall stability against uplift, and the section is relatively lightly loaded. It is possible to reduce the depth of the footing somewhat, but 18-in. thickness is quite reasonable for an 8-ft wide footing, and any possible material savings would be trivial. Figure 4.5 illustrates the final footing design. ▲

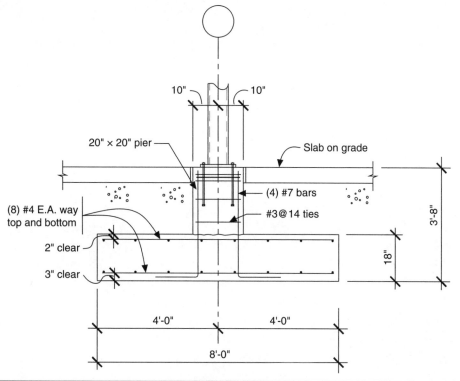

FIGURE 4.5 Isolated footing for Design Example 4.1.

References

2009 International Building Code® (IBC-09), International Code Council, Country Club Hills, IL, 2009.

ACI 318-08, *Building Code Requirements for Structural Concrete and Commentary*, American Concrete Institute, Farmington Hills, MI.

ACI Design Handbook SP-17, American Concrete Institute, Farmington Hills, MI, 1971-1997.

ASCE/SEI 7-05, *Minimum Design Loads for Buildings and Other Structures*, American Society of Civil Engineers, 2006.

CRSI Design Handbook, Concrete Reinforcing Steel Institute, Schaumburg, IL, 2002.

NAVFAC DM-7.2, *Foundations and Earth Structures*, Department of the Navy, Naval Facilities Engineering Command, Alexandria, VA, 1982.

Newman, Alexander, *Metal Building Systems: Design and Specifications*, 2d ed., McGraw-Hill, New York, 2004.

PCA Notes on ACI 318-08 Building Code Requirements for Structural Concrete with Design Applications, Kamara, Mahmoud E. et al, eds, Portland Cement Association, Skokie, IL, 2008.

CHAPTER 5
Foundation Walls and Wall Footings

5.1 The Basics of Design and Construction

5.1.1 Foundation Options for Support of Exterior Walls

The exterior walls in metal building systems (MBS) generally consist either of metal panels of various types or the so-called hard walls of masonry or concrete. Hard-wall systems include partial or full-height cast-in-place concrete, single-layer concrete masonry units (CMU), or masonry-veneer walls, as well as precast concrete panels. The weight of all of these needs to be supported at the foundation level. Metal panels weigh little, and the foundation demands they impose are minimal. Hard walls can be quite heavy and thick and require substantial foundations. The basic options for supporting the exterior building walls are:

1. Grade beams
2. Slab haunches built integrally with slabs on grade
3. Continuous foundation walls with footings

A basic grade beam without footing might suffice in a metal building system sheathed entirely with metal panels and located in the area with minimal frost-penetration requirements. The grade beam is a rectangular concrete element that either bears on the soil directly (Fig. 5.1a) or spans between the column foundations as a flexural member (Fig. 5.1b). The latter requires flexural and shear reinforcement to carry its own weight and the weight of the exterior building wall.

In the areas with deep frost lines some engineers prefer not to carry the flexural-type grade beams below the frost line and provide a frost cut (a sharply angled bottom surface) instead (Fig. 5.1b). The effectiveness of the frost cut can be supplemented by removing some of the soil from under the bottom of the grade beam. Both these steps attempt to minimize the upward pressure on the grade beam that might develop as a result of the soil frost heave. Flexural-type grade beams are particularly useful in combination with deep foundation piers discussed in Chap. 9.

A slab haunch is similar in function to a grade beam but is built integrally with the slab on grade (Fig. 5.1c). This design is commonly used in some areas of the United States where the frost penetration lines are located at shallow depths. A slab with haunch can be a foundation system by itself, and Chap. 8 covers it.

80 Chapter Five

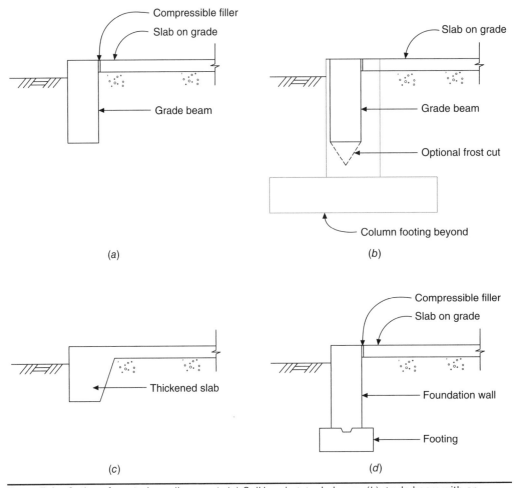

Figure 5.1 Options for exterior wall support: (a) Soil-bearing grade beam; (b) grade beam with an optional frost cut spanning between column foundations; (c) slab with haunch; (d) foundation wall with footing. (Reinforcement is not shown for clarity.)

The main subject of this chapter is the third option, continuous foundation walls with footings (Fig. 5.1d). These foundations are most cost-effective when they extend several feet below grade and carry significant gravity loads from the exterior walls above. They can also be indispensable in helping counteract the wind-uplift forces on exterior column foundations, such as moment-resisting footings discussed in Chap. 7.

5.1.2 Design and Construction Requirements for Foundation Walls

Specific provisions for design and construction of foundation walls are found in the model building codes, such as *2009 International Building Code®* (IBC-09), and in specialized concrete codes, such as ACI 318, *Building Code Requirements for Structural Concrete and Commentary*, published by the American Concrete Institute. In many cases plain-concrete or even masonry foundation walls are acceptable. Among other factors, design of foundation walls depends greatly on whether or not the walls are braced at the top. The bracing can be provided by the floor framing at crawl spaces and basements, or by slabs on grade connected to the foundation walls by dowels, as discussed later. Other important design considerations are the presence of unbalanced fill on both sides of the wall and the seismic design category of the building.

The thickness of the foundation wall depends on the type and thickness of the exterior superstructure wall it supports, as the foundation wall should be at least equal to that thickness (see IBC-09 Paragraph 1807.1.6). Some exceptions are granted to the foundation walls supporting brick veneer, which can overhang the foundation wall. (We recommend that the overhang does not exceed 1 in., for a variety of reasons.) The minimum thickness allowed for the plain-concrete foundation walls is 7.5 in.

The common thicknesses of the foundation walls used in metal building systems with metal siding range from 8 to 12 in. The walls supporting masonry-veneer cavity wall system with a backup of 8-in. CMU are usually 15 or 16 in. thick. Occasionally, lateral soil pressures (at basements, for example) control the design of the foundation walls and help determine their thickness.

In typical metal-building construction soil is placed at approximately the same elevation on each side of the foundation walls, although quite often the soil at the interior side is slightly higher than at the exterior. A 6-in. difference is usually sufficient to prevent the outside surface water from entering the building and to provide some modest barrier against the intrusion of the creeping wildlife (Fig. 5.2). Note that in the illustration the slab on grade is separated from the wall by compressible filler.

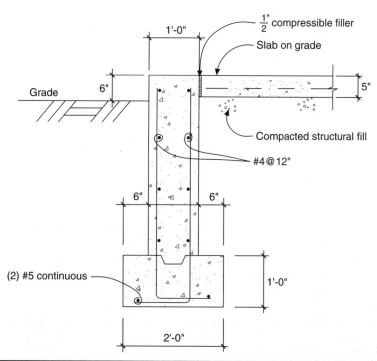

FIGURE 5.2 Reinforced-concrete foundation wall with a minimal difference of backfill elevations on two sides. The slab on grade is separated from the wall.

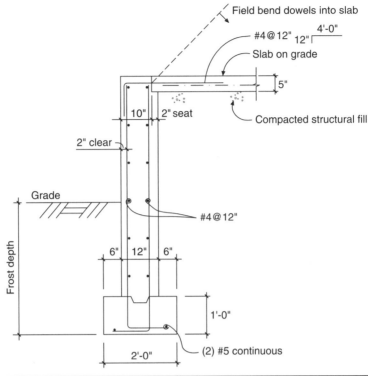

FIGURE 5.3 Reinforced-concrete foundation wall with significant difference of backfill elevations on two sides. The slab on grade is tied to the wall.

When the difference in the soil elevations on both sides of the foundation wall is significant, a retaining wall-like behavior is at hand, and the wall needs to be designed accordingly. A common example is the foundation wall at a loading dock, where the difference in the backfill elevations is typically substantial (a 4-ft drop is common). Depending on the configuration, loading, and soil conditions, such a wall can be designed as a reinforced wall with regular wall footing, similar to Fig. 5.3, or as a cantilevered concrete retaining wall. In both cases the wall is laterally supported at the top by the dowels extending into the floor slab.

The tables in IBC-09 Chapter 18 address the various combinations of the foundation wall materials (masonry or concrete), wall heights, maximum unbalanced backfill height, and the soil class or the associated design lateral soil loading. The tables indicate whether vertical steel reinforcement is needed, and, if it is, specify its minimum sizes. The tables assume that the walls are braced at the top and bottom; otherwise, custom design is required. (We should add that the walls unbraced at the top and supporting unbalanced fill need to be designed as pure cantilevered retaining walls.)

The IBC-09 Chapter 18 tables allow the use of 8-in.-thick solidly grouted foundation walls of unreinforced CMU for all the tabulated soil classes where the maximum wall height is 7 ft and the maximum unbalanced backfill height is 4 ft. The tables assume that CMU is placed in a running bond and its mortar is Type M or S. Similarly, the tables allow 7.5-in.-thick plain-concrete foundation walls to be used for all the tabulated design lateral soil loading where the maximum wall height is 7 ft and the maximum unbalanced backfill height is 5 ft. These conditions generally apply to a typical exterior foundation wall in the metal building system without a basement.

We reiterate that IBC-09 Chapter 18 tables are based on the assumption that the walls are braced at the top. In case of metal building systems, a slab on grade can provide such bracing. ACI 318-08 Commentary on Section 22.6 explains that lateral support of plain-concrete walls must "prohibit relative lateral displacement at top and bottom of individual wall elements." Walls without such lateral support must be designed as reinforced concrete members.

Even in cases where a plain-concrete foundation wall would be acceptable, some designers specify at least two No. 4 or No. 5 bars at the top and the bottom of the wall. This allows the wall to span over the soil areas with poor compaction and to reduce cracking owing to minor differential settlement. Others specify reinforced exterior wall in almost all cases.

Where required by analysis, code provisions, or designer's judgment, foundation walls are reinforced. To qualify as reinforced walls, the minimum reinforcement percentage of concrete walls, including foundation walls, is prescribed in ACI 318-08 as follows:

- Vertical: 0.0012 for No. 5 deformed bars and smaller, with $f_y \le 60{,}000$ psi; 0.0015 for other deformed bars
- Horizontal: 0.002 for No. 5 deformed bars and smaller, with $f_y \le 60{,}000$ psi; 0.0025 for other deformed bars

The maximum bar spacing in either direction is 18 in. Walls thicker than 10 in., except basement walls, need two layers of reinforcement. The exterior layer, with a clear cover of at least 2 in., must contain from one-half to two-thirds of the total required reinforcement area of in each direction. The interior layer, with a clear cover at least ¾ in., but not more than one-third the thickness of the wall, contains the remainder.

In cases where foundation walls help resist lateral loads on the column foundations, the walls should be designed as reinforced-concrete flexural members spanning horizontally (and vertically if needed) between the column piers. These designs are described in Chap. 7.

5.1.3 Construction of Wall Footings

The width of a wall footing depends on the thickness of the foundation wall it supports. When the foundation walls are made of cast-in-place concrete placed inside formwork, the minimum width of the footing ledge (the length of the projection beyond the face of the wall) should be ample to support the thickness of the formwork—typically 4 to 6 in. Therefore, the minimum width of the wall footing might be larger than required by the loading conditions alone. As discussed in Chap. 2, these circumstances might result in a differential settlement between the lightly loaded exterior foundation wall footings and the heavier-loaded isolated interior column footings.

The sides of wall footings are typically formed with dimensional lumber (see Fig. 2.2). Accordingly, some engineers prefer to specify the footing thicknesses equal to that of dimensional lumber (e.g., 11¼ in. for a 2 × 12 lumber, or 9¼ in. for a 2 × 10). Others prefer round numbers, such as 12 or 10 in. Still others specify the footing thickness to be equal to that of the foundation wall being supported. In any event, the minimum thickness of structural plain concrete footings is 8 in.

Because of their modest widths, continuous wall footings are usually made of plain concrete (*PCA Notes*). The applicability of plain-concrete footings depends on the Seismic Design Category (SDC) of the building, with slight variations among the various codes and their various editions. For example, ACI 318-08 Paragraph 22.10.1 stipulates that plain-concrete footings are allowed to carry plain-concrete foundation walls in SDCs A through C. In SDCs D, E, and F, plain-concrete foundation walls are also allowed, but with such caveats as:

- The walls must be of reinforced concrete or masonry.
- The wall footings must contain at least two continuous No. 4 or larger reinforcing bars, with the minimum area of the reinforcement of 0.002 times the gross cross-sectional area of the footing. The bars must be made continuous at corners and intersections.

However, IBC-09 Paragraph 1908.1.8 modifies the provisions of ACI 318-08 Section 22.10 and includes SDC C in the same group as SDCs D, E, and F. It also adds exceptions to those provisions. For instance, IBC-09 allows plain-concrete wall footings and plain-concrete "stemwalls," with only a single bar required at the top of the stemwall and the bottom of the footing.

Many engineers follow these ACI 318 provisions for all SDCs, making the details similar to Fig. 5.3 their standard practice. Note that the 12-by-24-in. footing is reinforced with two No. 5 bars, because the required area of the reinforcement is:

$$0.002 \times 12 \times 24 = 0.576 \text{ in.}^2$$

Two No. 5 bars provide 0.62 in.^{2} OK

5.1.4 Design of Wall Footings

Structural design of wall footings is straightforward and a bit simpler than the design of isolated column footings. The width of the footing is determined by the considerations outlined previously in Sec. 5.1.3, or, in the case of heavily loaded walls, on the basis of unfactored loads and allowable soil bearing pressures. The calculations for moment and shear follow the ultimate-design procedures. In most cases wall footings are designed as plain-concrete elements.

As discussed in Chap. 4, the maximum bending moment in a footing is found by passing a vertical plane through the footing at the critical section for flexure. For wall footings, the location of the critical section for moment depends on the wall material:

1. For concrete walls, the critical section for flexure is located at the face of the wall.
2. For masonry walls, the critical section for flexure is located halfway between the edge and the middle of the wall.

Design for shear is similar to the provisions used for isolated footings discussed in Chap. 4. The critical section for beam-action shear is located the distance h (footing thickness) from the critical section for moment. The cantilever projection distance, measured from the critical section for beam-action shear to the outer edge of the footing, is generally quite short. In a common scenario, the width of the footing ledge is less than the footing depth, and the critical section for beam-action shear occurs outside the footing. In this case beam-action shear need not even be checked.

Punching shear is not considered for walls. Design for uplift is similar to that for isolated footings, except that plain-concrete behavior is not questioned as much for this condition—it is the normal practice. As with unreinforced column footings, overall thickness of plain-concrete wall footings cast against the soil must be reduced by 2 in. for design.

Design Example 5.1: Wall Footing

Problem Check the wall footing design shown on Fig. 5.4. The footing is 10 in. thick and 20 in. wide. It supports 8-in. concrete foundation wall, which carries a fully grouted loadbearing masonry wall that is 16 ft tall 8 in. thick. The masonry wall serves as an endwall in the metal building system, with the tributary width of 12.5 ft. The following loads act on the roof: 3 pounds per square foot (psf) dead load and 30 psf design roof snow load. The unit weight of CMU is 80 psf. The frost depth of the footing is 3 ft 6 in. Use $f'_c = 3000$ psi and the allowable soil bearing capacity of 3000 psf. Assume the weight of soil 120 lb/ft³ and the weight of concrete 150 lb/ft³. The footing is placed on fill.

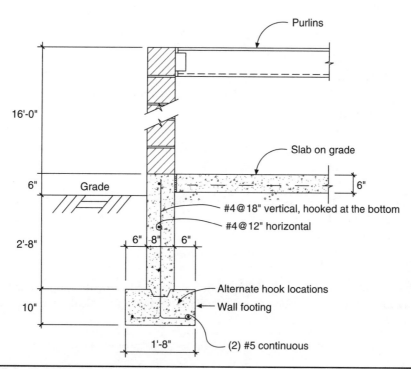

Figure 5.4 Wall footing for Design Example 5.1.

Solution Find the design loads on the footing. The design loads transmitted through the foundation wall (roof load + CMU + foundation wall) is:

Dead load D (not including the footing weight) = 3 psf × 12.5 ft + 80 psf × 16 ft + 100 psf × 3.17 ft = 1635 lb or 1.63 kips (1 kip = 1000 lb).

Snow load S = 30 × 12.5 = 375 lb or 0.375 kip.

Total load $D + S$ = 2010 lb = 2.01 kips.

Since the footing is placed on fill, subtract own weight of the footing as well as weight of soil and slab on its ledges from the allowable soil bearing capacity:

Weight of footing plus soil and slab on ledge = 125 + 120 × 2.67 + 150 × 0.5 = 520 psf = 0.52 ksf.

Total available soil pressure = 3 − 0.52 = 2.48 ksf.

Minimum required width of footing = 2.01/(2.48 × 1).

= 0.81 ft < 1.67 ft provided OK.

Factored load under load combination $1.2D + 1.6S$:

$$P_u = 1.2(1.63) + 1.6(0.375) = 2.556 \text{ kips}$$

Factored soil pressure

$$q_s = \frac{P_u}{A_f} = \frac{2.556}{1.67(1.0)} = 1.53 \text{ ksf}$$

$\phi M_n \geq M_u$
$M_u = 1.53 \times (0.5^2)/2 = 0.191$ kip-ft per lin. ft of footing
$t_{eff} = 10 - 2 = 8$ in.
$S_m = b(t_{eff})^2/6 = 12(8^2)/6 = 128$ in.^{3}
$\phi M_n = 0.6 \times 5\, \lambda \sqrt{f'_c}\, S_m = 0.6 \times 5 \times 1.0\, \sqrt{3000}(128) = 21{,}033$ lb-in. = 1.75 kip-ft
$\phi M_n = 1.75 > M_u = 0.191$ kip-ft   OK

Shear need not be considered since the critical section for shear occurs outside the footing ($t = 10$ in. $> l = 6$ in.).

∴ The footing is adequate.

As an added task, check whether the wall reinforcement complies with minimum requirements for reinforced walls. Minimum area of vertical bars spaced 18 in. o.c. (on center):

$$0.0012 \times 18 \times 8 = 0.17 \text{ in.}^2 < 0.2 \text{ in.}^2 \text{ provided (No. 4 bar)}$$

Minimum area of horizontal bars spaced 12 in. o.c.:

$$0.002 \times 12 \times 8 = 0.192 \text{ in.}^2 < 0.2 \text{ in.}^2 \text{ provided (No. 4 bar)}$$

Wall reinforcement shown complies with the minimum required for reinforced walls. Provide two No. 5 continuous longitudinal bars in the footing, as discussed earlier. ▲

References

2009 International Building Code® (IBC-09), International Code Council, Country Club Hills, IL, 2009.

ACI 318-08, *Building Code Requirements for Structural Concrete and Commentary*, American Concrete Institute, Farmington Hills, MI, 2008.

Notes on ACI 318-08 Building Code Requirements for Structural Concrete with Design Applications, Kamara, Mahmoud E., et al., eds., Portland Cement Association (PCA), Skokie, IL, 2008.

CHAPTER 6
Tie Rods, Hairpins, and Slab Ties

6.1 Tie Rods

6.1.1 The Main Issues

A typical rigid frame exerts both horizontal and vertical column reactions on its foundations, as discussed in Chap. 3. Under gravity loading a single-span gable rigid frame produces two equal horizontal forces acting outward in the opposite directions. The idea behind the tie-rod system is to interconnect the foundations under the frame columns and thus "extinguish" both horizontal reactions.

Tie rods are placed at or below the floor and are attached to the column foundations by hooks or mechanical devices. In one design they connect directly to the column base plates. Because the most reliable versions of the tie-rod system are relatively expensive to construct, they are commonly used for resisting large horizontal reactions, measured in tens of kips (1 kip = 1000 lb).

Being primarily tension members, the classic tie rods are not intended to resist the column reactions acting in the same direction, as from a wind load. If resistance to both tension and compression is required, as is often the case in metal building systems, tie rods are placed within substantial concrete grade beams designed as compression elements.

The typical tie-rod material is mild steel reinforcement, but high-strength post-tensioned steel can also be used in one system described later. Since the steel bars are placed in the vicinity of the soil, they are susceptible to corrosion. The cost of tie rod replacement is high, and it is prudent to protect them by multiple means. The first level of protection is encasement by concrete or grout; the second is hot-dip galvanizing or other corrosion-resisting coating. Some tie rods are placed within plastic sheaths that provide yet another protective barrier.

Should tie rods be classified as tension-tie members? The answer has practical implications. ACI 318-08 Section 12.15.6 requires splices in tension-tie members to be "full mechanical" or welded (lap splices alone are not sufficient). To help the users determine whether a tension-resisting element must comply with these provisions, the ACI 318-08 Commentary on Section 12.15.6 offers the following characteristics of tension-tie members:

- They are loaded with an axial tensile force that creates tension over the cross section.
- The level of stress requires each bar to be fully effective.
- Limited concrete cover exists on all sides.

Among the examples of tension-tie members given in the ACI Commentary are "arch ties." Accordingly, the tie rods consisting of one or two bars within a narrow encasement should indeed be classified as tension-tie members. For these elements, whether placed within a small thickened section of the slab on grade or in a separate narrow encasement, we recommend that full mechanical or welded splices be provided. If the tie-rod assembly consists of more than one bar, splices in the adjacent bars must be staggered by at least 30 in., another requirement of ACI 318-08 Section 12.15.6. Other tie rod systems, such as those with grade beams designed to resist both tension and compression, usually do not fall in the tension-tie category.

As discussed in Chap. 3, versatility of the tie-rod system is limited, because tie rods cannot be used in buildings with trenches, deep floor depressions, and pits. See Table 3.1 for a summary of the comparative reliabilities, degree of versatility, and costs for common foundation systems used in metal building systems.

6.1.2 Some Basic Tie-Rod Systems

The most basic tie-rod design consists of two deformed bars placed in a thickened slab (Fig. 6.1a). As just recommended, the reinforcement should be galvanized, and the splices made by mechanical means. The bars are usually anchored into the foundation by standard hooks. Alternatively, some engineers prefer to attach the bars directly to the column base plates by clevis-and-pin connections (Fig. 6.1b).

The only advantage of this design is the lowest possible construction cost, but there are at least two major disadvantages. The first is a total lack of survivability in case the slab is cut in the future. Slabs on grade are frequently cut during installation or maintenance of underground pipes, conduits, cables, and so on. When this happens, the tie-rod system would effectively cease to function, without any alternate load path being available.

The second disadvantage of this lowest-cost design relates to a common problem in slabs on grade: cracking resulting from drying shrinkage of concrete or from other causes. When a crack intersects tie-rod bars, it exposes them to localized corrosion and shortens their life span. A similar problem exists at the slab control and construction joints, if the tie-rod reinforcement continues through them.

A tie rod recessed below the slab represents a more reliable approach. One popular design has galvanized (sometimes epoxy-coated) bars placed in plastic pipes filled with grout (or grease in previous practice), as shown in Fig. 6.1c. This system certainly meets the conditions for tension-tie members, which means that mechanical splices are required here. In addition to the two levels of corrosion protection provided by bar coating and grout encasement, the plastic sheath provides the third.

A major advantage of this design over simply placing the bars into a thickened concrete slab is better survivability. While a thickened slab with a couple of bars is unlikely to alarm a worker determined to cut it, workers are trained not to cut cables. Naturally, fitting the bars with mechanical splices into pipes and then filling the pipes with grout increases the cost of this system.

A less reliable variation on this theme is shown in Fig. 6.2, where a simple concrete encasement replaces the grouted plastic sheath. The encasement is likely to crack in service from concrete shrinkage, bar elongation under load, and sagging under its own weight. Once the cracks appear, corrosion-protection effectiveness of the encasement is greatly reduced.

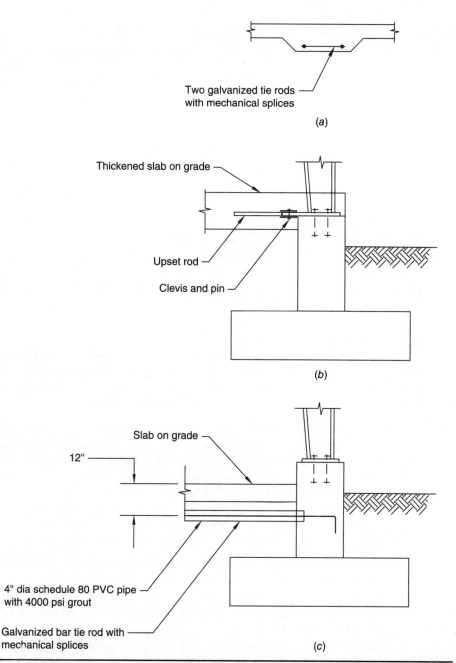

Figure 6.1 Basic tie-rod designs: (a) Tie rod in thickened slab; (b) tie rod directly attached to column base plate; (c) tie rod encased in plastic sheath filled with grout.

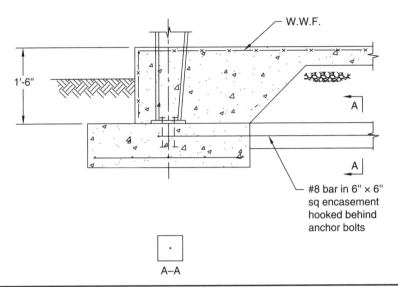

Figure 6.2 Tie rod in plain-concrete encasement.

6.1.3 A Reliable Tie-Rod Design

A much more reliable tie-rod design than those just described is a deep grade beam that extends from one end of the frame to another (Fig. 6.3). The horizontal reinforcement terminates in standard hooks embedded into the column piers or other column foundations. There are several significant advantages to this design:

- In lieu of a single tie rod, at least four reinforcing bars fulfill its function here, providing a measure of redundancy. The cross-section of the grade beam could be quite substantial, which might avoid classifying it as a "tension-tie member" and therefore allowing for less-expensive lap splices in the reinforcement instead of mechanical or welded splices.

- In addition to resisting tension, the grade beam can also resist compression. It could also act as a horizontal seismic tie between the pile caps or column footings subjected to seismic loading (see ACI 318-08 Section 21.12.3).

- Owing to the smaller bar diameters of the horizontal reinforcement, the hook development lengths become more reasonable (see Sec. 6.1.4).

- Because grade beams are typically placed in formwork, horizontal reinforcement can be readily supported at close intervals, removing a problem of bar sagging discussed in Sec. 6.1.6.

- For large horizontal reactions, many reinforcing bars might be needed, as demonstrated in Design Example 6.1. The grade-beam system makes placement of multiple bars practical.

There are advantages to locating the grade beams some distance below the bottom of the slab on grade. This allows the slab to settle without loading the grade beams and limits the damage from any future slab cutting. Alternatively, some engineers prefer to place the grade beams in grillagelike configuration, where they run in both directions integral with the floor, essentially making a stiffened mat system.

Tie Rods, Hairpins, and Slab Ties 93

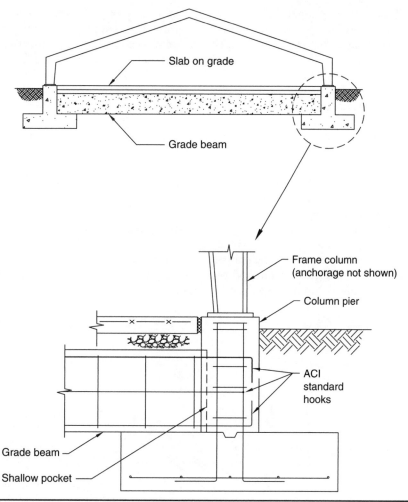

Figure 6.3 Grade-beam tie rod.

What are the minimum reinforcing requirements for a grade-beam tie rod? The codes are silent on the minimum reinforcement of concrete tension members, but since the grade beams can also be used as compression elements, it is wise to reinforce them accordingly. The longitudinal reinforcement consists of the tie-rod bars, and their area should be no less than 1 percent of the gross concrete area and no more than 8 percent (see ACI 318-08 Section 10.9.1).

Lateral ties (stirrups) in these grade beams are typically made of No. 4 bars, although No. 3 bars are allowed for longitudinal bars No. 10 and smaller (see ACI 318-08, Paragraph 7.10.5). The tie spacing should not exceed:

- 16 longitudinal bar diameters (d_b)
- 48 tie-bar diameters
- Least dimension of the grade beam

If there are more than four longitudinal bars, additional ties may be needed to ensure that none of the bars is farther than 6 in. from a bar supported by a tie (measured as a clear distance between the bars). The minimum bar spacing in compression members is $1.5d_b$, but not less than 1.5 in., a requirement that also applies to the distances between the lap splices and the adjacent splices or bars (ACI 318-08, Section 7.6).

When the tie rod of the grade beam design also acts as a seismic tie in Seismic Design Categories D, E, and F, it must meet certain design requirements, such as those of ACI 318-08 Section 21.12.3. Each of its cross-sectional dimensions should be at least equal to 1/20th of the clear spacing between the connected concrete piers, but need not exceed 18 in. Closed ties are required; their spacing must not exceed the lesser of one-half of the smallest cross-sectional dimension and 12 in. (For pre-engineered buildings with clear spans between the column piers of at least 30 ft, the 18 in. minimum dimension controls, with a corresponding maximum tie spacing of 9 in.) Also, standard hooks are not considered effective in developing bars in compression, and straight-bar embedment or other means of load transfer may be needed.

If the tie rod of this design is narrower than the column pedestal, it can be embedded into a shallow pocket in the pier (say, 2-in. deep). This provides some lateral bracing for the end of the grade beam and protects the joint between the two elements. The joint between the column pedestal and grade-beam is shown in Fig. 6.4.

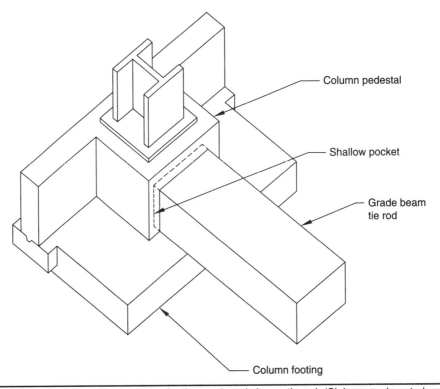

Figure 6.4 Assembly of column pier, footing, and grade-beam tie rod. (Slab on grade not shown for clarity.)

6.1.4 Development of Tie Rods by Standard Hooks

As already stated, in many tie-rod systems the rods are anchored into the column foundations by standard hooks. ACI 318-08 Section 7.1 defines a standard hook either one of the following:

- 90° bend plus $12d_b$ (the bar diameter) extension at free end of bar
- 180° bend plus $4d_b$ extension, but ≤ 2.5 in. at free end of bar

According to ACI 318-08 Section 12.5.2, the development length of standard hooks in tension l_{dh} is a function of several variables:

$$l_{dh} = \left(\frac{0.02\psi_e f_y}{\lambda\sqrt{f'_c}}\right)d_b$$

Here, f'_c is 28-day strength of concrete; f_y is yield strength of reinforcement; $\lambda = 1.0$ for normal-weight concrete (0.75 for lightweight concrete); $\psi_e = 1.0$ for uncoated bars (1.2 for epoxy-coated); and d_b is bar diameter. Figure 6.5 illustrates the development length of a hook with a 90° bend. The development length computed by this formula may be multiplied by the following modification factors:

- 0.7 for 90° hooks of bars ≤ No. 11 with side cover ≥ 2.5 in. and cover on bar extension beyond the hook ≥ 2 in.
- 0.8 for 90° and 180° hooks of bars ≤ No. 11 enclosed within ties or stirrups of specified designs.
- Excess area of reinforcement (A_s) is accounted for by the ratio of (A_s required) to (A_s provided), assuming that anchorage for full yield strength is not needed.

In any event, l_{dh} should not be less than the larger of $8d_b$ or 6 in.

The modification factors are of great help when reinforcing bars of large diameters are used as tie rods. Even so, the typical sizes of column pedestals (piers) might be insufficient to develop the standard hooks of very large bars, and through-the-pier anchorage might be needed. In some situations proprietary inserts cast into the column foundation can be used, with tie-rod reinforcement mechanically attached to the inserts.

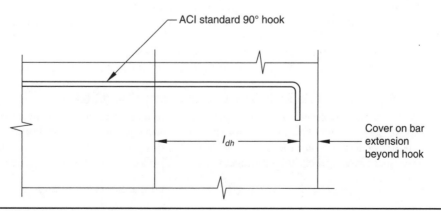

FIGURE 6.5 Hook development length l_{dh}.

Bar Size No.	$f'_c = 3000$ psi	$f'_c = 4000$ psi
5	9.6	8.3
6	11.5	10
7	13.4	11.6
8	15.3	13.3
9	17.3	14.9
10	19.2	16.6
11	21.1	18.3

The table is based on the following assumptions:
(1) Equation from ACI 318-08 Section 12.5.2.
(2) $\psi_e = 1.0$ for uncoated bars Grade 60 bars ($f_y = 60,000$ psi).
(3) $\lambda = 1.0$ for normal-weight concrete.
(4) Modification factor of 0.7 used for 90° hooks of bars ≤ No. 11 with side cover ≥ 2.5 in. and cover on bar extension beyond the hook ≥ 2 in.
(5) For bars not listed, the following equations may be used: $15.34d_b$ for $f'_c = 3000$ psi and $13.28d_b$ for $f'_c = 4000$ psi.
(6) The listed values may be reduced by the ratio (A_s required)/(A_s provided), but should not be less than the larger of $8d_b$ or 6 in.

TABLE 6.1 Development Length L_{DH} of Standard Hooks, inches

Table 6.1 lists the development lengths of standard hooks for common sizes of reinforcing bars used in tie rods. The table uses 28-day compressive strengths of concrete commonly used in foundation construction. Note that, without taking advantage of the excess-area reduction (A_s required)/(A_s provided), bars larger than No. 11 cannot be developed in the 24-by-24-in. and smaller piers. The solution for very high horizontal forces is to use a large amount of smaller bars in the grade beam, or at least to use more large bars than required by strength considerations alone. As discussed in Sec. 6.1.5, this might be necessary for other reasons as well.

Another code provision dealing with the development of hooked bars (ACI 318-08 Section 12.5.4) applies to cases where bars are developed by hooks "at discontinuous ends of members with both side cover and top or bottom cover over hook less than 2½ in." In this situation, stirrups or ties of the prescribed design must enclose the hooked ends. To avoid unnecessarily triggering these provisions, it is wise to provide a minimum cover of at least 2½ in. at the sides, top, and bottom of the hooked bars.

6.1.5 Design of Tie Rods Considering Elastic Elongation

As further discussed in Chap. 4, concrete design for tension is relatively rare, and the codes address tension resistance of concrete only for some specific situations. For example, ACI 318-08 Section 11.8 provides an equation for the design strength of tension reinforcement in brackets and corbels, which must exceed the applied factored tensile force N_u:

$$\phi A_n f_y \geq N_u, \text{ where } \phi = 0.75 \text{ and } A_n \text{ is the area of bars in tension}$$

The maximum *factored* tension stress is then $0.75 \times 60 = 45$ kips/in² (ksi).

However, the tie rods proportioned on the basis of this rather high stress value will tend to undergo significant elastic elongation, as demonstrated in Design Example 6.1. The elongation can be computed by the standard formula:

$$\Delta_{\text{rod}} = \frac{PL}{AE}$$

Here, P is the unfactored force in the tie rod; L, its length; A, the area; and E is modulus of elasticity.

A tie rod that stretches a lot under load will allow the frame columns to move apart and will result in additional stresses in the frames. These stresses are typically not considered in the design of metal building frames. Frame movements might also result in damage to cladding and finishes, as well as in floor cracking. The magnitude of these deleterious effects will depend on the building design. For example, a building clad in metal siding and having a large eave height might be able to tolerate larger deformations than a stocky building clad in masonry.

It is wise to decrease the stresses in the tie rods to limit their elastic elongation to some maximum value, but this limit should be left to the foundation designer's discretion. Some building manufacturers feel that the column spread of up to ½ in. is not harmful to the primary frames in most circumstances. Accordingly, in the absence of definitive research that establishes another value, it might make sense to limit the *total* tie-rod elongation under service (unfactored) loads to ½ in. This limit corresponds to the maximum movement at each end of the frame of ¼ in. If the building manufacturer permits a different limit for the frame spread, that limit could be used instead.

6.1.6 Post-Tensioned Tie Rods

Even a relatively strict limit on the elastic elongation does little to alleviate another common problem with tie rods: sagging under their own weight. A tie rod that is 100 or 150 ft long and not placed in a grade beam will sag a significant amount. The sagging will not totally disappear even when the rods are supported by the ground or formwork at regular intervals and when the turnbuckles are used to lightly tighten the bars.

The vertical deflection of the rods will result in a slack that will have to be overcome when the building frame applies tension to the tie rod. It will permit the foundations—and the frame columns on them—to spread apart somewhat, unless another mechanism exists to restrain the movement. This column spreading is in addition to the elastic elongation discussed in Sec. 6.1.5. The amount of slack in tie rods, and the eventual lateral movement of the foundations, is difficult to quantify. Many factors are involved, including the degree of soil compaction and the amount of rod tension induced by the turnbuckles.

The problem is minimized in the grade-beam tie rod system described in Sec. 6.1.3. Another system, post-tensioned tie rods, goes even further. In this design, a high-strength steel tendon is placed in the middle of a concrete grade beam reinforced with mild steel (Fig. 6.6). The grade beam extends from one column pedestal to another and is cast integrally with the pedestals. The post-tensioned grade beam is a compression element surrounded and laterally supported by the soil.

The tendon is post-tensioned after the concrete in the grade beam attains its 28-day compressive strength and the interior slab on grade is placed. The end anchors are protected by appropriate means, such as by a nonshrink grout in combination with a waterproofing membrane.

98 Chapter Six

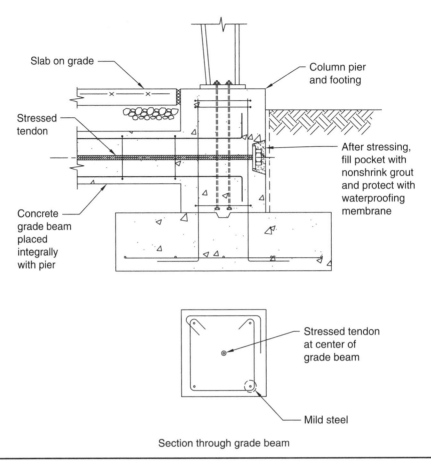

FIGURE 6.6 Post-tensioned grade-beam tie rod.

To reduce the upward pressure on the grade beam caused by the inevitable foundation settlement under load (see the discussion in Chap. 2), the soil underneath can be loosened after concrete in the grade beam reaches its design strength. Alternatively, the grade beam can be placed on continuous compressible filler or on cardboard. However, these steps invite a problem of grade-beam deflection under its own weight and should be used cautiously.

But why not simply post-tension the basic tie rods in Fig. 6.1? For post-tensioning to work, a compression element is needed. The grade beam is precompressed, and the tension forces from the building frame applied in the future will tend to stretch it. With proper design, the outward-acting column reactions will be less than the amount of the precompression, and the grade beam will stay compressed. But an attempt to post-tension a tie rod *without* a grade beam will simply result in overloading and perhaps fracturing the rods, as the post-tensioning stresses and those caused by the service loads would be additive.

The grade beam and the column pier in Fig. 6.6 are placed together to facilitate load transfer from the column to the grade beam. Otherwise, the load-transfer dowels or other devices would be needed to carry the column reaction to the grade beam across the horizontal construction joint above it.

Since the post-tensioned tie rod is placed some distance below the floor (for the same reasons the tie rod in a plastic sheath is), an eccentricity exists between the point of the load application (the top of the pier) and the center of the tie rod. In essence, the column pier and footing function as a cantilevered beam. The load is applied outward at the top of the pier—the end of the cantilever—while the tie rod and the bottom of the footing act as supports. Accordingly, the load on the tie rod is greater than the column reaction. The bottom of the footing should be capable of resisting the balance of the reactions by soil friction. If insufficient, soil friction can be augmented by passive soil pressure of the pier and the foundation walls. This mechanism of resisting lateral loading is discussed in Chap. 7.

6.1.7 Tie-Rod Grid

Yet another reliable approach to using tie rods in metal building systems is to place them as a grid of grade beams running in two directions. In this system, the tie-rod grade beams are placed integrally with the slab on grade and are spaced at relatively close intervals—from 10 to 15 ft on centers. For the typical primary-frame spacing of 25 ft, the grade beams can be spaced at 12.5 or 25 ft on centers. The grade beams are reinforced at top and bottom and contain closed stirrups (Fig. 6.7).

At first glance, the approach appears rather expensive, but a similar design has been used throughout the United States in the areas of expansive soils. One such design (UFC 3-220-07) has grade beams that are 24 in. deep, 12 in. wide, and spaced from 11 to 12 ft on centers. The grade beams are reinforced with two No. 5 bars at top and bottom and contain No. 2 stirrups (rarely available today). Sometimes, these grade beams are post-tensioned with ½-in high-strength tendons.

This design system has been widely used in Texas and many Southern states where expansive soils are prevalent. It is considered one of the most economical solutions for lightly loaded single-family houses, as well as pre-engineered buildings in those areas. Why not expand the use of this system to the rest of the country?

A grid of tie-rod grade beams would be especially economical in the areas where soils allow for placing concrete in the excavated trenches integrally with the slab, thus avoiding the expense of using the formwork in the interior areas. The *exterior* edges of the grade beams should of course be formed for aesthetic appearance. Because the grade beams are placed relatively close to one another, the system should provide significant redundancy even if one of the tie rods is accidentally severed. The design of the longitudinal reinforcement follows the basic procedure outlined in Sec. 6.1.5.

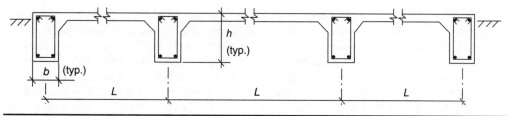

FIGURE 6.7 Tie-rod grid.

One of the questions about the grid of grade beams concerns the need to provide the grade beams in the direction *perpendicular* to the primary tie rods. While these grade beams do not serve as tie rods per se, they provide lateral bracing to the primary grade beams and stiffen the floor structure in general. It is certainly possible to reduce their spacing, but not eliminate them totally.

For example, if the primary grade beams are spaced 12.5 ft on centers, the perpendicular grade beams might perhaps be spaced at 25 to 50 ft on centers. Since little empirical guidance exists on the effects of the beam spacing in the transverse direction, the spacing becomes a matter of the engineering judgment.

One intriguing advantage of placing the perpendicular grade beams at close intervals is a possibility of engaging the passive soil pressure in resisting horizontal frame reactions, a subject explored in Chap. 7. Where the passive soil pressure on the perpendicular grade beams can be developed, at the very least the redundancy of the floor system greatly increases; it might even be possible to make this the primary mechanism for resisting lateral frame forces.

6.1.8 Which Tie-Rod Design Is Best?

Which of the designs described in this chapter is the best? The answer depends on a number of variables, including the frame span, the magnitude of the design loading, the type of soil, and the sensitivity of the frame and of the building cladding to foundation movement.

For small, lightly loaded buildings, where the possibility of a future slab cutting is negligible, the tie rods placed integrally with the slab might be acceptable. But as stated at the beginning of the chapter, tie rods are not necessarily the best solution for those circumstances anyway, and the hairpins might work just as well.

Indeed, the main advantage of tie rods is that they can "extinguish" very large horizontal reactions at the opposite ends of the frame. This typically occurs when the frame spans are very large and the loading is significant. It is in this situation that the issues of elastic elongation and sagging become acute. Using tie rods placed in grade beams, with or without post-tensioning, helps overcome these challenges.

Post-tensioned tie rods hold the promise of minimizing both problems, but their construction requires specialized equipment and expertise. When those are not readily available, the grade-beam tie rods of Sec. 6.1.3 might be as effective and probably less expensive. The system of the grade-beam tie rods deserves our nod as the most reliable performer in the tie-rod family. When the soil conditions allow placing a grid of the tie-rod grade beams integrally with the slab on grade (Sec. 6.1.7), a use of this system should certainly be explored.

Design Example 6.1: Tie Rod in Isolated Grade Beam

Problem Design a grade-beam tie rod for a large single-story single-span rigid frame spanning 150 ft. The spacing between the frames is 25 ft; the eave height is 20 ft. The building manufacturer supplied the column reactions. The following load combination was found to produce the maximum outward horizontal load: dead (14 kips) + snow (76 kips). The depth of the footing must be at least 4 ft below the floor. The column is supported by a 24-by-24-in. concrete pedestal, the top of which is flush with the floor slab. Use $f'_c = 3000$ psi and $f_y = 60,000$ psi. Assume the grade beam will *not* act as a seismic tie. Limit the total tie-rod elongation under service loads to ½ in. (1 kip = 1000 lb)

Solution Find the maximum factored and service (unfactored) design loads on the tie rod. The service design tension force from dead and snow loads is:

$$N = 14 + 76 = 90 \text{ kips}$$

The factored design tension force from dead and snow loads is:

$$N_u = 14 \times 1.2 + 76 \times 1.6 = 138.4 \text{ kips}$$

Find the minimum required bar area from the factored tension force:

$$A_{rq} = \frac{138.4}{0.75(60)} = 3.08 \text{ in.}^2$$

This requires three No. 9 bars ($A_s = 2.72$ in.²) or four No. 7 bars ($A_s = 3.5$ in.²).

However, to limit the total tie-rod elongation to ½ in., the minimum bar area should be:

$$A_{min} = \frac{PL}{0.5E} = \frac{90(150)12}{0.5(29,000)} = 11.17 \text{ in.}^2$$

This can be achieved with 12 No. 9 bars ($A_s = 12.0$ in.² > 11.17 in.²).

Find the minimum and maximum gross cross-sectional area of the grade beam A_g.

The maximum $A_g = 12.0/0.01 = 1200$ in.²

The minimum $A_g = 12.0/0.08 = 150$ in.²

The minimum clear distance between the bars, using No. 9 bars and assuming contact lap splices are used, is $1.5d_b + d_b = 2.5d_b$ or $2.5 \times 9/8 = 2.812$ in.

The minimum center-to-center bar spacing is then $2.812 + 9/8 = 3.94$ in., say 4 in.

Choosing a 16-in.-wide grade beam for easy constructability, and 2 ½-in. clear cover to the longitudinal bars on all sides (see discussion in Sec. 6.1.4), the minimum depth is $2.5 + 9/8 + 4 \times 5 + 2.5 = 26.125$ in. Use 16×30 in. grade beam with $A_g = 16 \times 30 = 480$ in.²

$$150 < A_g < 1200 \text{ in.}^2 \quad \text{OK}$$

Setting the top of the grade beam 12 in. below top of slab, the top of column footing needs to be at least 3.5 ft below top of slab.

Bar spacing = $(30 - 2.5 \times 2 - 9/8)/5 = 4.78$ in. < 6 in., so only one additional tie is required. Other longitudinal bars will be close enough to the laterally supported bars.

Determine tie spacing:

16 longitudinal bar diameters = $16 \times 9/8 = 18$ in.

48 tie bar diameters = $48 \times 4/8 = 24$ in.

Least dimension of grade beam = 16 in.   Controls

Space two sets of ties 16 in. on centers.

Note that if the grade beam also had to act as a seismic tie in the Seismic Design Category D, E, or F, its minimum size would be 18 in. and the tie spacing would be 9 in.

Since the grade beam is narrower than the pier, a 2-in. embedment pocket can be provided. The No. 9 longitudinal bars are developed into the column by standard hooks. Good practice dictates maximizing the hook's development length l_{dh}, as long as at least 2 in. clear cover is provided on bar extension beyond the hook. The available development length is (24 – 2 – 2) = 20 in. > 17.3 in. from Table 6.1, even before using the excess reinforcement adjustment factor. For consistency with other clear cover requirements, specify 2½ in. cover beyond the hook.

The cross-section through the grade-beam tie rod is shown in Fig. 6.8.

Notes
1. We remind the reader that this substantial grade beam has become necessary as a result of very heavy loading on a very large (150 ft) frame, and because of our intent to limit the elastic elongation of the tie rods to a strict limit of ½ in. For many other situations the cross-section will be smaller.

2. The design example considers only the resistance to the horizontal column reactions acting outward. A comprehensive foundation design would also consider downward and uplift loading on the column footing (see Design Example 4.1), as well as inward-acting horizontal column reactions (see Design Example 7.1). An added bonus of having the substantial grade beam is that it helps resist uplift loading and lateral sliding. It also helps transfer the inward-acting horizontal column reactions to the opposite side of the building, where they are resisted by passive pressure. ▲

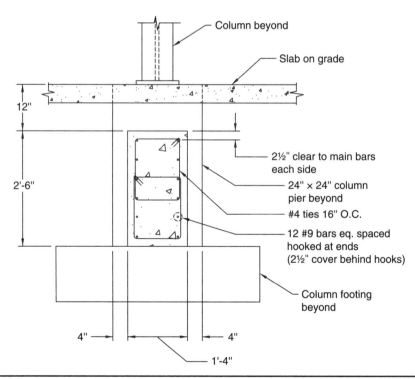

FIGURE 6.8 Cross-section through the grade-beam tie rod in Design Example 6.1.

6.2 Hairpins and Slab Ties

6.2.1 Hairpins: The Essence of the System

The basic principle behind the hairpin system and the tie-rod system is the same: Two ends of the frame are tied together to "extinguish" the opposing horizontal column reactions. The critical difference lies in the way the load is transferred from one end of the frame to another. In the hairpin system the tension force is resisted by distributed steel reinforcement in the floor slab, rather than by discrete tie rods.

Here is how the hairpin system is supposed to work. Hairpins are deformed reinforcing bars bent into a V-shape behind the frame column and extending some distance into the floor slab (Fig. 6.9). The slab is reinforced with either deformed bars or welded-wire fabric (WWF). The hairpins engage a certain number of slab reinforcing bars or welded-wire fabric wires and transfer the horizontal force to them. The bars or wires continue through the slab and eventually transfer their loading to the column foundations at the opposite side.

The hairpin system represents the least expensive solution for resisting horizontal column reactions, both in terms of design and construction costs. Hairpins can be effective when used in elevated structural slabs reinforced with properly spliced deformed bars that extend through the joints. However, when used in slabs on grade, the single advantage of this system—its low cost—must be balanced against a host of significant weaknesses, as discussed next.

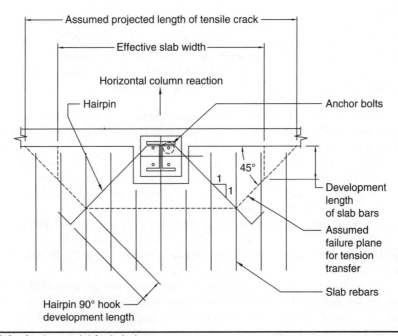

Figure 6.9 Design model for hairpins.

6.2.2 Hairpins in Slabs on Grade

The idea of relying on slabs on grade to transfer significant tension forces raises many questions. Unlike the elevated structural slabs reinforced with deformed bars, slabs on grade are frequently unreinforced, or are lightly reinforced with welded-wire fabric. The typical welded-wire fabric used in floors on ground has a modest cross-sectional area that is not sufficient even for meeting the minimum shrinkage reinforcement requirements of ACI 318. For this reason some call it "skin reinforcement." (Further information on various types of reinforcement used for designing slabs on ground can be found in ACI 360R-06.)

An unreinforced slab on grade requires close spacing of control or construction joints to avoid cracking caused by concrete drying shrinkage. For example, a 6-in.-thick slab needs a joint spacing of about 14 ft on centers, according to ACI 360R-06 Figure 5.6. To be effective in the hairpin system, reinforcing bars or welded-wire fabric wires must be continuous and properly spliced throughout the slab. They cannot be stopped at the slab construction joints, or weakened (e.g., by cutting every other wire) at control joints, as is common practice. If the reinforcement or welded-wire fabric wires are allowed to continue through the joints, they compromise the very function of the joints. States ACI 360R-06:

> As a general rule, the continuation of larger percentages of deformed reinforcing bars should not be used across sawcut contraction joints or construction joints because they restrain joints from opening as the slab shrinks during drying, and this will increase the probability of out-of-joint random cracking.

Therefore, a design that requires the slab reinforcement to be continuous from one end of the building to another is likely to facilitate slab cracking (and, we should add, potentially increase the designers' professional liability, should any claim be brought against them for a cracked slab).

From a practical standpoint, if continuity of steel reinforcement is needed across the slab on grade, the author prefers to use deformed reinforcing bars rather than welded-wire fabric. The bars are typically lap spliced in a code-prescribed manner, but splicing welded-wire fabric is not always done as reliably in the field. Needless to say, an unreinforced slab, or a slab reinforced only with fibers, does not provide the needed continuity.

Yet another disadvantage of a slab-based hairpin system is the same caveat we have identified for the tie rods located in thickened slabs: These systems do not work when large pits, trenches, or floor depressions are present.

In any event, ACI 318-08 Section 1.1.7 excludes slabs on ground from its scope, "unless the slab transmits vertical or lateral forces from other portions of the structure to the soil." Essentially, slabs on ground are not considered structural concrete, unless they transmit forces from the other elements to the soil. In the author's experience, the quality standards related to their construction are not followed as strictly on the construction site as those for elevated slabs. In some cases, even meeting the minimum slab thickness requirements could be a challenge, and the floor thickness ends up fluctuating widely (see Figure 6.9 in Newman, 2001).

If the designer intends to have the slab on grade comply with ACI 318 requirements, the slab must be designed and constructed with greater care than the prevalent practices. At the very least, it should be reinforced more substantially than with a layer of light welded-wire fabric.

When the hairpin system is intended to transmit seismic forces in seismic Design Categories D, E, and F, special provisions of ACI 318-08 Section 21.12.3.4 apply. These provisions require that the slab on ground be designed as a structural diaphragm per ACI 318-08 Section 21.11 and be clearly identified as such on structural drawings. To act as a diaphragm the slab needs to be positively attached to the perimeter foundation walls with dowels, rather than be separated from them by slab expansion joints. We should note that the diaphragm action would also be needed for resisting the horizontal column reactions acting inward.

Another reason *not* to separate the floor slab from the foundation walls by compressible filler or rigid insulation (see Fig. 5.1) is a potential for localized hairpin corrosion within a short hairpin length not embedded in concrete. Unlike tie rods, hairpins are usually made of uncoated steel reinforcement and are vulnerable to corrosion in that area, which can result in their fracture under load.

But perhaps the biggest problem with the hairpin system is its total lack of survivability, as discussed in Chap. 3. Were the slab to be cut at any time (for pipe, cable, or conduit installation and maintenance, among other reasons), the system would be destroyed. Some additional discussion of this system can be found in Newman (2004).

Because of the serious weaknesses of the hairpin system, we recommend limiting its use to small utility buildings only, where a possibility of the slab being cut in the future is remote. We further recommend placing deformed bar reinforcement rather than welded-wire fabric in the slabs on grade when this system is used.

The minimum suggested practical size of deformed bars is No. 4, although closely spaced No. 3 bars could be used as well. The maximum spacing of primary reinforcement in structural slabs per ACI 318-08 Section 7.6 is three times the slab thickness or 18 in. o.c. (on centers). Therefore, the maximum bar spacing should be:

In 4-in. slabs: 12 in. o.c.

In 5-in. slabs: 15 in. o.c.

In 6-in. and thicker slabs: 18 in. o.c.

The reinforcement in the slabs on grade conforming to ACI 318 should provide at least the required area to act as shrinkage and temperature reinforcement. Following the provisions of ACI 318-08 Section 7.12.2 for deformed bars with f_y = 60,000 psi and slab thickness t, the minimum bar area A_s per ft of slab width becomes:

$$A_s = 0.0018 \times t \times 12$$

Table 6.2 provides the area of minimum shrinkage and temperature reinforcement for various common slab thicknesses, with the bars placed in single layer. Both hairpins and slab bars are typically placed at least 1.5 in. from the top of the slab. ACI 360R-06 states that "common practice is to specify that the steel have 1.5 to 2 in. cover below the top surface of the concrete."

6.2.3 Hairpins: The Design Process

The design of hairpins is uncomplicated. It consists of two separate steps:

1. Determination of the hairpin's length
2. Finding the hairpin's bar size (diameter)

Slab Thickness	A_s Required, in.² per lin. ft	Bar Size and Spacing
4	0.086	No. 4 @ 12"
5	0.108	No. 4 @ 15"
6	0.130	No. 4 @ 18"
8	0.173	No. 4 @ 12"

The table is based on assuming the bars to be primary reinforcement in structural slabs, following the provisions of ACI 318-08 Section 7.6.

TABLE 6.2 Minimum Shrinkage and Temperature Reinforcement for Some Common Slab Thicknesses

The first step determines the minimum required tensile capacity of the slab reinforcement; the second, the capacity of the hairpin itself. The design model for hairpins is given in Fig. 6.9. The hairpins shown there have a 90° included angle between their legs, which makes for a 45° angle between the legs and the direction of the outward-acting horizontal force. Some engineers prefer to use a smaller included angle, such as 60°.

The minimum required length of each leg of a V-shaped hairpin bar depends on the number of slab bars or wires that need to be engaged. The total area of those should be sufficient to resist the applied horizontal column reaction. The failure plane shown in Fig. 6.9 represents an idealized location of the tensile crack in the slab that is assumed to form at the theoretical ends of the hairpins. The longer the hairpins, the farther this assumed plane extends into the slab and the more of the slab reinforcement is engaged.

As a minimum, the number and spacing of the slab bars should satisfy the provisions for shrinkage and temperature reinforcement, as discussed in Sec. 6.2.2 and as shown in Table 6.2. Each slab bar or wire that intersects the assumed failure plane and extends past it for a certain distance is considered effective for tension transfer. This distance should be at least equal to the straight-bar development length.

The total number of the effective slab bars is approximately equal to the projected length of the tensile crack divided by the bar spacing minus two times the development length. (This excludes the bars that extend less than the straight-bar development length past the assumed tensile crack.) The short horizontal bar length at the apex of the "V" may be conservatively neglected.

The simplified assumed projected length of the tensile crack L_{slab} can then be expressed via the length of each hairpin leg L_{hair} as:

$$L_{slab} = L_{hair} \sqrt{2}(2) \approx 2.82\, L_{hair}$$

The *effective* slab width for engaging slab bars in tension transfer $L_{slab,\,eff}$ reduces this length by the development length of the slab bars on each side:

$$L_{slab,\,eff} = L_{hair} \sqrt{2}(2) - 2l_d \approx 2.82\, L_{hair} - 2l_d$$

Each hairpin leg can be developed past the assumed failure plane by a standard hook, with the appropriate hook development length l_{dh} (see Table 6.1). It can also be developed as a straight bar (see Sec. 6.2.4). Thus the actual length of each hairpin is equal to L_{hair} plus the development length of hook or straight bar.

The lack of design provisions relating to tension in deformed bars used as hairpins, and our desire to reduce the effects of elastic elongation in the bars, both argue for assigning a conservative tension capacity to hairpins. The traditional way of doing so is to use the allowable-stress design method for their design. In this venerable approach, still widely used in masonry design, the allowable tension in deformed reinforcement is taken as 24 ksi. (The allowable tension in welded-wire fabric is 20 ksi.) Multiplying this stress by the effective area of the bars or wires gives the allowable capacity of the hairpin system. This provides the solution for the first step of the design.

For the second step, we need to find the required hairpin's bar size—a separate effort from finding the hairpin's length. For a 90° included angle between hairpin legs, the force F in each leg can be found from the following equation, as illustrated in Fig. 6.10:

$$F = \frac{T\sqrt{2}}{2}$$

Here, T is the horizontal force on the hairpin (in kips). Using the allowable tension stress of 24 ksi gives the following equation for the required hairpin area:

$$A_{hair} = \frac{T\sqrt{2}}{2(24)}$$

We recommend using hairpin bars of at least for No. 5, to provide some additional corrosion resistance for these critical elements.

6.2.4 Development of Straight Bars in Slabs

As just stated, hairpins and slab bars can developed past the assumed failure plane by one of two methods. In the first, standard hooks are used; in the second, straight bars

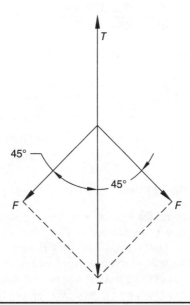

Figure 6.10 Determination of forces in hairpins with an included angle of 90°.

are developed. Development of standard ACI hooks is addressed in Sec. 6.1.4, and the equations for straight-bar development length in footings are given in Example 4.1. Here, were focus on the development length of deformed reinforcement l_d in slabs. The applicable formula of ACI 318-08, Section 12.2 is:

$$l_d = \left[\frac{3}{40}\frac{f_y}{\lambda\sqrt{f'_c}}\frac{\psi_t\psi_e\psi_s}{\left(\dfrac{c_b + K_{tr}}{d_b}\right)}\right]d_b$$

The variables for this equation, assuming common slab conditions, are listed as follows. The assumptions are: Bar sizes d_b are No. 6 and smaller; uncoated bars; bar cover of at least 1.5 in.; and $f'_c = 3000$ psi.

$\psi_t = 1.0$ (< 12 in. of concrete below bars)

$\psi_e = 1.0$ (uncoated bars)

$\psi_s = 0.8$ (bars less than No. 7)

$\psi = 1$ (normal weight concrete)

$c_b = 1.5 + 6/16 = 1.875$ in. (distance from top of concrete to center of No. 6 bar, this rules over the criterion of ½ bar spacing)

$K_{tr} = 0$ (no transverse reinforcement)

Check $\dfrac{c_b + K_{tr}}{d_b} = \dfrac{1.875 + 0}{0.75} = 2.5$ (same as 2.5 in. max. value) $\therefore$ use 2.5 in.

$$l_d = \left(\frac{3}{40}\frac{60{,}000}{1\sqrt{3000}}\frac{1\times 1\times 0.8}{2.5}\right)d_b = 26.3d_b$$

When $f'_c = 4000$ psi concrete is used, the formula becomes:

$$l_d = \left(\frac{3}{40}\frac{60{,}000}{1\sqrt{4000}}\frac{1\times 1\times 0.8}{2.5}\right)d_b = 22.8d_b$$

The straight-bar development lengths in slabs for common bar sizes are given in Table 6.3.

Design Example 6.2: Hairpin Bars

Problem Design hairpins for a small single-story single-span rigid frame spanning 60 ft. The spacing between the frames is 20 ft; the eave height is 18 ft; the roof pitch is 1:12. The roof live load (RLL), reduced for primary frames, is 12 pounds per square foot (psf). The building manufacturer supplied the column reactions. The following load combination produces the maximum outward horizontal load: Dead (1 kip) + RLL (4 kips). Use $f'_c = 3000$ psi and $f_y = 60{,}000$ psi. (1 kip = 1000 lb). The slab on grade is 5 in. thick and reinforced with deformed bars for temperature and shrinkage.

Bar Size No.	$f'_c = 3000$ psi	$f'_c = 4000$ psi
4	13.2	11.4
5	16.4	14.3
6	19.7	17.1

The table is based on the assumptions listed in the text. For bars not listed, the following equations may be used: $l_d = 26.3 d_b$ for $f'_c = 3000$ psi; $l_d = 22.8 d_b$ for $f'_c = 4000$ psi.

TABLE 6.3 Straight-Bar Development Length L_d in Slabs for Some Common Bars, in.

Solution The total outward-acting horizontal force on the hairpin T is $1 + 4 = 5$ kips. Using the allowable bar tension stress of 24 ksi, find the hairpin size:

$$A_{\text{hair}} = \frac{T\sqrt{2}}{2(24)} = \frac{5\sqrt{2}}{2(24)} = 0.147 \text{ in.}^2$$

One No. 4 or No. 5 bar is sufficient. As noted previously, we recommend a minimum bar size of No. 5 for increased corrosion resistance. To determine the length of the hairpins, use No. 4 slab bars at 15 in. o.c. (see Table 6.2), with $A_s = 0.2/1.25 = 0.16$ in.^{2}/ft. For concrete with $f'_c = 3000$ psi and No. 4 bars $l_d = 13.2$ in. (Table 6.3).

Using $F_t = 24$ ksi, the reinforced slab can resist the following service load:

$$0.16 \text{ in.}^2/\text{ft} \times 24 \text{ ksi} = 3.84 \text{ kips/ft of slab width}$$

The total required width of slab to be engaged is:

$$5 \text{ kips}/3.84 \text{ kips/ft} = 1.3 \text{ ft}$$

With this trivial length only two slab bars need to be engaged, so even adding the development length of a hook (9.6 in. from Table 6.1), a nominal hairpin length of 3 ft is more than sufficient.

Conclusion No. 5 hairpins with 90° included angle can be used, with each leg 3 ft long, standard hooks at ends.

As a side note, simple slab ties could have been sufficient in this situation. These are described in the next section. ▲

6.2.5 Slab Ties (Dowels)

For small horizontal loads, such as those in Design Example 6.2, slab ties might provide sufficient resistance to the horizontal column reactions. Slab ties are bent bars or dowels that extend from the foundation piers and walls into the slab. In plan, they are perpendicular to the outside edge of the wall. Figure 5.3 is one example of such dowels extending into the slab; Fig. 6.11 nearby is another.

One difference between the two illustrations is that the slab in Fig. 6.11 is thickened at its bearing area on the foundation. The added thickness allows the slab to better

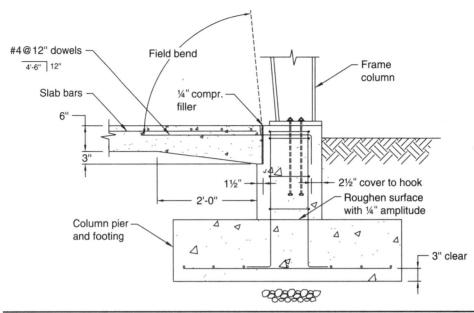

Figure 6.11 Slab ties (dowels).

withstand any concentrated loads, such as wheels of a forklift, in the vicinity of the foundation wall, where soil compaction is likely to be less than ideal. Various practices exist on the recommended amount of the added thickness and taper; the author has successfully used the design shown.

Because of the difficulty with soil compaction near exterior foundation walls and piers, as well as around any horizontal dowels extending into the slab, the dowels in Fig. 6.11 are field bent into the slab *after* the soil is compacted. The length of the dowels should be sufficient to overlap the point where the slab taper starts by at least 6 to 12 in. Why? If both the thickened part of the slab and the dowels stop at the same location, a plane of relative weakness in the slab would be created, inviting shrinkage cracking there. By continuing the reinforcement past the point where the taper begins, the effects of change in slab strength and stiffness are reduced.

Slab ties essentially act as distributed hairpins. The dowels are intended to overlap the slab reinforcement, and their horizontal length should be sufficient to provide for the required lap-splice length of the slab bars. In Design Example 6.2, only two of those ties would have been needed to develop the required horizontal resistance.

Slab ties are typically placed throughout the entire building perimeter at close intervals, such as 12 in. o.c. Because of their large numbers slab ties are much more redundant than a few hairpin bars, making each tie much less critical. Accordingly, the additional bar thickness that is prudent to specify for heavily loaded hairpins subject to corrosion, might not be needed for the ties, and their size may be kept is common No. 4 bars. As with hairpins, the effectiveness of this system depends on the continuity of the slab reinforcement, and we still recommend using deformed bars in the slab whenever slab ties are specified for transfer of horizontal loads.

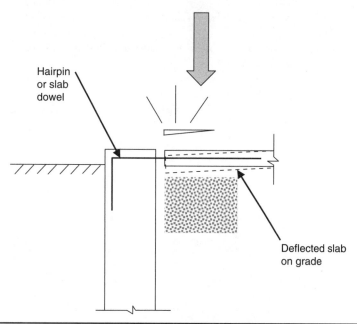

Figure 6.12 Potential failure at slab on grade separated from the foundation caused by settlement or deflection.

6.2.6 Using Foundation Seats

The last point regards the need for a slab seat at the interface of the slab on grade and the foundation wall or column pedestal. Whenever slab ties or hairpins are used, the slab should bear on a continuous seat, typically 1.5 or 2 in. wide, such as that illustrated in Figs. 6.11 and 5.3. The purpose of the seat is to prevent the slab from deflecting relative to the foundation wall or pier.

The slab on grade might deflect in this fashion because of concentrated loading on the slab, caused by wheeled equipment, for example, or because of long-term soil settlement. When the slab starts to deflect, the hairpins or dowels extending from the foundation will try to restrain the movement and will likely fail or cause the slab to crack (Fig. 6.12). The larger the separation between the edge of the slab and the foundation, the bigger the problem. Recall that in some designs the floor on ground is separated from the perimeter walls by compressible filler (see Figs. 6.3 and 5.2) or even by rigid insulation. Whenever those boundary conditions occur, the slab should not be tied to the foundation walls or pedestals, and other foundation systems should be used instead.

References

2009 International Building Code® (IBC-09), International Code Council, Country Club Hills, IL, 2009.

ACI 318-08, *Building Code Requirements for Structural Concrete and Commentary*, American Concrete Institute, Farmington Hills, MI.

ACI 360R-06, *Design of Slabs-on-Ground*, American Concrete Institute, Farmington Hills, MI.

Newman, Alexander, *Metal Building Systems: Design and Specifications*, 2d ed., McGraw-Hill, New York, 2004.

Newman, Alexander, *Structural Renovation of Building: Methods, Details, and Design Examples*, McGraw-Hill, New York, 2001.

UFC 3-220-07 (US Army TM 5-818-7), Foundations on Expansive Soils, 2004.

CHAPTER 7
Moment-Resisting Foundations

7.1 The Basic Concept

Is there a foundation system designed to resist both vertical and horizontal column reactions that does not rely on the contribution of the slab on grade? For reasons explained in Chaps. 3 and 6, the slabs are not the best building elements to rely on for transfer of lateral loads. The slabs typically contain joints, are liable to be cut or partly removed during the service life of the building, and are often lightly—if at all—reinforced. The reliability of any foundation system that does not have to rely on floors on ground would be rather high. Fortunately, at least one such system exists. Moment-resisting foundations are intended to resist both the vertical and horizontal column reactions exerted by metal building frames.

7.1.1 A Close Relative: Cantilevered Retaining Wall

The basic idea behind this system is far from new: It follows the decades-old approach for designing cantilevered retaining walls made of cast-in-place concrete. These walls are designed to resist the overturning and sliding effects of the soil behind them by using the weight of that same soil.

The weight of the soil carried on top of the wall *heel* (the back part of the wall base) acts as ballast, which helps resist both the sliding and the overturning on the wall. Accordingly, the heel of the cantilevered concrete retaining wall is rather long, while the *toe*—the front part of the base—is relatively short (Fig. 7.1). The proportions of the base are established by trial and error, depending on whether the soil surface behind the wall is horizontal or sloped.

With horizontal backfill, the width of the base might be approximately one-half of the wall height, measured from the top of the soil to the bottom of the base. With sloping backfill, the width of the base might be two-thirds of the wall height. These rules of thumb provide only a rough idea about the proportions of the walls, since the proportions are heavily influenced by the depth of the wall, the type of the backfill, and the required factors of safety. The longer the heel, the larger the weight of the soil on top of it and the larger the resisting moment against the overturning. In some circumstances the base of the wall might equal or even exceed the height.

To increase the sliding resistance of the cantilevered retaining wall, its depth can be increased, or a shear key can be provided at the base. A detailed procedure for the design of cantilevered concrete retaining walls is provided in *CRSI Design Handbook*.

114 Chapter Seven

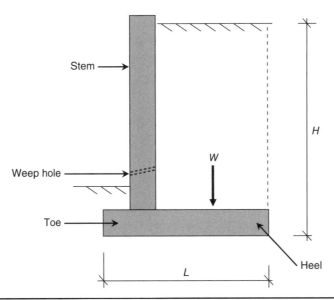

FIGURE 7.1 Cantilevered concrete retaining wall.

One key difference exists between the cantilevered retaining wall and the moment-resisting foundation supporting the metal building: The latter benefits from the vertical reaction of the building frame, while the former rarely support columns. In an effort to utilize the effects of the frame vertical reaction P to our advantage, the proportions of the heel and toe are reversed in moment-resisting foundations vis-à-vis those in cantilevered retaining walls. The base of a moment-resisting foundation typically has a long toe and a short heel (Fig. 7.2)—the exact opposite of the situation in a cantilevered retaining wall.

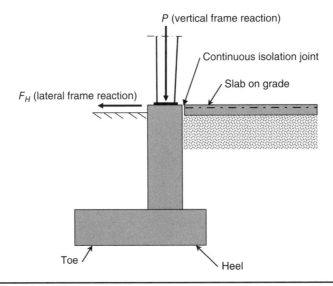

FIGURE 7.2 Moment-resisting foundation.

7.1.2 Advantages and Disadvantages

The main advantage of the moment-resisting foundation system has already been stated: a complete independence of the slab on grade. It means that the integrity of the foundation system will not suffer when the slab is cut, repaired, or partly removed. As discussed in Chap. 6, these operations often accompany building renovations or installation of the under-slab pipes, conduits, and so on. The freedom from the floor also means that any number of floor pits, depressions, and trenches can be easily accommodated. Another advantage of this system is that horizontal frame reactions that act both inward and outward can be resisted by the same foundation.

The moment-resisting foundation has its disadvantages. The design process is lengthy and cumbersome; the resulting foundation is often large and therefore more expensive than a conventional foundation designed only for vertical loading. Also, as discussed next, these types of foundations, as their cantilevered concrete retaining wall cousins, must rotate slightly under load to develop the passive pressure of the soil. This rotation might become problematic when the foundation supports brittle wall materials and finishes.

7.2 Active, Passive, and At-Rest Soil Pressures

To understand how the moment-resisting foundation resists horizontal and vertical frame reactions, the concept of active and passive soil pressures must be discussed first. The easiest way to illustrate the topic is to examine the behavior of the cantilevered retaining walls.

7.2.1 The Nature of Active, Passive, and At-Rest Pressures

When a cantilevered retaining wall retains the unbalanced soil (that is, the soil elevation is higher on one side of the wall than on the other), the soil on the high side produces lateral pressure on the wall called the *active pressure*. The retaining wall rotates slightly under load until it compresses the soil on the opposite side, which pushes back on the wall and develops the *passive pressure*.

The total force P_a exerted by active pressure can cause the wall to fail in two ways: overturning and sliding. In addition to the passive pressure force P_p, sliding is resisted by the force of soil friction (Fig. 7.3). If the combined resistance of the passive pressure and soil friction is sufficient to counteract the overturning and sliding effects of the active pressure, a stable equilibrium is achieved. Otherwise, the wall could topple over and fail or move horizontally.

Moment-resisting foundations are subjected to the same active and passive soil pressures, but there are two key differences. First, the soil elevations on both sides of the foundation are generally similar, although the top of the soil at the exterior is generally somewhat lower than the soil at the interior of the building. Second, unlike in a cantilevered retaining wall, there is also a horizontal column reaction to resist. The active and passive soil pressures, as well as other forces acting on a moment-resisting foundation, are shown in Fig. 7.4.

When the retaining wall or the moment-resisting foundation is restrained at the top by the floor or other structure, the active pressure coefficients do not apply. In this case, a so-called *at-rest coefficient* of soil pressure should be used. The at-rest coefficient has a higher value than the active-pressure coefficient, as discussed in the next section.

116 Chapter Seven

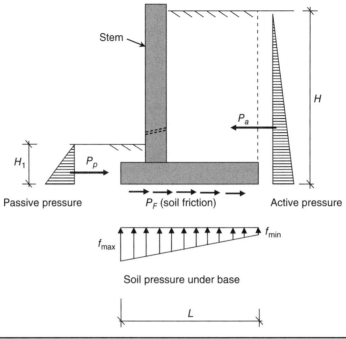

Figure 7.3 Active and passive soil pressures in a cantilevered retaining wall.

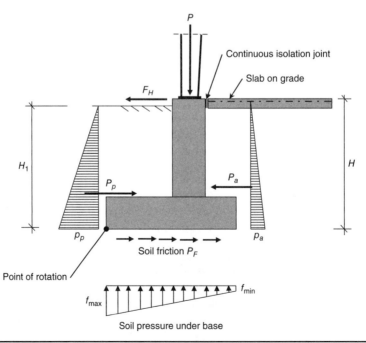

Figure 7.4 Active and passive soil pressures in a moment-resisting foundation.

7.2.2 How to Compute Active, Passive, and At-Rest Pressure

The magnitude of active and passive soil pressures depends on the type and characteristics of the soil. There are various methods of establishing these pressures. Perhaps the most widely used is the Rankine theory, where the active and passive soil pressures are approximated by comparing the soil pressures to those of a liquid, with triangular stress distribution that linearly increases with increasing depth.

According to the Rankine theory, both active and passive pressures depend on the angle of internal friction (ϕ), which can be found from the Mohr's failure envelope for shear strength. In practice, the angle of internal friction is generally determined by the geotechnical consultants, although there are some sources, including *CRSI Design Handbook*, that provide their values directly for some typical soils. The Rankine liquid-analogy formulas for active pressure (p_a) and passive pressure (p_p) are as follows:

$$p_a = K_a \gamma (H), \text{ and } K_a = \tan^2 \phi \ (45° - \phi/2)$$
$$p_p = K_p \gamma (H_1), \text{ and } K_p = \tan^2 \phi \ (45° + \phi/2)$$

where γ = soil density in lb/ft³ (pounds per cubic foot [pcf])
 K_a = active pressure coefficient
 K_p = passive pressure coefficient
 H and H_1 = heights of active and passive pressure areas (see Fig. 7.4)

Once the active and passive pressure values are determined, the active (P_a) and passive (P_p) pressure forces can be found from the following equations:

$$P_a = \tfrac{1}{2} K_a (H^2)$$
$$P_p = \tfrac{1}{2} K_p (H_1^2)$$

These forces are located at the centroids of the respective triangles, that is, the force P_a is located a distance $1/3H$ and the force P_p a distance $1/3H_1$ from the bottom of the foundation's base. We should note that with level backfill the direction of the force P_a is horizontal, while with sloped backfill the force vector is parallel to the slope. The direction of the force P_p is normally assumed to be horizontal. Unlike in cantilevered retaining walls, the backfill in moment-resisting foundations is generally horizontal.

As mentioned in the previous section, the at-rest soil pressure coefficient K_o (sometimes pronounced "Kay-not") applies when the retaining wall or the moment-resisting foundation is restrained at the top. (A good example is a basement wall laterally braced at the top by the first-floor framing.) The value of K_o is usually determined by geotechnical analysis or taken directly from the building codes and authoritative publications, as further discussed later.

The active, passive, and at-rest pressures represent the loading experienced by the soil in the moist condition, while compacted to the optimum density level. If the soil is saturated or submerged, and additional loading component is added—the hydrostatic pressure of groundwater. In this case the lateral soil loads are reduced by the force of buoyancy, but another force of the lateral groundwater pressure is added. As active pressure, the lateral pressure of water increases linearly downward in a triangular fashion.

7.2.3 Typical Values of Active, Passive, and At-Rest Coefficients

As just stated, the values of K_a (active pressure coefficient), K_p (passive pressure coefficient) and K_o (at-rest pressure coefficient) are generally determined by the geotechnical analysis.

However, some reference sources list these coefficients for some idealized typical soils. For example, *CRSI Design Handbook* provides the following coefficients:

- For clean drainable fill, such as sand and gravel free of fines:

$$\phi = 30°, K_a = 0.33, K_p = 3.00$$

- For mixed grain sizes with fines, dense enough for relatively low permeability:

$$\phi = 35°, K_a = 0.271, K_p = 3.69$$

In addition, some building codes and authoritative publications directly provide the design lateral soil loads rather than the coefficients. ASCE/SEI 7-10 Table 3.2-1 (and ASCE/SEI 7-05 Table 3-1) *Design Lateral Soil Load* lists the lateral pressures for various backfill materials. For example, for well-graded and poorly graded clean gravel and gravel/sand mixes (soil types GW and GP) and for silty gravels the design lateral soil load is 35 pounds per square foot (psf)/ft of depth.

As for at-rest soil pressures, a footnote to the tables calls for increasing the lateral soil loads "for relatively rigid walls, as when braced by floors" as follows:

- For sand and gravel-type soils, use 60 psf/ft of depth.
- For silt and clay-type soils, use 100 psf/ft of depth.

An exception from these requirements is given to the basement walls extending not more than 8 ft below grade and supporting light floors.

Another source is the *2009 International Building Code®* (IBC) Table 1610.1, which provides the values of the design active pressure and design at-rest pressure for various soils, as summarized in Table 7.1. As in ASCE 7, the IBC provisions allow the retaining walls that can move and rotate under lateral soil loading to be designed for active pressures. The walls restricted against movement at the top must be designed for at-rest pressures, except the basement walls extending not more than 8 ft below grade and

Backfill Material (Classification per ASTM D 2487)	Unified Soil Classification	Design Active Pressure (psf/ft of Depth)	Design At-Rest Pressure (psf/ft of Depth)
Well- and poorly graded clean gravels or sands; gravel-sand mixes	GW, GP, SW, SP	30	60
Silty sands, poorly graded sand-silt mixes	SM	45	60
Clayey sands, poorly graded sand-clay mixes; mixture of inorganic silt and clay; inorganic clays of low to medium plasticity	SC, ML-CL, CL	60	100

Notes
1. Partial table derived from IBC-06 and IBC-09 Table 1610.1.
2. Design loads are for optimum densities; add hydrostatic loading for saturated soil.

TABLE 7.1 Representative Soil Lateral Loads per *International Building Code®*

Material	Lateral Bearing (psf/ft below Natural Grade)	Lateral Sliding Coefficient of Friction	Cohesion (psf)
Sandy gravel and/or gravel	200	0.35	—
Sand, silty sand, clayey sand, silty gravel, clayey gravel	150	0.25	—
Clay, sandy clay, silty clay, clayey silt, silt, sandy silt	100	—	130 × contact area

Note: Partial table derived from IBC-09 Table 1806.2 (IBC-06 Table 1804.2).

TABLE 7.2 Representative Soil Passive Resistance per *International Building Code®*

laterally supported by flexible diaphragms may be designed for active pressures. A separate IBC table (IBC-09 Table 1806.2 and IBC-06 Table 1804.2) provides the values for soil passive pressures and lateral-sliding coefficients of friction for a variety of soils, as summarized in Table 7.2. The tabulated values for lateral sliding resistance can be increased linearly up to a maximum value of 15 times the tabular value. The tabulated lateral bearing pressures can be increased by one-third when used with the IBC alternative basic load combinations that include wind or seismic loads.

7.3 Lateral Sliding Resistance

7.3.1 The Nature of Lateral Sliding Resistance

As the name suggests, lateral sliding resistance refers to the frictional forces that develop between the soil and the bottom of the retaining wall or the moment-resisting foundation under lateral loading. Lateral sliding resistance is expressed in terms of the force P_F developed at the bottom of the foundation and acting in the horizontal direction (see Figs. 7.3 and 7.4). The magnitude of the force P_F depends on two main factors, the vertical loading W_o and the lateral sliding coefficient of friction μ between the soil and concrete:

$$P_F = \mu(W_o)$$

The vertical loading W_o is a sum of the weights of all the foundation elements, the soil on top of the base, and the vertical frame reaction. A number of approaches can be taken to establish the lateral sliding coefficient of friction μ. First, the coefficient can be determined from the project-specific geotechnical analysis. Second, for typical soils, some of the authoritative technical publications listed previously can help.

For example, *CRSI Design Handbook* uses two values of μ in its tables for cantilevered retaining walls: 0.45 and 0.55. These values are used for backfill having the densities γ of either 115 or 130 lb/ft³ (pcf) and the angles of internal friction ϕ of 30°, 35°, and 40°.

Finally, the default IBC values may also be used. As can be seen from Table 7.2, the values of the lateral sliding coefficient of friction listed there are very low: 0.35 for sandy gravel and/or gravel, and 0.25 for sand, silty sand, clayey sand, silty gravel, and clayey gravel. Essentially, the listed IBC default values are approximately one-half of what *CRSI Design Handbook* uses. Why such a disparity?

The reason IBC default values are so low is that the code is intended to be used in a wide variety of locations, including those outside the United States. Some allowance is evidently made for a possible misidentification of the soil and for field errors. Incidentally, the presumptive load-bearing values listed in IBC-09 Table 1806.2 are also very conservative. Among other authoritative sources, NAVFAC DM-7.2, *Foundations and Earth Structures*, lists much higher allowable load-bearing values for soils. For example, for medium-to-compact coarse-to-medium sand NAVFAC DM-7.2 lists the "recommended value for use" as 3 tons per sq. ft, or 6000 psf. By contrast, the IBC presumptive load-bearing value for sand is only 2000 psf—three times less!

It should be clear from the foregoing discussion that using the default IBC values is rarely the first option to consider. Chances are, the values obtained from a geotechnical investigation would be higher. There are certainly circumstances when using the IBC values is justified, such as when a small project is involved, or when the geotechnical expertise is not available for some reason.

The design approach using the lateral sliding coefficient of friction works well with cohesionless soils, such as sands and gravels. For cohesive soils, such as clay, sandy clay, silty clay, clayey silt, silt, and sandy silt, IBC-09 Table 1806.2 (IBC-06 Table 1804.2) stipulates another approach. For these soils, lateral sliding resistance is determined as the listed value of cohesion (130 psf) times the contact area, but not more than one-half dead load.

7.3.2 Combining Lateral Sliding Resistance and Passive Pressure Resistance

Can the passive pressure resistance force P_p discussed in Sec. 7.2 be combined with the lateral sliding resistance force P_F? Some have questioned the common practice of combining the two, pointing out that these two forces are dissimilar in nature. To develop the passive-pressure resistance, some rotation and movement of the foundation is necessary under load. To develop the soil frictional force, a constant concrete-to-soil contact is necessary. Thus the conceptual objection has some theoretical merit, but the successful long-term practice of combining the two argues otherwise.

In any event, the code provisions specifically allow combining the two lateral-resistance mechanisms. According to IBC-09 Paragraph 1806.3.1, Combined Resistance:

> The total resistance to lateral loads shall be permitted to be determined by combining the values derived from the lateral bearing pressure and the lateral sliding resistance specified in Table 1806.2.

Curiously, the previous edition of IBC (2006) includes an even more definitive statement on the subject. IBC-06 Paragraph 1804.3, Lateral Sliding Resistance states:

> The resistance of structural walls to lateral sliding shall be calculated by combining the values derived from the lateral bearing and the lateral sliding resistance shown in Table 1804.2 unless data to substantiate the use of higher values are submitted for approval.

Accordingly, the forces P_p and P_F are typically combined, so that the total exceeds the applied lateral reaction P_a times the factor of safety:

$$P_p + P_F \geq \text{S.F.} (P_a)$$

The factors of safety against overturning and sliding are discussed next.

7.4 Factors of Safety against Overturning and Sliding

7.4.1 No Explicit Factors of Safety in IBC Load Combinations

Factors of safety can be either explicitly stated or implicitly accounted for in the load combinations. The IBC does not include explicit factors of safety for the foundations subjected to overturning, sliding, and uplift. Instead, the code provides implicit factors of safety built into the load combinations where the dead load counteracts the effects of wind or live load. (We should also note that IBC-09 Paragraph 1806.1 also permits a one-third stress increase in vertical foundation pressure and lateral bearing pressure in these the load combinations when wind load is present.)

For example, the basic IBC load combination of $0.6D + W + H$ for allowable stress design (ASD) has an implied safety factor for the dead-load effects counteracting wind (W) and lateral earth pressure (H) of $1/0.6 = 1.67$.

For alternative basic load combinations, such as $D + L + \omega W$, where the weight of the foundation counteracts the effects of wind load, the load D should include "only two-thirds of the minimum dead load likely to be in place", which corresponds to the implied safety factor of $1/(2/3) = 1.5$. (See Sec. 4.2 for additional discussion on IBC load combinations.)

Accordingly, the minimum safety factors against uplift caused by wind should be taken as either 1.67 or 1.5, depending on whether basic or alternative basic IBC load combinations are used.

7.4.2 Explicit Factors of Safety for Retaining Walls

Section 1807.2.3, Safety Factor, of the *2009 International Building Code* does include the explicit factors of safety of 1.5 against overturning and sliding of retaining walls. Since moment-resisting foundations closely resemble retaining walls, it follows that this IBC section should apply to their design as well. When the explicit factors of safety of 1.5 are employed, the load combinations listed in IBC-09 Section 1605 are not used. Instead, nominal loads should be considered without any multipliers.

An exception applies to earthquake loading, which is taken as 0.7 times the design earthquake load. Also, the minimum safety factor against overturning and sliding of retaining walls designed for earthquake loading is 1.1.

IBC-09 Section 1807.2.3 states that the safety factor against lateral sliding of retaining walls is determined by "the available soil resistance at the base of the retaining wall foundation divided by the net lateral force applied to the retaining wall." The terms "resistance at the base" and "net lateral force" suggest a possible design procedure to determine the safety factor against lateral sliding of the moment-resisting foundation: One first reduces the combination of the applied lateral forces and active pressure by the counteracting effects of the passive pressure resistance and then checks if the available lateral sliding resistance divided by the remainder equals or exceeds 1.5. Of course, a simpler approach would be to consider the total available resistance to lateral sliding (the sum of passive pressure and lateral-sliding resistance forces) vis-à-vis the applied horizontal forces (the sum of F_H and P_a in Fig. 7.4).

We should note that different factors of safety against overturning and sliding of the retaining walls are included in other authoritative publications. For example, *CRSI Design Handbook* recommends a factor of safety of 1.5 against sliding, but 2.0 against overturning 2.0 at service loads.

7.4.3 How to Increase Lateral Sliding Resistance

If the safety factors against overturning and sliding of a moment-resisting foundation are insufficient, which steps can be taken to increase its lateral sliding resistance? The following options are available:

1. Make the foundation longer and heavier. For a substantial lateral force, this often means that the footing becomes rectangular in plan, with the longer side parallel to the direction of the force F_H. A longer footing will weigh more and carry more soil on top of its ledges thus increasing the dead load. The larger the dead load, the larger the lateral sliding resistance.

2. Make the foundation deeper. The deeper the foundation, the larger the passive-pressure resistance that can be engaged (up to a depth of 15 ft, as discussed earlier).

3. Provide keys at the bottom of the footing.

The last point requires some elaboration. Keys are sometimes employed in retaining walls, but their use comes with a few caveats. First, construction of keys involves some precision in excavation and in reinforcement placement, which may be difficult to achieve reliably. Obviously, added costs for the excavation, formwork, and concrete are involved. According to *CRSI Design Handbook*, its tables for retaining wall designs attempt to avoid the keys whenever possible.

Second, the keys are typically placed in the undisturbed soil rather than in the fill (NAVFAC DM-7.2). Since some contractors prefer to overexcavate the site and bring the compacted fill up to the required elevation, the keys might end up in the filled area, even when the foundation designer had assumed otherwise.

For those interested in the design procedures for the keys, *CRSI Design Handbook* provides a design example. Essentially, the key extends the depth of the passive-pressure resistance triangle. The key is designed as a reinforced-concrete cantilever subjected to the applicable passive-pressure loading. The keys used in moment-resisting foundations are illustrated in Fig. 7.5.

7.5 The Design Procedures

7.5.1 Design Input

The design procedure for moment-resisting foundations depends on the eccentricity of the applied load relative to the center of the footing. The eccentricity e is determined by dividing the applied bending moment (M) by the vertical load applied to the footing (P):

$$e = \frac{M}{P}$$

The force P is generally the same as the sum of weights (W_o) discussed earlier.

For pin-base columns, the applied bending moment is found by multiplying the horizontal column reaction (F_H) by the distance to the bottom of the footing (H):

$$M = F_H(H)$$

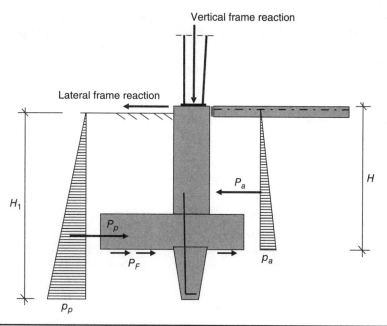

Figure 7.5 Using a key in moment-resisting foundation.

For fixed-base columns, the moment of fixity (M_{fix}) is added:

$$M = F_H(H) + M_{fix}$$

Depending on the magnitude of the load eccentricity e, one of the design methods discussed next can be used.

Consistent with our discussion in the previous chapters, we use the service (unfactored) loads in proportioning the foundations for stability against the overturning and sliding, as well as for determination of the soil pressures. Once the size of the foundation is established, the design of concrete and reinforcement is performed using the factored (ultimate) loading.

If one nevertheless attempts to use the factored forces to determine the eccentricity of load, the ultimate load factors would result in the value of the eccentricity that differs from that found by using the service-load method. This would affect all other calculations that follow.

7.5.2 Design Using Combined Stresses Acting on Soil

Since the footing of the moment-resisting foundation resists the combined effect of the vertical load (P) and the applied bending moment (M), the soil is subjected to the following maximum and minimum pressures:

$$f_{p,\max,\min} = \frac{P}{A} \pm \frac{M}{S}$$

In this equation A is the area and S the section modulus of the footing. The section modulus, computed for the direction of the applied moment, is computed as

$$S = \frac{bL^2}{6}$$

where b is the width and L the length of the footing (measured in the direction of the applied moment). There are three possible relationships between the terms P/A and M/S:

1. $P/A > M/S$. The downward force P predominates, and a trapezoidal distribution of the soil pressure under the footing exists.
2. $P/A = M/S$. The effects of the downward force P and the applied moment M are equal. A triangular distribution of the soil pressure under the footing exists.
3. $P/A < M/S$. The applied moment M predominates. However, since soil cannot resist tension, a double triangle of the soil pressure under the footing cannot be developed. As a result, this design procedure cannot be used.

These three possible relationships between the terms P/A and M/S can be expressed in terms of the position of the resulting eccentricity e relative to the *kern limit* of the footing. The kern limit is a diamond-shaped area in the center of the footing, with the length of the diamond oriented parallel to the direction of the applied moment. The length of the diamond is equal to one-third the length of the footing; the width is equal to one-third the width of the footing (Fig. 7.6). The outer edges of the kern limit are thus located one-sixth of the footing length and width from its center. There relationships are expressed in Fig. 7.7:

1. $P/A > M/S$. The eccentricity e falls within the kern limit.
2. $P/A = M/S$. The eccentricity e falls exactly at the outer edge of the kern limit.
3. $P/A < M/S$. The eccentricity e falls outside the kern limit.

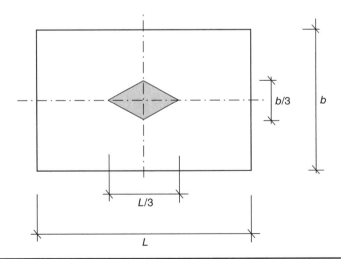

FIGURE 7.6 The kern limit of the footing.

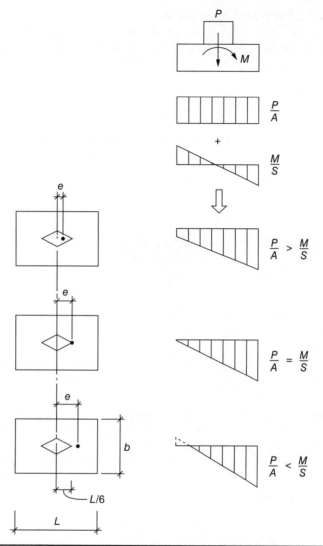

FIGURE 7.7 Three possible relationships between the terms P/A and M/S.

Working with the kern limit of the footing simplifies the decision-making process involved in the selection of the design method. In general, when the first two relationships exist (the eccentricity e falls either within or at the outer edge of the kern limit of the footing), the design procedure that uses combined stresses on the soil can be used. Design Example 7.1 illustrates this procedure.

When the eccentricity e falls outside the kern limit of the footing—but still within the footing's footprint—the pressure wedge method described next can be used instead. When the eccentricity e is so large that it falls outside the footing's footprint, neither of the methods can be used. In that case, the footing must be made larger, or some other method of resisting the applied external moment employed, such as using mats or soil anchors.

7.5.3 The Pressure Wedge Method

As just discussed, the pressure wedge method is useful when the eccentricity of load falls outside the kern limit of the footing, but still within the footing's footprint. In this method the soil is loaded on less than the whole area of the footing, part of which is assumed not to touch the soil. (As stated in the preceding section, soil cannot resist tension, and a double triangle of the soil pressure under the footing cannot be developed.)

From the force equilibrium, the resultant of the soil pressure R must align with the position of the vertical load P, and also be at the centroid of the triangular soil-pressure profile—the pressure wedge (Fig. 7.8). Therefore, the distance a from the outer edge of the footing to the location of the load P is equal to one-half of the footing length L minus the eccentricity e:

$$a = \frac{L}{2} - e$$

Since the centroid of the triangular soil-pressure profile is located at a distance of one-third the length of the triangle (l), the latter can be found as

$$l = 3a$$

Once the length of the pressure wedge is determined, the following relationship should be satisfied to describe the volume of the wedge, including the width of the footing b:

$$R = P = \tfrac{1}{2} b(l) f_{p,\,max}$$

The maximum soil pressure $f_{p,\,max}$ determined from this relationship is then compared to the maximum allowable soil-bearing capacity F_p. If the calculated soil pressure is excessive, the footing length can be increased until the soil pressure is within the allowable. Another alternative is to increase the width of the footing—or increase both the length and the width.

Increasing the length is generally preferable, as this increases the part of the footing that is in contact with the soil. This occurs because the distance to the edge a increases with the increased footing length, which in turn increases the loaded length l.

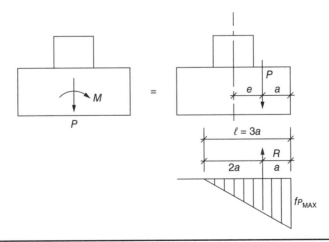

Figure 7.8 The pressure wedge method.

The longer the loaded length, the less rotation the footing is likely to undergo as a result of uneven distribution of the soil pressure. The larger the local pressure spike at the edge of the footing, the more the footing rotates under load. The smoother the pressure distribution under the footing, the less rotation occurs.

This is one of the reasons we prefer to proportion the moment-resisting foundations so that their footings are long enough to allow the eccentricity e to fall within the kern limit, which results in a trapezoidal distribution of the soil pressure under the footing and less movement under load.

7.5.4 General Design Process

The general process for foundation design is outlined in Chap. 3. In the design example that follows we assume that the basic determination of loads, the frost depth, the allowable bearing capacity of the soil, the concrete strength, and so on, have been already made.

Two questions require an answer: First, should the collateral load be considered in the design of moment-resisting foundations? While it should certainly be considered in the design of axially loaded foundations, such as those used to support the intermediate columns of multiple-span rigid frames, the situation is more complex here.

As will become evident in the design example, the proportions of moment-resisting foundations are typically controlled by overturning, sliding, and uplift, rather than the maximum downward-acting loads. The foundation sizes are much larger than required for the resistance of the gravity loads alone. The additional gravity load, such as collateral load, tends to *help* resist the effects of overturning, sliding, and uplift on the foundation, rather than make the loading worse.

It is certainly appropriate to consider the load combinations with and without collateral load, if time permits and computer assistance is available, and use the most critical effects for the design of moment-resisting foundations. However, in the interest of brevity, the design examples that follow do not include the collateral load.

The second question: Should the IBC load combinations be used, or should we use the service (nominal) loads and explicit safety factors? While both approaches can be followed, for moment-resisting foundations we prefer using the service (nominal) loads with explicit safety factors. The latter are based on IBC-09 Section 1807.2.3 factors of safety of 1.5 against the overturning and sliding for the retaining walls. For the load combinations involving wind uplift, we also use a safety factor of 1.5, following the discussion in Sec. 7.4.1 for alternative basic load combinations.

One of the disadvantages of moment-resisting foundations is that their design generally requires a lengthy trial-and-error process. The basic configuration of the foundation needs to be estimated first and then checked for stability against at least two load combinations (with the horizontal frame reaction acting away and toward the building). The trial configuration includes not only the sizes of the footing and the column pedestal (pier), but also the location of the pedestal on the footing.

Once the stability requirements are satisfied, the soil pressures are checked next. If the first trial produces either inadequate or overly conservative results, the configuration is changed and rechecked.

7.5.5 Moment-Resisting Foundations in Combination with Slab Dowels

As mentioned already, one disadvantage of moment-resisting foundations lies in the fact that they tend to rotate under load. If the rotation is large enough, it could increase the stresses in the metal building frame and damage some brittle finishes. One way of avoiding or greatly reducing the rotation is to tie the top of the foundation pier and the perimeter foundation walls to the slab on grade, as discussed in Sec. 6.2.5 and shown in Fig. 6.11.

As long as one designs the moment-resisting foundation independently and does not count on the ties to resist the applied loading, the presence of the ties does not necessarily require that the foundation be designed for at-rest lateral pressures, although not everyone will agree with this statement. Essentially, the assumption is made that the slab ties would not provide the sufficient resistance against rotation under the design level of loading and would be effective only at low levels of load.

Once the design loading takes place—a record-breaking snow accumulation, for example—the moment-resisting foundation will become effective, with all the accompanying cosmetic damage its rotation might cause. Such an occurrence is quite rare, of course, and at that time some minor distress would be judged against the performance of the surrounding structures that might not even survive the event. During the vast majority of the building's service life the foundation rotation under load will not occur, prevented by the slab ties.

Design Example 7.1: Moment-Resisting Foundation

Problem Design the moment-resisting foundation for an industrial building framed with a metal building system. The primary framing consists of single-span rigid frames with pin-base columns. The frames have an eave height of 18 ft, span 80 ft, and are spaced 25 ft on centers. The frost depth is 3.5 ft. Each column is supported by a 24-in. square pedestal (pier), the top of which is 6 in. above the adjacent soil. A continuous foundation wall, of the same depth as the column footing, exists between the piers. The 6-in. slab on grade covers the interior of the building. The column vertical loads are applied at the center of the pier. The roof snow load is 30 psf. Use the frame reactions from the tables in the Appendix. Assume the following material properties:

- Soil weight: 120 lb/ft^3
- Concrete unit weight: 150 lb/ft^3
- Allowable bearing pressure of soil: 2 ksf
- Concrete 28-day compressive strength: $f'_c = 4000$ psi

Assume soil is clean sand free of fines and use parameters from *CRSI Design Handbook*:

$$\phi = 30°, K_a = 0.33, K_p = 3.00, \mu = 0.55$$

Solution

Determine the Design Column Reactions The following column reactions are listed in the Appendix table for an 80-ft-wide frame:

Vertical: dead 4.8 kips, snow 30.9 kips

Horizontal: dead 2.9 kips, snow 21.8 kips

Wind reactions on the right-side column, wind from right to left: Horizontal: 13.6 kips (inward); vertical: 12.2 kips (uplift)

Wind reactions on the left-side column, wind from right to left: Horizontal: 3.1 kips (outward); vertical: 8.4 kips (uplift)

Design the foundation for the right-side column for the following controlling load combinations.

Case 1. Dead + Snow Load Vertical: 4.8 + 30.9 = 35.7 kips (downward); horizontal: 2.9 + 21.8 = 24.7 kips (acting outward).

Case 2. Dead + Wind Load from Right Vertical: 4.8 − 12.2 = − 7.4 kips (uplift); horizontal: 2.9 −13.6 = −10.7 kips (inward).

The direction of the external forces applied on the foundations by gravity loads is shown in Fig. 7.9a and by wind loads in Fig. 7.9b. (Note that these forces act in the opposite direction to the frame reactions.)

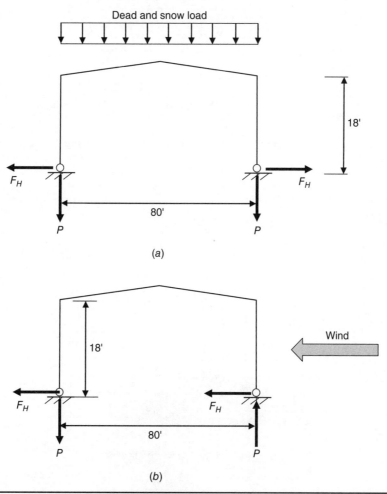

FIGURE 7.9 The direction of the column reactions acting on the foundation in Design Example 7.1: (a) Case 1, gravity loads; (b) Case 2, wind loads.

Proportion the Foundation Establish the foundation size to resist overturning, sliding, and uplift. The foundation size is determined by trial and error.

Case 1: Dead + Snow Load

$$P = 35.7 \text{ kips (downward)}, F_H = 24.7 \text{ kips (outward)}$$

This case provides the largest horizontal force on the foundation. The force acts in the direction away from the building. Try a footing 9 ft long, 4 ft wide and 2 ft thick, with 2 ft by 2 ft column pier. The weights and restoring moments are tabulated as follows. The applied horizontal force (F_H) attempts to overturn the foundation by causing it to rotate about Point A (Fig. 7.10).

Check resistance to overturning first and then check the resulting soil pressures.

Weight	Distance to Point A	Restoring Moment M_R
$W_1 = (0.5 \times 0.15 + 1.5 \times 0.12) 3 \times 4$	= 3.06 kips × 7.5 ft	= 22.95 kip-ft
$W_2 = 2 \times 2 \times 2[(0.15) + (0.12)]$	= 2.16 × 5 ft	= 10.80 kip-ft
$W_3 = 0.15 \times 2 \times 4 \times 9$	= 10.8 × 4.5 ft	= 48.60 kip-ft
$W_4 = 0.12 \times 1.5 \times 4 \times 4$	= 2.88 × 2 ft	= 5.76 kip-ft
P	= 35.7 × 5 ft	= 178.5 kip-ft
	W_o = 54.6 kip	ΣM_R = 266.61 kip-ft

Applied overturning moment $M_{OT} = 24.7 \text{ kips} \times 4 \text{ ft} = 98.8 \text{ kip-ft}$ (clockwise):

$$M_{OT} < \Sigma M_R \quad \text{OK}$$

Factor of safety against overturning:

$$\text{S.F.}_{OT} = \frac{266.61}{98.8} = 2.70 > 1.5 \quad \text{OK}$$

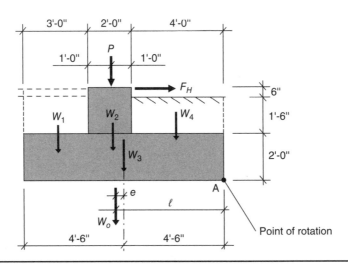

FIGURE 7.10 Forces acting on foundation for Design Example 7.1, Case 1.

The location of the resultant force, measured from point A:

$$l = \frac{266.61 \text{ kip-ft}}{54.6 \text{ kip}} = 4.88 \text{ ft}$$

The resultant of vertical loads acts with an eccentricity with respect to the footing centerline of

$$e = 4.88 - 4.5 = 0.38 \text{ (ft) left of the footing centerline}$$

Then the overall eccentricity of load is

$$e_o = \frac{M_{OT}}{W_o} - e = \frac{98.8 \text{ kip-ft}}{54.6 \text{ kip}} - 0.38 \text{ ft} = 1.43 \text{ ft}$$

The kern limit of the 9-ft footing is

$$\frac{9 \text{ ft}}{6} = 1.5 > 1.43 \text{ ft}$$

The resultant is barely but still within the kern limit of the footing, so the proportion of the foundation is so far satisfactory. The soil pressure can then be determined from the equation:

$$f_{p,\max,\min} = \frac{P}{A} \pm \frac{M}{S}$$

where $P = W_o = 54.6$ kips
$A = 9 \text{ ft} \times 4 \text{ ft} = 36 \text{ ft}^2$
$M = 54.6 \text{ kips} \times 1.43 \text{ ft} = 78.05 \text{ kip-ft}$

$$S = \frac{4 \times 9^2}{6} = 546 \text{ ft}^3$$

$$f_{p,\max,\min} = \frac{54.6}{36} \pm \frac{78.05}{546} = 1.52 \pm 0.14$$

$$f_{p,\max} = 1.66 \text{ ksf} < 2.0 \text{ ksf} \quad \text{OK}$$

$$f_{p,\min} = 1.38 \text{ ksf}$$

Check resistance to sliding, as a combination of passive pressure and soil frictional resistance. From the input data, $K_a = 0.33$, $K_p = 3.00$, $\mu = 0.55$.
One of two approaches can be taken to determine the passive pressure resistance:

1. Count only the passive pressure acting on an inverted "T" section, representing the projected vertical area of the pier and the footing.
2. Since a continuous foundation wall of the same depth as the column footing exists between piers, we can count the passive pressure acting on a 4-ft-wide strip of soil behind the combined area of the footing, pier and the wall on top of the column footing.

For this design example, select the second approach.

The combined soil friction and the passive minus active pressure on a 4-ft-wide strip of soil:

$$F_R = 54.6 \text{ kips} \times 0.55 + \frac{1}{2}(3.00 - 0.33) \times 0.12 \times 3.5^2 \times 4 \text{ (ft)}$$
$$= 30.03 + 7.85 = 37.88 \text{ kips} > 24.7 \text{ kips}$$

Factor of safety against lateral sliding = 37.88/24.7 = 1.53 > 1.5   OK

The factors of safety against overturning and sliding are adequate, even without reliance on the passive pressure of soil.

Case 2: Dead + Wind from Right

$$P = -7.4 \text{ kips (uplift)}, F_H = -10.7 \text{ kips (toward the building)}$$

This case combines the inward-acting horizontal force with uplift. The applied horizontal force (F_H) attempts to overturn the foundation by causing it to rotate about Point B (Fig. 7.11).

Check resistance to overturning first and then check the resulting soil pressures.

Weight	Distance to Point B	Restoring Moment M_R
$W_1 = (0.5 \times 0.15 + 1.5 \times 0.12)3 \times 4$	= 3.06 kips × 1.5 ft	= 4.59 kip-ft
$W_2 = 2 \times 2 \times 2[(0.15) + (0.12)]$	= 2.16 × 4 ft	= 8.64 kip-ft
$W_3 = 0.15 \times 2 \times 4 \times 9$	= 10.8 × 4.5 ft	= 48.60 kip-ft
$W_4 = 0.12 \times 1.5 \times 4 \times 4$	= 2.88 × 7 ft	= 20.16 kip-ft
P	= −10.7 × 4 ft	= −42.8 kip-ft
	W_o = 8.2 kips	ΣM_R = 39.19 kip-ft

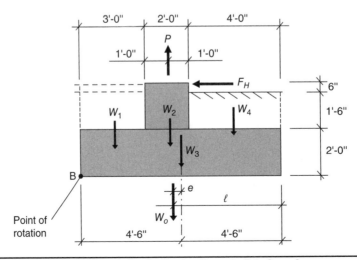

FIGURE 7.11 Forces acting on foundation for Design Example 7.1, Case 2.

The location of the resultant force, measured from point B, using the table of weights shown earlier:

$$l = \frac{39.19 \text{ kip-ft}}{8.2 \text{ kip}} = 4.78 \text{ ft}$$

The resultant of vertical loads acts with an eccentricity with respect to the footing centerline of

$$e = 4.78 - 4.5 = 0.28 \text{ (ft) to the right of the footing centerline}$$

Applied overturning moment $M_{OT} = 10.7$ kips $\times$ 4 ft = 42.8 kip-ft (counterclockwise):

$$M_{OT} = 42.8 \text{ kip-ft} > \Sigma M_R = 39.19 \text{ kip-ft} \quad \text{NG}$$

The weight of the foundation alone is insufficient to resist the overturning.

Check if the passive pressure can help overcome the overturning. (Note: A reliance on passive pressure is rarely done in the analysis of retaining walls, simply because the difference in the heights of the active and passive pressure triangles is quite significant and the passive pressure does not add much resistance. In the case of moment-resisting foundations, passive pressure is more useful.) As is the calculations for Case 1, using for simplicity $h = h_1 = 3.5$ ft:

$$P_p = \frac{1}{2}(3.00 - 0.33) \times 0.12 \times 3.5^2 \times 4 = 7.85 \text{ (kips)}$$
$$M_{R,P} = 7.85 \times 1/3(3.5) = 9.16 \text{ kip-ft}$$

Combined $\Sigma M_R + M_{R,P} = 48.35$ kip-ft $> M_{OT} = 42.8$ kip-ft, but the safety factor is obviously insufficient.

Additional resistance to overturning can be provided by the weight of the continuous foundation wall running between the column foundations. Assume the foundation wall is 12 in. thick and 3 ft deep, bearing on the wall footing that is 1 ft thick and 2 ft wide (Fig. 7.12). Our intent is not to engage the passive pressure on the wall

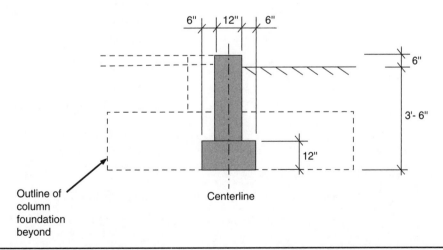

Figure 7.12 Continuous foundation wall running between the primary-frame moment-resisting foundations (the reinforcement shown in Fig. 7.16).

(although this avenue could also be pursued), but simply to use the wall as an additional dead load. The weight of the wall with its foundation plus the soil on its ledges is, conservatively counting only the wall outside the edges of the 4-ft column footings:

$$w_{wall} = [(3 \times 1 + 2 \times 1) 0.15 + (0.5 \times 2.5 + 0.5 \times 3)0.12] = 1.08 \text{ (kip-ft)}$$
$$W_{wall} = 1.08 \times (25 \text{ ft} - 4 \text{ ft}) = 22.68 \text{ (kips)}$$

Note that the centerline of the wall aligns with the center of the column footing. Then

$$W_o = 8.2 + 22.68 = 30.88 \text{ (kips)}$$
$$M_R = 39.19 + 22.68 \times 4.5 = 141.25 \text{ (kip-ft)} > M_{OT} = 42.8 \text{ kip-ft}$$
$$S.F._{OT} = \frac{141.25}{42.8} = 3.3 > 1.5 \quad \text{OK}$$

Sliding is OK by observation, since it was OK for Case 1, with a much higher lateral force. The foundation pressure is also OK by observation, considering the net weight of the moment-resisting foundation itself being only 8.2 kips.

Design the Concrete Elements of the Foundation Once the overall foundation sizes are established, the concrete design can start. As already mentioned, the service loads used so far need to be converted into the factored (ultimate) loads.

1. *Design the Column Pier* The maximum lateral force on the pier acts in Case 1.

$$F_u = 1.2D + 1.6S = 1.2 \times 2.9 + 1.6 \times 21.8 = 38.36 \text{ (kips)}$$
$$M_u = 38.36 \text{ kips} \times 2 \text{ ft} = 76.72 \text{ kip-ft}$$

Check flexure: For a 2- × 2-ft concrete pier (Fig. 7.13) with a minimal vertical reinforcement consisting of three No. 7 hooked bars along two faces placed in the direction parallel to the wall, No. 4 ties and 3-in. clear cover:

$$d = 24 - 3 - \frac{1}{2} - 7/16 = 20.06 \text{ (in.)}$$
$$A_s = 1.80 \text{ (in.}^2\text{)}$$
$$\rho = \frac{1.80}{20.06 \times 24} = 0.00374$$
$$a_u = 4.35*$$
$$\phi M_n = 1.80 \times 4.35 \times 20.06 = 157 \text{ kip-ft} > 76.72 \text{ kip-ft} \quad \text{OK}$$

Check shear:

$$V_u = 38.36 \text{ kips}$$
$$\phi V_n = \phi V_C = 0.75 \times 2 \ (1)\sqrt{4000}(b)(d) = 0.0949(b)(d) \text{ (kips)}$$
$$\phi V_n = 0.0949 \times 20.06 \times 24 = 45.69 \text{ (kips)} > V_u \quad \text{OK}$$

---

*Using the tables in the older versions of *ACI Design Handbook*, as discussed in Chap. 4. See Table 4.1.

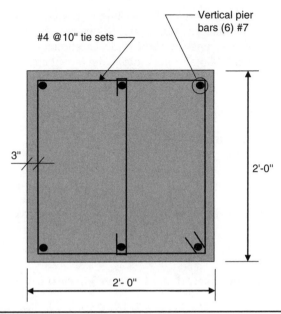

FIGURE 7.13 Concrete pier used in Design Example 7.1.

  2. *Design the Column Footing*   Case 1 controls. The profile of the soil pressure under the footing is shown in Fig. 7.14.
  As determined for Case 1,

$$f_{p,\,max} = 1.66 \text{ ksf}$$
$$f_{p,\,min} = 1.38 \text{ ksf}$$

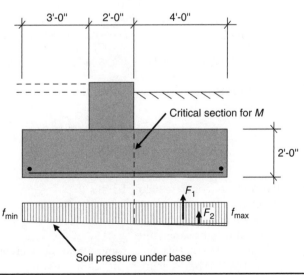

FIGURE 7.14 Soil pressures under the footing for Design Example 7.1, Case 1.

Chapter Seven

The critical section for flexure for the footing is located at the right side of the column pier. To simplify calculations, it is useful to determine the rate of change for the soil pressure, determine the pressure at the critical section for flexure, and find the forces F_1 and F_2 shown in Fig. 7.14. For this example, the rate of change in the soil pressure is

$$1.66 - 1.38 = 0.28/9 = 0.031 \text{ ksf/ft}$$

The pressure at the location underneath the right side of the pier is

$$1.66 - 0.031 \times 4 = 1.536 \text{ ksf}$$
$$F_1 = 1.536 \times 4 \times 4 = 24.58 \text{ (kips)}$$
$$F_2 = \tfrac{1}{2}(1.66 - 1.536) \times 4 \times 4 = 0.99 \text{ (kips)}$$
$$M_{max} = 24.58 \text{ kips} \times 2 \text{ ft} + 0.99 \text{ kips} \times 4(2/3) \text{ ft} = 51.8 \text{ kip-ft}$$

Conservatively using an overall load factor of 1.6 to avoid recomputing the soil pressures,

$$M_u = 51.8 \times 1.6 = 82.89 \text{ kip-ft}$$

For a 24-in.-thick footing with No. 7 bars and 3-in clear cover:

$$d = 24 - 3 - 7/16 = 20.56 \text{ (in.)}$$

Try minimum reinforcement in the long direction, using six No. 7 bars

$$A_s = 3.61 \text{ (in.}^2)$$
$$\rho = \frac{3.61}{20.56 \times 48} = 0.0036$$
$$a_u = 4.35$$
$$\phi M_n = 3.61 \times 4.35 \times 20.56 = 322.9 \text{ (kip-ft)}$$
$$\phi M_n > M_u = 82.89 \text{ kip-ft} \quad \text{OK}$$

Check development length of the reinforcement in the long direction. Center-to-center spacing of for six No. 7 bars with 3 in. clearance at the footing sides is

$$[4 \times 12 - 2 \times 3 - 7/8]/5 = 8.225 \text{ in.} < 18 \text{ in.} \quad \text{OK}$$

The development length of reinforcement for No. 7 bars per ACI 318-08 formula 12-1:

$$l_d = \left[\frac{3 f_y}{40 \lambda \sqrt{f'_c}} \frac{\psi_t \psi_e \psi_s}{\left(\dfrac{c_b + K_{tr}}{d_b}\right)}\right] d_b$$

where $\psi_t = 1.0$ (< 12 in. of concrete below bars)
$\psi_e = 1.0$ (uncoated bars)
$\psi_s = 1.0$ (bars No. 7 and larger)
$\lambda = 1$ (normal weight concrete)
$c_b = 3 + 7/16 = 3.438$ in. (distance from bottom of concrete to center of bar, controls over ½ bar spacing of ½(8.225) = 4.11 in.
$K_{tr} = 0$ (no transverse reinforcement)

Check

$$\frac{c_b + K_{tr}}{d_b} = \frac{3.438 + 0}{0.875} = 3.93 > 2.5 \therefore \text{use } 2.5 \text{ in.}$$

$$l_d = \left[\frac{3}{40} \frac{60,000}{1\sqrt{4000}} \frac{1 \times 1 \times 1}{2.5}\right] 0.875 = 24.9 \text{ in.} > 12.0 \text{ in.}$$

$l_d = 24.9$ in. < available embedment length of $(48 - 3) = 45$ in.   OK

Note that l_d is adequate for the left side of the footing as well, where the available embedment length is $(36 - 3) = 33$ in.

∴ Use six No. 7 bars in the long direction.

The bars in the short direction can be given the minimum shrinkage reinforcement

$$A_{min} = 0.0018 \times 24 \times 9 \times 12 = 4.66 \text{ (in.}^2\text{)}$$

Try 11 No. 6 bars ($A_o = 4.84$ in.^{2})

Bar spacing is

$$[9 \times 12 - 2 \times 3 - 6/8]/10 = 10.12 \text{ in.} < 18 \text{ in.}   \text{OK}$$

Check the development length of reinforcement in the short direction.

$$l_d = \left[\frac{3}{40} \frac{f_y}{\lambda\sqrt{f'_c}} \frac{\psi_t \psi_e \psi_s}{\left(\frac{c_b + K_{tr}}{d_b}\right)}\right] d_b$$

where $\psi_t = 1.0$ (< 12 in. of concrete below bars)
$\psi_e = 1.0$ (uncoated bars)
$\psi_s = 0.8$ (bars less than No. 7)
$\lambda = 1$ (normal weight concrete)
$c_b = 3 + 6/16 = 3.375$ in. (distance from bottom of concrete to center of bar, controls over ½ bar spacing of ½(10.12) = 5.06 in.
$K_{tr} = 0$ (no transverse reinforcement)

Check

$$\frac{c_b + K_{tr}}{d_b} = \frac{3.375 + 0}{0.75} = 4.5 > 2.5 \therefore \text{use } 2.5 \text{ in.}$$

$$l_d = \left[\frac{3}{40} \frac{60,000}{1\sqrt{4000}} \frac{1 \times 1 \times 0.8}{2.5}\right] 0.75 = 17.1 \text{ in.} > 12.0 \text{ in.}$$

$l_d = 17.1$ in. < available embedment length of $(24 - 12 - 3) = 9$ in.   NG

Since the footing cantilever in the short direction is only 12 in., it may be difficult to develop straight reinforcement bars in that direction. Some might attempt to reduce the development length by the "excess reinforcement" ratio of ACI 318-08 Sec. 12.2.5. However, as the commentary to that section states, the reduction factor *does not* apply for development of shrinkage and temperature reinforcement.

Instead, try using either the bars of a smaller diameter or hooked bars. Specifying the smallest commonly used No. 4 bars would require

$$4.66/0.2 = 23.3 \approx 24 \text{ bars, probably an excessive number}$$

Using hooked No. 6 bars would require a basic development length l_{dh} of

$$l_{dh} = \left(\frac{0.02\psi_e f_y}{\lambda\sqrt{f'_c}}\right) d_b$$

where $\psi_e = 1.0$ (uncoated bars)
$\lambda = 1$ (normal weight concrete)

$$l_{dh} = \left(\frac{0.02(1)(60,000)}{1\sqrt{4000}}\right) 0.75 = 14.23 \text{ (in.)}$$

Using modification of 0.7 for concrete cover normal to the plane of the hook ≥ 2.5 in. and cover on bar extension beyond hook ≥ 2 in.:

$$l_{dh} = 0.7 \times 14.23 = 9.96 \text{ in.} > 9 \text{ in.}$$

The required development length in the short direction is still excessive.
Try 15 No. 5 hooked bars instead ($A_o = 4.65$ in.²).

$$l_{dh} = \left(\frac{0.02(1)(60,000)}{1\sqrt{4000}}\right) 0.625 = 11.86 \text{ (in.)}$$

Using modification of 0.7 for concrete cover normal to the plane of the hook ≥ 2.5 in. and cover on bar extension beyond hook ≥ 2 in.:

$$l_{dh} = 0.7 \times 11.86 = 8.3 \text{ in.} < 9 \text{ in.} \quad \text{OK}$$

Check also for minimum l_{dh} of $8d_b$ or 6 in.

$$l_{dh} = 8.3 \text{ in.} > 8 \times 5/8 = 5 \text{ in.} \quad \text{OK}$$
$$l_{dh} = 8.3 \text{ in.} > 6 \text{ in.} \quad \text{OK}$$

Bar spacing is

$$[9 \times 12 - 2 \times 3 - 5/8]/14 = 7.2 \text{ in.} < 18 \text{ in.} \quad \text{OK}$$

∴ Use 15 No. 5 hooked bars in the short direction. The reinforcement is shown in Fig. 7.15.
Check shear in the footing, conservatively considering it at the right face of the pier:

$$V_{max} = 24.58 + 0.99 = 25.57 \text{ (kips)}$$

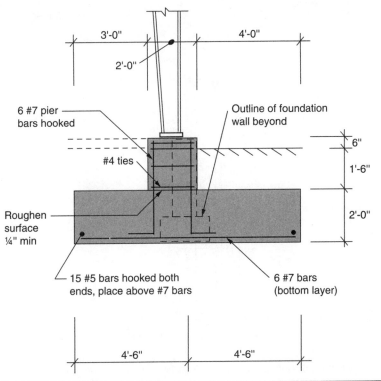

FIGURE 7.15 Moment-resisting foundation designed in Design Example 7.1.

Conservatively using an overall load factor of 1.6 to avoid recomputing the soil pressures,

$$V_u = 25.57 \times 1.6 = 40.91 \text{ (kips)}$$

$$\phi V_c = 0.75 \times 2(1)\sqrt{4000}(b)(d) = 0.0949(b)(d) \text{ (kips)}$$

$$\phi V_n = 0.0949 \times 20.56 \times 48 = 93.62 \text{ (kips)} > V_u \quad \text{OK}$$

3. *Design the Foundation Wall* As mentioned earlier, we have used the 12-in. continuous foundation wall running between the column foundations as additional "ballast." The only check this wall requires is against carrying its own weight suspended in the air. The weight of the wall with foundation plus the soil on its ledges is found earlier:

$$w_{wall} = [(3 \times 1 + 2 \times 1)\,0.15 + (0.5 \times 2.5 + 0.5 \times 3)0.12] = 1.08 \text{ (kip-ft)}$$
$$w_u = 1.08 \times 1.2 = 1.30 \text{ (kip-ft)}$$

Maximum negative moment (at exterior face of first interior support, per ACI 318 Section 8.3 formulas, assuming more than two spans):

$$M_u^- = \frac{1.3 \times 25^2}{10} = 81.25 \text{ (kip-ft)} \quad \text{Controls}$$

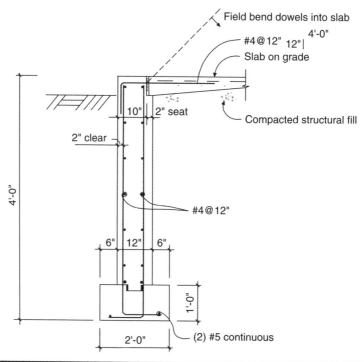

Figure 7.16 Steel reinforcement in the foundation wall designed in Design Example 7.1.

Maximum positive moment (at end spans, discontinuous end unrestrained, per ACI 318 Section 8.3 formulas):

$$M_u^+ = \frac{1.3 \times 25^2}{11} = 73.86 \text{ (kip-ft)}$$

Check if two No. 4 bars at the top and two No. 5 bars at the bottom of the stem are acceptable (see Fig. 7.16).

$$d^- = 48 - 2 - 1/4 = 45.75 \text{ (in.)}$$

$$A_s = 0.4 \text{ (in.}^2)$$

$$\rho = \frac{0.4}{45.75 \times 12} = 0.0007$$

$$a_u = 4.43$$
$$\phi M_n = 0.4 \times 4.43 \times 45.75 = 81.07 \text{ (kip-ft)} \approx 81.25 \text{ OK, since the contribution of other horizontal bars was neglected}$$

The wall can carry its own weight and the weight of the soil on its ledges suspended in the air. This was, of course, a conservative analysis, since no net uplift exists on the wall.

A Final Note   This design example assumed that the column base was pinned. For fixed-base columns, the moment of fixity is added to the overturning moment produced by the horizontal frame reaction. The general procedure would be the same as in this design example. We should note that the moment-resisting foundations designed for the fixity moment tend to become excessively large. This illustrates the suggestion we have made that the specifiers avoid fixed-base columns in metal building frames. The frames with fixed-base columns might provide some relatively minor cost savings for the metal building manufacturer, but the significant increase in the foundation cost will more than offset any such savings.   ▲

References

2009 International Building Code® (IBC), International Code Council, Country Club Hills, IL, 2009.

ACI 318-08, *Building Code Requirements for Structural Concrete and Commentary*, American Concrete Institute, Farmington Hills, MI.

ASCE/SEI 7-05, *Minimum Design Loads for Buildings and Other Structures*, American Society of Civil Engineers, 2006.

ASCE/SEI 7-10, *Minimum Design Loads for Buildings and Other Structures*, American Society of Civil Engineers, 2011.

CRSI Design Handbook, Concrete Reinforcing Steel Institute, Schaumburg, IL, 2002.

NAVFAC DM-7.2, *Foundations and Earth Structures*, Department of the Navy, Naval Facilities Engineering Command, Alexandria, VA, 1982.

Newman, Alexander, *Metal Building Systems: Design and Specifications*, 2d ed., McGraw-Hill, New York, 2004.

CHAPTER 8

Slab with Haunch, Trench Footings, and Mats

8.1 Slab with Haunch

In this chapter we examine three types of shallow foundations used in metal building systems: slab with haunch, trench footings, and mats. The general design principles established in the previous chapters apply for all three. Perhaps the most commonly used of the three is the slab with haunch, and it is examined first.

8.1.1 General Issues

Instead of placing a separate foundation at the exterior of the building, why not simply deepen the edge of the slab and support the building structure on that? The slab with haunch, also known as the downturned slab (Newman, 2004), has been widely used for the support of residential structures in the areas with shallow frost lines. Many contractors familiar with this rather basic system tried to adopt it for supporting the frames in metal building systems. As the following discussion demonstrates, this is not as easy as some had hoped.

The depth and width of the haunch commonly used in residential applications in the warm areas of the United States range from 12 to 18 in. In certain competent soils these modest foundation sizes might be sufficient for the support of lightweight structures, such as single-story houses. The house structure applies distributed reactions from gravity loading on the foundation. However, narrow and shallow foundations of the type represented by the slab with haunch are inappropriate for the support of heavily loaded discrete columns of metal building systems.

Even the larger versions of this system, where the haunch is 24-in. wide and 24- or 30-in. deep, are clearly inadequate: They generally cannot provide either the requisite "ballast" for wind uplift or the passive-pressure resistance against lateral forces. Unfortunately, these haunch sizes are rather commonly used in pre-engineered buildings in the southern regions. Why?

The situation is confused by the presence of the concrete floor slab placed integrally with the haunch. If somebody proposed to separate the haunch from the slab and attempted to bear a large and heavily loaded pre-engineered building column on, say, a 24-in. square footing that was also 24 in. deep—and bear it eccentrically—the folly of this idea would immediately become evident even to a layman (Fig. 8.1).

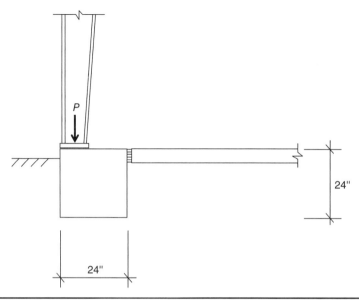

FIGURE 8.1 A small shallow haunch separated from the slab does not inspire confidence.

However, when the slab is present, some people seem to assume that the slab helps in carrying the load. In what fashion can that possibly take place? The point of transition between the haunch and the slab will most likely develop a crack, as the haunch becomes a point of restraint for thermal expansion and contraction of the slab (Fig. 8.2a). Once the crack forms, the continuity disappears and the joint will behave at best as a hinge. It is possible, of course, to provide a gradual transition between the haunch and the slab by using curved formwork (Fig. 8.2b), but this design introduces additional costs and is rarely used.

8.1.2 The Role of Girt Inset

The exact location of the metal building column on the foundation depends on the girt inset. As discussed in Chap. 1, girts can be positioned relative to the columns in three possible ways (insets): bypass, flush, and semiflush. A bypass girt is placed wholly outside of the column's outer flange, while a flush girt is framed in a way that its outside flange is approximately even with the column's.

The column with flush girts will typically be located close to the edge of the footing, while the column with bypass girts will be located farther away. Obviously, using bypass girts helps reduce the load eccentricity on the foundation (Figs. 8.3a and b). In general, whenever slab with haunch is used, we recommend that bypass girts be specified.

8.1.3 Resisting the Column Reactions

The basic behavior of the slab with haunch is similar to that of the moment-resisting foundation. The eccentricity of the vertical column reaction is resisted by the restoring moment provided by the haunch. The overturning and sliding caused by the horizontal column reactions are prevented by the passive pressure at the edge of the haunch and the soil friction under the haunch and the slab. The wider and heavier the haunch, the better it can develop these load-resisting mechanisms.

Slab with Haunch, Trench Footings, and Mats

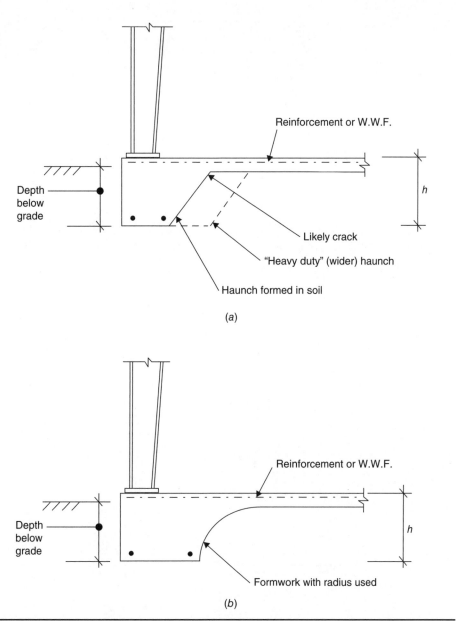

Figure 8.2 Slab with haunch: (*a*) As commonly used; (*b*) using formwork with a radius.

Using a wide haunch introduces a significant amount of dead load, which can greatly lessen the eccentricity of the vertical column reaction. Ideally, the eccentricity will fall within the kern limit of the haunch (see Chap. 7), which will keep the soil pressures under the footing in a trapezoidal pattern (Fig. 8.4). Otherwise, the eccentricity will tend to fall outside the footing's kern limit and result in a spike of high soil pressures at the exterior edge under the column.

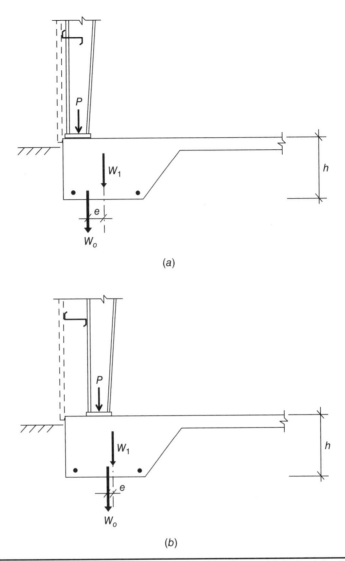

FIGURE 8.3 The influence of column inset on the overall load eccentricity in slab with haunch: (a) With flush girts; (b) with bypass girts.

In addition to the slab haunch at the column location, a perimeter grade beam—the slab with haunch running under the exterior walls between the column foundations—also helps resist the frame reactions. The plan view of slab with haunch with perimeter grade beams is shown in Fig. 8.5.

The perimeter haunch provides additional dead load and a vertical face that can develop passive soil pressure. The basic mechanism of resisting vertical and horizontal frame reactions is shown in Fig. 8.6.

For an outward-acting horizontal frame reaction, the weights of the haunch under the column and the grade beam, as well as the vertical frame reaction, will develop a

Slab with Haunch, Trench Footings, and Mats

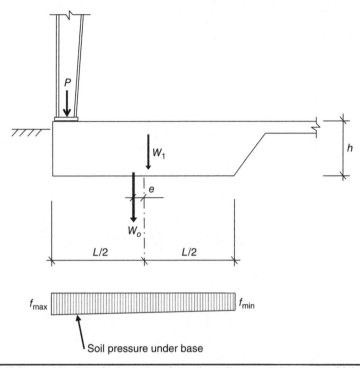

FIGURE 8.4 Using a wide slab with haunch to keep the soil pressure in a trapezoidal pattern.

restoring moment against rotation about point B (Fig. 8.6a). For an inward horizontal reaction, the same weights (minus the vertical column reaction that now acts upward) will develop a restoring moment against rotation about point A (Fig. 8.6b). But what about the slab?

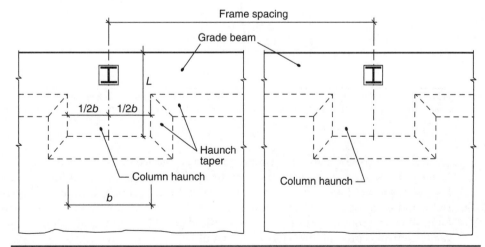

FIGURE 8.5 Plan of slab with haunch with perimeter grade beams.

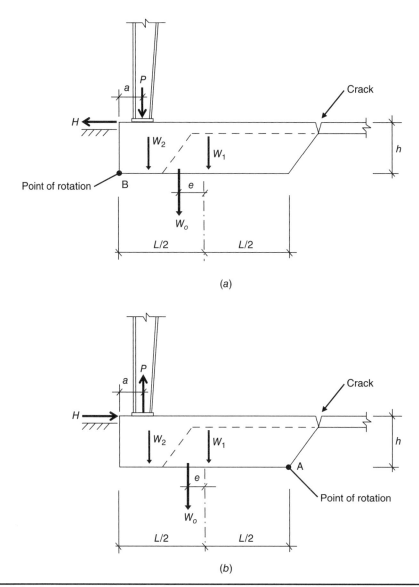

Figure 8.6 Overturning and restoring forces acting on slab with haunch with perimeter grade beams and a crack at the end of the haunch: (a) At maximum downward load; (b) at wind load acting from left.

The degree of the slab's contribution to the resistance of the haunch against overturning and sliding depends on the slab design. If deformed steel reinforcement or at least welded-wire fabric extends from the slab into the haunch, the joint between the two could be considered a hinge. It means that the slab will not be able to stabilize the haunch against rotation, but the weight of the slab could help resist sliding. If no slab reinforcement is present, the haunch area would have to rely only on its own resistance to sliding when the horizontal frame reaction acts outward.

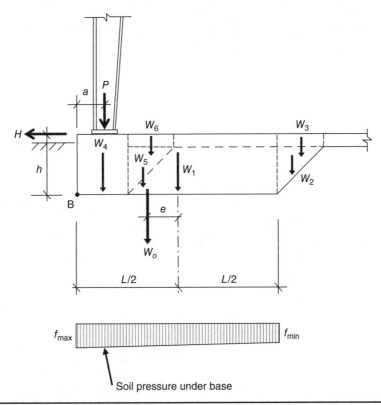

FIGURE 8.7 Overturning and restoring forces acting on slab with haunch for the load combination that produces the maximum downward load.

Since neither the haunch under the column nor the grade beam is rectangular in cross section, the weights of the trapezoidal transition areas between the slab and the haunch could also be included in the system's weight. The bottom of the transition area is sloped, and for this reason it is often conservatively assumed not to bear on the soil at all. The designation of the various components of the weight that exist under the outward or inward-acting horizontal frame reactions are shown in Figs. 8.7 and 8.8.

However carefully the weights are computed, the necessarily shallow depth of the haunch means that this foundation system can resist quite limited horizontal and vertical column reactions. Accordingly, this system is best suited for small buildings with light loading. As discussed in Chap. 3, the moderate cost of the slab with haunch has to be balanced against its low reliability and versatility (see Table 3.1).

The process of designing a slab with haunch for a small building without any slab reinforcement is illustrated in Design Example 8.1.

Design Example 8.1: Slab with Haunch

Problem Design the slab-with-haunch foundation for a small metal building system with single-span rigid frames and pin-base columns. The frames have the eave height of 18 ft, span of 40 ft, and are spaced 25 ft on centers. A continuous foundation grade beam,

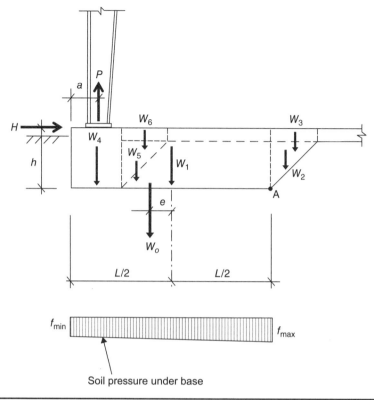

FIGURE 8.8 Overturning and restoring forces acting on slab with haunch for a wind load acting from left.

also a slab with haunch, of the same depth as the column haunch, runs between the column foundations. The 6-in. slab on grade covers the interior part of the foundation.

The column vertical loads are applied 12 in. away from the outside edge (bypass girts are used). No minimum frost depth is specified, and the depth of the haunch for column foundation and the grade beam is 2.0 ft below grade. The top of the slab is 6 in. above grade.

The roof live load is 20 pounds per square foot (psf). Use the frame reactions from the tables in the Appendix. Assume the following material properties:

- Soil weight: 120 lb/ft^3
- Concrete unit weight: 150 lb/ft^3
- Allowable bearing pressure of soil: 4 ksf
- Concrete 28-day compressive strength: $f'_c = 4000$ psi

Assume soil is clean sand, free of fines, and use the following parameters from *CRSI Design Handbook*:

$$\phi = 30°, K_a = 0.33, K_p = 3.00, \mu = 0.55$$

Slab with Haunch, Trench Footings, and Mats

Solution

Determine the Design Column Reactions The following column reactions are listed in the Appendix table for a 40-ft-wide frame:

Vertical: dead 1.9 kips, roof live 10.3 kips

Horizontal: dead 0.5 kip, live 3.3 kips

Wind reactions on the left-side column, wind from left to right: Horizontal: 7.4 kips (inward); vertical: 7.7 kips (uplift)

Wind reactions on the right-side column, wind from left to right: Horizontal: 2.2 kips (outward); vertical: 2.3 kips (uplift)

Design the foundation for the left-side column for the following controlling IBC basic load combinations, but using explicit factors of safety rather than the IBC reduction factors (see a related discussion in Sec. 7.4).

Case 1: Dead + Roof Live Load Vertical: 1.9 + 10.3 = 12.2 kips (downward); horizontal: 0.5 + 3.3 = 3.8 kips (acting outward).

Case 2: Dead + Wind Load from Left on Left-Side Column Vertical: 1.9 − 7.7 = − 5.8 kips (uplift); horizontal: 0.5 − 7.4 = − 6.9 kips (acting inward).

The direction of the external forces applied on the foundations by gravity loads is shown in Fig. 8.9a and by wind loads in Fig. 8.9b. (Note that these forces act in the opposite direction to the frame reactions.)

Proportion the Foundation Establish the foundation sizes to resist overturning, sliding and uplift. The foundation sizes are determined by trial and error.

Case 1: Dead + Roof Live Load

$$P = 12.2 \text{ kips (downward)}, F_H = 3.8 \text{ kips (outward)}$$

This case provides the largest vertical column force on the foundation, applied eccentrically. The horizontal force acts in the direction outward the building. Try a footing 4 ft long, 4 ft wide, and 2.5 ft thick, with a grade beam 2 ft wide. The weights and restoring moments are tabulated as follows. The applied horizontal force (F_H) attempts to overturn the foundation by causing it to rotate about Point B (Fig. 8.10).

Check resistance to overturning first. For the first try, neglect the contribution of the grade beam.

Weight	Distance to Point B	Restoring Moment M_R
$W_1 = 4 \times 4 \times 2.5 \times 0.15$	= 6.00 kips × 2.0 ft	= 12.0 kip-ft
$W_2 = 2 \times 2 \times 0.5 \times 4 \times 0.15$	= 1.2 × 4.67 ft	= 5.6 kip-ft
$W_3 = 0.15 \times 0.5 \times 4 \times 2$	= 0.6 × 5.0 ft	= 3.0 kip-ft
P	= 12.2 × 1.0 ft	= 12.2 kip-ft
	W_o = 20.0 kips	ΣM_R = 32.8 kip-ft

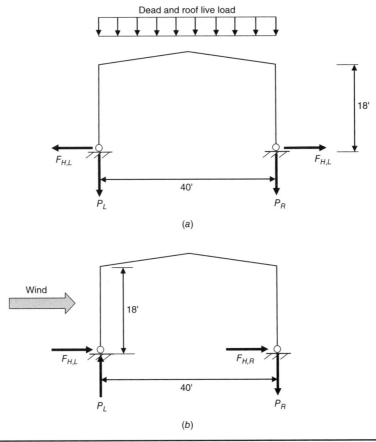

Figure 8.9 Direction of the column reactions acting on the foundation in Design Example 8.1: (a) Case 1, gravity loads; (b) Case 2, wind loads.

Applied overturning moment $M_{OT} = 3.8$ kips $\times$ 2.5 ft = 9.5 kip-ft (counterclockwise).

$$M_{OT} < \Sigma M_R \quad \text{OK}$$

Factor of safety against overturning:

$$\text{S.F.}_{OT} = \frac{32.8}{9.5} = 3.45 > 1.5 \quad \text{OK}$$

The location of the resultant force, measured from point B:

$$l = \frac{32.8 \text{ kip-ft}}{20.0 \text{ kip}} = 1.64 \text{ ft}$$

The resultant of vertical loads acts with an eccentricity with respect to the footing centerline of

$$e = 2.0 - 1.64 = 0.36 \text{ (ft) left of the footing centerline}$$

Slab with Haunch, Trench Footings, and Mats

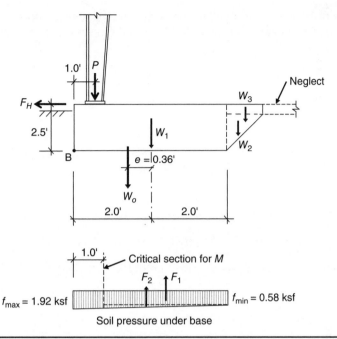

FIGURE 8.10 Overturning and restoring forces acting on the slab with haunch for Design Example 8.1, Case 1, without relying on the grade beam for stability.

The kern limit of the footing is

$$\frac{4 \text{ ft}}{6} = 0.67 > 0.36 \text{ ft}$$

The resultant is within the kern limit of the footing, so the proportion of the foundation is so far satisfactory. The soil pressure is trapezoidal in shape, and the maximum and minimum pressures can be determined from the equation:

$$f_{p,\max,\min} = \frac{P}{A} \pm \frac{M}{S}$$

where
$P = W_o = 20$ kips
$A = 4 \text{ ft} \times 4 \text{ ft} = 16 \text{ ft}^2$
$M = 20 \text{ kips} \times 0.36 \text{ ft} = 7.2 \text{ kip-ft}$
$S = \dfrac{4 \times 4^2}{6} = 10.67 \text{ ft}^3$

$$f_{p,\max,\min} = \frac{20}{16} \pm \frac{7.2}{10.67} = 1.25 \pm 0.67$$

$f_{p,\max} = 1.92 \text{ ksf} < 4.0 \text{ ksf}$ OK
$f_{p,\min} = 0.58 \text{ ksf} > 0 \text{ ksf}$ OK

Check resistance to sliding. Even neglecting the passive pressure and friction under the slab, soil frictional resistance under the foundation alone is sufficient:

$$F_R = 20 \text{ kips} \times 0.55 = 11 \text{ kips} > 3.8 \text{ kips}$$

Factor of safety against lateral sliding = 11/3.8 = 2.89 > 1.5   OK

Case 2: Dead + Wind Load from Left on the Left-Side Column

$$P = -5.8 \text{ kips (uplift)}, F_H = -6.9 \text{ kips (inward)}$$

This case combines the inward-acting horizontal force with uplift. The applied horizontal force (F_H) attempts to overturn the foundation by causing it to rotate about Point A (Fig. 8.11).

Check resistance to uplift first. For the first try, neglect the contribution of the grade beam. From the table of weights for Case 1 the weight of the column haunch alone is

$$6 + 1.2 + 0.6 = 7.8 \text{ (kips)} > 5.8 \text{ kips}$$

However, the factor of safety against uplift is insufficient:

$$\text{S.F.}_{UP} = \frac{7.8}{5.8} = 1.34 < 1.5 \quad \text{NG}$$

The following two options are available to increase the safety factor against uplift:

1. The haunch foundation could be made larger, which would involve placing additional concrete.

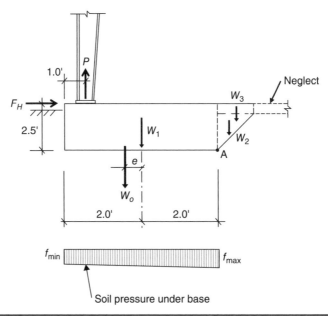

FIGURE 8.11 Overturning and restoring forces acting on the slab with haunch for Design Example 8.1, Case 2, without relying on the grade beam for stability.

2. The weight of the grade beam could be considered in addition to the weight of the haunch at the column. In this case, the grade beam would require flexural reinforcement to allow it to carry its own weight suspended in the air.

Looking into the first option, the required weight of concrete to provide a factor of safety of 1.5 against uplift would be

$$5.8 \times 1.5 = 8.7 \text{ (kips)}$$

which corresponds to a footing of the following width:

$$b_{min} = \left(\frac{8.7}{7.8}\right) 4 = 4.46 \text{ ft}$$

Increasing the footing width from 4 ft to 4.5 ft provides the needed factor of safety against uplift, while preserving the overall cross-section of the haunch. At this stage, such a minor increase in the haunch dimension seems more economical than designing the grade-beam haunches as flexural members.

Now, check resistance to overturning about point A (see Fig. 8.11). Again, for the first try neglect the grade beam's contribution to overturning. Note that the transition areas (weights W_2 and W_3) increase the overturning moment, rather than help reduce it.

Weight	Distance to Point A	Restoring Moment M_R
$W_1 = 4 \times 4.5 \times 2.5 \times 0.15$	$= 6.75$ kips $\times 2$ ft	$= 13.5$ kip-ft
$W_2 = 2 \times 2 \times 0.5 \times 4.5 \times 0.15$	$= 1.35 \times (-0.67)$ ft	$= (-0.9)$ kip-ft
$W_3 = 0.5 \times 2 \times 4.5 \times 0.15$	$= 0.675 \times (-1.0)$ ft	$= (-0.675)$ kip-ft
P	$= (-5.8) \times 3$ ft	$= (-17.4)$ kip-ft
	$W_o = 2.98$ kips	$\Sigma M_R = (-5.48)$ kip-ft

As can be seen, there is still no positive resistance to overturning in this load combination without considering some other resistance mechanism, such as adding the weight of the grade beam to the total "ballast" resisting overturning.

Accordingly, for the next step the design will consider the added weight of the grade beam, as shown in Fig. 8.12. The width of the column haunch can now be reduced back to the original 4.0 ft.

Weight	Distance to Point A	Restoring Moment M_R
$W_1 = 4 \times 4 \times 2.5 \times 0.15$	$= 6.00$ kips $\times 2$ ft	$= 12.0$ kip-ft
$W_2 = 2 \times 2 \times 0.5 \times 4 \times 0.15$	$= 1.2 \times (-0.67)$ ft	$= (-0.80)$ kip-ft
$W_3 = 0.15 \times 0.5 \times 4 \times 2$	$= 0.6 \times (-1.0)$ ft	$= (-0.60)$ kip-ft
$W_4 = 2 \times 2.5 \times 21 \times 0.15$	$= 15.75 \times 3$ ft	$= 47.25$ kip-ft
$W_5 = 2 \times 2 \times 0.5 \times 21 \times 0.15$	$= 6.3 \times 1.33$ ft	$= 8.38$ kip-ft
$W_6 = 0.15 \times 0.5 \times 2 \times 21$	$= 3.15 \times 1.0$ ft	$= 3.15$ kip-ft
P	$= (-5.8) \times 3$ ft	$= (-17.4)$ kip-ft
	$W_o = 27.2$ kips	$\Sigma M_R = 51.98$ kip-ft

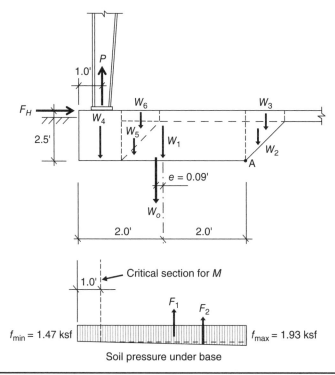

FIGURE 8.12 Overturning and restoring forces acting on the slab with haunch, and soil pressures, for Design Example 8.1, Case 2, relying on the grade beam for stability.

The overall system's safety factor against the uplift force of 5.8 kips is OK by observation.
Applied overturning moment from horizontal force F_H is

$$M_{OT} = 6.9 \text{ kips} \times 2.5 \text{ ft} = 17.25 \text{ kip-ft (clockwise)}$$
$$M_{OT} < \Sigma M_R \quad \text{OK}$$

Factor of safety against overturning:

$$\text{S.F.}_{OT} = \frac{51.98}{17.25} = 3.01 > 1.5 \quad \text{OK}$$

The weight of the column haunch plus the grade beam is sufficient to resist overturning.
Check resistance to sliding. Even neglecting the passive pressure and friction under the slab, soil frictional resistance under the foundation and the grade beam is sufficient:

$$F_R = 27.2 \text{ kips} \times 0.55 = 14.96 \text{ kips} > 6.9 \text{ kips}$$

Factor of safety against lateral sliding = $14.96/6.9 = 2.17 > 1.5$ OK

A determination of the soil pressures under the various components of the system is rather complex, as it involves apportioning the uplift force among the grade beam and the column haunch. To simplify calculations for practical design, an assumption can be made that the grade beam does not bear on the soil and its weight is included in the dead load of the column haunch. (If desired, this assumption can be followed through in the actual foundation construction by removing the soil from under the grade beam after its concrete reaches its 28-day compressive strength.) With this assumption, the resultant force W_o, which includes the weight of the grade beam, is applied at the following distance measured from point A:

$$l = \frac{51.98 \text{ kip-ft}}{27.2 \text{ kip}} = 1.91 \text{ ft}$$

The overall eccentricity of load e with respect to the centerline of the column haunch is

$$e = 2.0 - 1.91 = 0.09 \text{ (ft) to the right of the haunch centerline}$$

The resultant is approximately in the middle of the footing and certainly within the kern limit. The soil pressure is then trapezoidal in shape. Similar to Case 1, the maximum and minimum pressures can be determined from the equation:

$$f_{p,\text{max,min}} = \frac{P}{A} \pm \frac{M}{S}$$

where
$P = W_o = 27.2$ kips
$A = 4 \text{ ft} \times 4 \text{ ft} = 16 \text{ ft}^2$
$M = 27.2 \text{ kips} \times 0.09 \text{ ft} = 2.45 \text{ kip-ft}$
$S = \frac{4 \times 4^2}{6} = 10.67 \text{ ft}^3$

$$f_{p,\text{max,min}} = \frac{27.2}{16} \pm \frac{2.45}{10.67} = 1.7 \pm 0.23$$

$f_{p,\text{max}} = 1.93 \text{ ksf} < 4.0 \text{ ksf}$ OK
$f_{p,\text{min}} = 1.47 \text{ ksf} > 0 \text{ ksf}$ OK

The column haunch can be kept as 4 × 4 ft, if the contribution of the grade beam is considered.

Design the Concrete Elements of the Foundation Once the overall foundation sizes are established, the concrete design can proceed. For concrete design the service loads used so far need to be converted into the factored (ultimate) loads.

1. Design the Footing of the Column Slab Haunch

CASE 1 The profile of the soil pressure under the footing is shown in Fig. 8.10. As determined for Case 1,

$$f_{p,\text{max}} = 1.92 \text{ ksf}$$
$$f_{p,\text{min}} = 0.58 \text{ ksf}$$

Since the size of the column was not given, conservatively assume the critical section for flexure being at the centerline of the column, or 1 ft away from the outside edge of concrete. To simplify the calculations, it is useful to compute the rate of change for the soil pressure, determine the pressure at the critical section for flexure, and find the forces F_1 and F_2 shown in Fig. 8.10. For this example, the rate of change is

$$(1.92 - 0.58) \text{ ksf}/4 \text{ ft} = 1.07/4 = 0.335 \text{ ksf/ft}$$

The soil pressure at the centerline of the column is

$$1.92 - 0.335 \times 1 = 1.585 \text{ (ksf)}$$
$$F_1 = 0.58 \times 3 \times 4 = 6.96 \text{ (kips)}$$
$$F_2 = \tfrac{1}{2}(1.585 - 0.58) \times 3 \times 4 = 6.03 \text{ (kips)}$$
$$M_{max} = 6.96 \text{ kips} \times 1.5 \text{ ft} + 6.03 \text{ kips} \times 3(1/3) \text{ ft} = 16.47 \text{ kip-ft}$$

Find the average load factor:

$$\text{Ave. L.F.} = \frac{10.3(1.6) + (20 - 10.3)1.2}{20} = 1.41$$

$$M_u = 16.47 \times 1.41 = 23.22 \text{ kip-ft}$$

Assume that grade-beam reinforcement, which runs parallel to the exterior edge of the foundation, will consist of No. 6 bars top and bottom. These bars will be placed outside the bars in the column haunch.

For 30-in.-thick footing with No. 4 haunch bars and 3-in. clear cover to the No. 6 bars grade-beam bars:

$$d^{(+)} = 30 - 3 - 6/8 - 4/16 = 26.0 \text{ (in.)}$$

Following ACI 318 Section 10.5.3, the required area of steel reinforcement in the long direction is approximately

$$A_{rq}^{(+)} \approx \frac{(4/3)M_u}{4d} = \frac{23.22(4/3)}{26.0 \times 4} = 0.30 \text{ (in.}^2\text{)}$$

Two No. 4 bars would be sufficient to meet the flexural demand ($A_o = 0.4$ in.^{2}).

However, ACI 318 Section 10.5.4 requires the minimum reinforcing ratio for footings of uniform thickness to be $0.0018 A_g$, which results in a much higher value for the area of reinforcement:

$$A_{s,\,min} = 0.0018 \times 30 \times 48 = 2.33 \text{ (in.}^2\text{)}$$

This area can be satisfied with 12 No. 4 bars ($A_o = 2.4$ in.^{2}) or 8 No. 5 bars ($A_o = 2.48$ in.^{2}). One-half of the bars could be placed at top and bottom of the column haunch. The bar size will likely be controlled by the development length.

Check the development length of No. 4 bars first per ACI 318-08 formula 12-1.

1. Center-to-center spacing of for six No. 4 bars with 3 in. clearance at the footing sides is

$$[4 \times 12 - 2 \times 3 - 1/2]/5 = 8.3 \text{ in.} < 18 \text{ in.} \quad \text{OK}$$

The development length for No. 4 bars is

$$l_d = \left[\frac{3}{40}\frac{f_y}{\lambda\sqrt{f'_c}}\frac{\psi_t\psi_e\psi_s}{\left(\frac{c_b+K_{tr}}{d_b}\right)}\right]d_b$$

where $\psi_t = 1.0$ for bottom bars; $\psi_t = 1.3$ for top bars
$\psi_e = 1.0$ (uncoated bars)
$\psi_s = 0.8$ (bars smaller than No. 7)
$\lambda = 1$ (normal weight concrete)
$c_b = 3 + 6/8 + 1/4 = 4.0$ in. [distance from bottom of concrete to center of bar, controls over ½ bar spacing of ½(8.3) = 4.15 in.]
$K_{tr} = 0$ (no transverse reinforcement)

Check

$$\frac{c_b + K_{tr}}{d_b} = \frac{4+0}{0.5} = 8 > 2.5 \quad \therefore \text{use } 2.5 \text{ in.}$$

For bottom bars:

$$l_d = \left[\frac{3}{40}\frac{60{,}000}{1\sqrt{4000}}\frac{1\times 1\times 0.8}{2.5}\right]0.5 = 11.4 \text{ in} < 12.0 \text{ in.}$$

Use $l_d = 12$ in.
For top bars:

$$l_d = \left[\frac{3}{40}\frac{60{,}000}{1\sqrt{4000}}\frac{1\times 1\times 1}{2.5}\right]0.5 = 14.2 \text{ in.} > 12.0 \text{ in.}$$

Use $l_d = 14.2$ in.
For the right side of the haunch both are less than the available embedment length of

$$36 - 3 = 33 \text{ in.} \quad \text{OK}$$

However, the development length for straight No. 4 bars is insufficient for the left side, where the available embedment length is

$$12 - 3 = 8 \text{ in.} \quad \text{NG}$$

Try six No. 4 bars at top and bottom, hooked at exterior end. The basic development length l_{dh} of No. 4 bars is

$$l_{dh} = \left(\frac{0.02\psi_e f_y}{\lambda\sqrt{f'_c}}\right)d_b$$

where $\psi_e = 1.0$ (uncoated bars)
$\lambda = 1$ (normal weight concrete)

$$l_{dh} = \left(\frac{0.02(1)(60,000)}{1\sqrt{4000}}\right)0.5 = 9.5 \text{ (in.)}$$

Using modification of 0.7 for concrete cover normal to the plane of the hook ≥ 2.5 in and cover on bar extension beyond hook ≥ 2 in.:

$$l_{dh} = 0.7 \times 9.5 = 6.64 \text{ in.} < 8 \text{ in.} \quad \text{OK}$$

The bars in the other direction, parallel to the exterior edge, have the same sizes and spacing, but do not require hooks at the ends, since the available embedment length is (24 − 3) = 21 in. > 14.2 in. for top bars (OK).

Some designers consider negative bending in the haunch caused by the eccentricity between the column load P and the centroid of soil pressure W_o:

$$e_{col} = 2.0 - 1.0 - 0.36 = 0.64 \text{ (ft)}$$
$$P_u = 1.6 \times 10.3 + 1.2 \times 1.9 = 18.76 \text{ (kips)}$$
$$M_u^{(-)} = 18.76 \times 0.64 = 12.01 \text{ (kip-ft)}$$

For the top No. 4 haunch bars, if used, with 2-in. clear cover to the No. 6 bars grade-beam bars:

$$d^{(-)} = 30 - 2 - 6/8 - 4/16 = 27.0 \text{ (in.)}$$

Following ACI 318 Sec. 10.5.3, the required area of steel reinforcement in the direction parallel to that shown in Fig. 8.11 is approximately

$$A_{rq} \approx \frac{(4/3)M_u}{4d} = \frac{12.01(4/3)}{27.0 \times 4} = 0.15 \text{ (in.}^2\text{)}$$

Even a single No. 4 bar would be sufficient to meet this flexural demand ($A_o = 0.2$ in.²), and six bars are more than sufficient. The top bars can also be used to reinforce the transition between the haunch at the column and the slab. Specify their length to extend 12" beyond the theoretical transition point, for a total of 7 ft (see Fig. 8.13).

Check beam-type shear of the column haunch, conservatively taking the critical section at the location of the force P:

$$V_u = (F_1 + F_2)1.41 = (6.96 + 6.03)1.41 = 18.32 \text{ (kips)}$$
$$\phi V_c = 0.75 \times 2(1)\sqrt{4000}(b)(d)(lb) = 0.0949(b)(d) \text{ (kip)}$$
$$\phi V_n = \phi Vc = 0.0949 \times 26 \times 48 = 118.44 \text{ (kips)} > V_u \quad \text{OK}$$

CASE 2   The profile of the soil pressure under the footing is shown in Fig. 8.12. As determined for Case 2,

$$f_{p,\text{max}} = 1.93 \text{ ksf}$$
$$f_{p,\text{min}} = 1.47 \text{ ksf}$$

Slab with Haunch, Trench Footings, and Mats

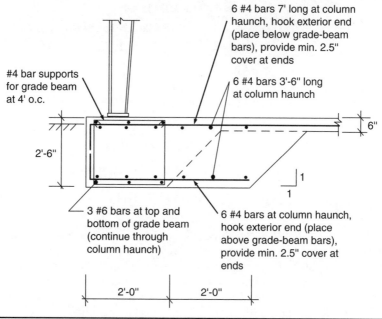

FIGURE 8.13 Reinforcement for the slab with haunch in Design Example 8.1.

As in Case 1, conservatively assume the critical section for flexure being at the centerline of the column, or 1 ft away from the outside edge of concrete. The rate of change in soil pressure is

$$(1.93 - 1.47) \text{ ksf}/4 \text{ ft} = 0.115 \text{ ksf/ft}$$

The soil pressure at the centerline of the column is

$$1.93 - 0.115 \times 3 = 1.585 \text{ (ksf)}$$
$$F_1 = 1.585 \times 3 \times 4 = 19.02 \text{ (kips)}$$
$$F_2 = \tfrac{1}{2}(1.93 - 1.585) \times 3 \times 4 = 2.07 \text{ (kips)}$$
$$M_{max} = 19.02 \text{ kips} \times 1.5 \text{ ft} + 2.07 \text{ kips} \times 3(2/3) \text{ ft} = 32.67 \text{ kip-ft}$$

Find the average load factor:

$$\text{Ave. L.F.} = \frac{5.8(1.6) + (27.2 - 5.8)1.2}{27.2} = 1.29$$

$$M_u = 32.67 \times 1.29 = 42.14 \text{ kip-ft}$$

Check the flexural capacity of the minimum reinforcement designed for Case 1 (six No. 4 hooked bars top and bottom), using the same $d = 26.0$ in.:

$$A_s = 1.2 \text{ in.}^2$$

$$\rho = \frac{1.20}{26.0 \times 48} = 0.00096$$

$$a_u = 4.45 \text{ (See Table 4.1)}$$
$$\phi M_n = 1.20 \times 4.45 \times 26.0 = 138.84 \text{ (kip-ft)}$$
$$138.84 > (4/3)42.14 = 56.19 \text{ (kip-ft)} \quad \text{OK}$$

The development length of the standard ACI hook for No. 4 bar is 6.64 in., as determined for Case 1, which is adequate for the haunch dimensions.

Check beam-type shear, conservatively taking the critical section at the location of the force P:

$$V_u = (F_1 + F_2)1.29 = (19.02 + 2.07)1.29 = 27.21 \text{ (kips)}$$
$$\phi V_c = 0.75 \times 2(1)\sqrt{4000}(b)(d) = 0.0949(b)(d) \text{ (kip)}$$
$$\phi V_n = \phi V_c = 0.0949 \times 26.0 \times 48 = 118.44 \text{ (kips)} > V_u \quad \text{OK}$$

2. Design the Grade Beam Since the perimeter grade beam running between the columns was used to counteract the effects of uplift and overturning, the grade beam needs to be designed to carry its own weight (plus the weight of any metal building components supported by the grade beam) suspended in the air. The span of the grade beam is assumed to be the frame spacing, that is, 25 ft. Its cross-section is 2 ft wide by 2.5 ft deep. The weight of the grade beam plus its haunch is

$$w_{gb} = (2 \times 2.5 + 2 \times 2 \times 0.5 + 0.5 \times 2)\,0.15 = 1.20 \text{ (kip-ft)}$$

Assume the weight of the metal siding and wall girts supported by the grade beam is 2 psf. The eave height is 18 ft, thus the superimposed dead load on the grade beam is

$$(18 \times 2)/1000 = 0.036 \text{ (kip-ft), say } 0.04 \text{ kip-ft}$$

Since the dead load in combination with wind is checked here, the load factor of 1.2 is used for dead load (ACI 318-08 Equation 9-4). The total factored dead load on the grade beam is

$$w_u = 1.2(1.2 + 0.04) = 1.49 \text{ (kip-ft)}$$

The maximum negative moment (at exterior face of first interior support, per ACI 318 Section 8.3 formulas, assuming more than two spans):

$$M_u^{(-)} = \frac{1.49 \times 25^2}{10} = 93.12 \text{ (kip-ft)} \quad \text{Controls}$$

The maximum positive moment (at end spans, discontinuous end unrestrained, per ACI 318 Section 8.3 formulas):

$$M_u^{(+)} = \frac{1.49 \times 25^2}{11} = 84.66 \text{ (kip-ft)}$$

For a 30-in.-deep grade beam with No. 6 bars and 3-in. clear cover (these bars are placed outside the bars for the column haunch to maximize the flexural capacity of the grade beam):

$$d^{(+)} = 30 - 3 - 6/16 = 26.62 \text{ (in.)}$$

Slab with Haunch, Trench Footings, and Mats

At the top, the clear cover is 2 in., and

$$d^{(-)} = 30 - 2 - 6/16 = 27.62 \text{ (in.)}$$

For the first try, the required area of negative steel reinforcement in the grade beam is approximately given by

$$A_{rq} \approx \frac{(4/3)M_u}{4d} = \frac{93.12(4/3)}{4 \times 27.62} = 1.12 \text{ (in.}^2\text{)}$$

Try three No. 6 bars top and bottom ($A_o = 1.32$ in.² at top and bottom, for a total of 2.64 in.²). Check is this satisfies the minimum reinforcing ratio for footings of uniform thickness of $0.0018 A_g$:

$$A_{s,\,min} = 0.0018 \times 30 \times 24 = 1.3 \text{ (in.}^2\text{)} < A_o = 2.64 \text{ in.}^2 \quad \text{OK}$$

For negative bending:

$$\rho = \frac{1.32}{27.62 \times 24} = 0.002 < \rho_{min}$$

$a_u = 4.42$ (from Table 4.1)

$$\phi M_n^{(-)} = 1.32 \times 4.42 \times 27.62 = 161.15 > 93.12(4/3) = 124.16 \text{ (kip-ft)} \quad \text{OK}$$

For positive bending:

$$\rho = \frac{1.32}{26.62 \times 24} = 0.002 < \rho_{min}$$

$a_u = 4.42$

$$\phi M_n^{(+)} = 1.32 \times 4.42 \times 26.62 = 155.31 > 84.66(4/3) = 112.88 \text{ (kip-ft)} \quad \text{OK}$$

Check shear in the grade beam at the distance d from the assumed support at the center of the column (alternatively, this could have been taken at the distance d from the edge of the column haunch):

$$V_u = 1.49(25/2 - 27.62/12) = 15.20 \text{ (kips)}$$

$$\phi V_c = 0.75 \times 2(1)\sqrt{4000}(b)(d) = 94.9(24)(27.62) = 62{,}907 \text{ (lb)} = 62.91 \text{ kips}$$

$$\phi V_n > V_u \quad \text{OK}$$

Stirrups are not needed by analysis, and minimum shear reinforcement is not needed per ACI 318 Section 11.4.6, because

$$V_u = 15.20 \text{ (kips)} < 0.5 \phi V_c = 0.5 \times 62.91 = 31.45 \text{ (kips)}$$

However, since the top bars will need support, specify some No. 4 stirrups for that purpose; a reasonable spacing of 4 ft could be used. The overall reinforcement of the system is shown in Fig. 8.13. Note that 6 No. 6 bars in the grade beam are designed slightly conservatively to account for heavier-than-assumed wall weights in some spots. ▲

8.2 Trench Footings

The trench footings, also known as mass foundations or formless footings, can be thought of as the slab-with-haunch foundations where the haunches are separated from the slabs and are much, much deeper. The idea of a separate haunch might not work well with the haunch of a modest size (see Fig. 8.1), but if the haunch is wide and deep enough the system becomes viable (Fig. 8.14a).

The main advantage of the trench footing is an almost complete elimination of the formwork ordinarily needed for constructing the foundation: A trench excavated with a backhoe is all that's needed. Quite often, the trench footing contains no steel reinforcement,

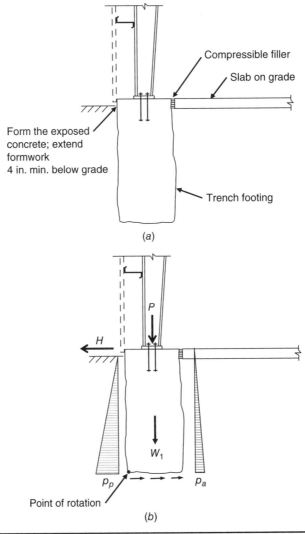

FIGURE 8.14 Trench footing: (a) Basic design; (b) resistance to applied forces.

another way to save money. To give the foundation a more refined appearance, the exposed area of the footing is typically finished by using formwork in that area. For example, the visible edge of the exterior concrete plus a distance of 4 in. below the top of the soil might be finished.

Trench footings could be economical in the areas with cohesive soils. As long as a trench can be safely excavated and kept from collapsing until concrete is placed, as determined by engineering analysis and confirmed by prior successful experience, the system might work. In most locations with cohesionless or mixed soils, however, the possibility of a trench collapse rules out this system. Unfortunately, trench collapses are among the common causes of construction accidents, and the foundation designer should think twice before specifying trench footings without an exhaustive analysis.

Even if a stable trench can be cut without using any excavation support systems, the cost-effectiveness of the formless footings should be carefully considered. While the expense of the concrete formwork is avoided, the amount of concrete used in construction of the trench footings will typically exceed that used in other systems. Also, many engineers, the author included, are uncomfortable with using plain concrete foundations for the applications involving wind uplift and seismic loading. At the very least, we recommend using a couple of continuous reinforcing bars placed near the top and bottom of the trench footings, to provide a measure of ductility.

The trench footing resists vertical and lateral forces applied by the building columns in the same fashion as the moment-resisting foundation and the slab with haunch—by developing the passive-pressure soil resistance and the friction at the base (Fig. 8.14b). As described earlier in this chapter, the weight of the foundation W_1 counteracts the overturning moment caused by the lateral column reactions. Similarly, the resultant of the combined vertical forces should preferably fall within the kern limit of the foundation. To achieve this, the column location should align as much as possible with the center line of the foundation. As discussed earlier, using bypass girts will help achieve this objective.

As with the slab with haunch, the trench footing can be constructed with the same dimension throughout the perimeter of the building, or a wider trench can be excavated at the column locations. Making the trench wider at the columns increases the width and the weight of the foundation where it is needed the most. (The construction complexity increases too.) A design that includes local widening at the column locations and continuous steel reinforcement placed near the top and bottom is shown in Fig. 8.15.

8.3 Mats

8.3.1 Common Uses

Mat foundations could be useful in situations where the load-bearing capacity of the soil is uniform but rather poor. The individual isolated footings designed to bear on these soils tend to become inordinately large. When the combined area of isolated footings covers more than 50% of the building footprint, mat foundations often become economical. By spreading the load over a wide area, a mat makes it possible to avoid the cost of forming many large isolated footings. Depending on the loading, soil specifics, and design details, mats could be competitive with the alternative foundation systems used in poor soils, such as deep foundations.

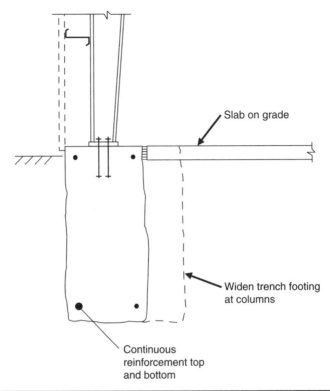

FIGURE 8.15 Possible improvements to the basic design of trench footing.

Mat foundations tend to be most cost-effective when they support a closely spaced grid of similarly loaded columns, such as multiple-span rigid frames. Mat foundations are less useful for supporting widely spaced metal building columns with heavy and unequal loads. Cohesionless soils are generally considered to be the most suitable for mat support. As discussed in Chap. 2, silts and clays tend to undergo long-term deformations under sustained loading. When such deformations occur locally, under the heaviest loaded areas of the mat, the mat could become overloaded and fail.

One particular type of mat foundation could be of use in the buildings with basements. The so-called floating mat is designed not to exert any more pressure on the soil than had existed originally, prior to the basement excavation. Since the soil does not receive any additional pressure, even the soils otherwise considered unsuitable might become viable in this scenario.

For example, suppose the soil weighs 120 lb/ft³, the mat is 18 in. thick, the concrete weight is 150 lb/ft³ and the basement is 10 ft deep, measured from the top of the original soil level to the bottom of the mat. The total pressure on the soil caused by its own weight 10 ft below the surface was originally

$$120 \times 10 = 1200 \text{ lb/ft}^2 \text{ (psf)}$$

The maximum allowable pressure under the floating mat is then

$$1200 - 1.5(150) = 975 \text{ (psf)}$$

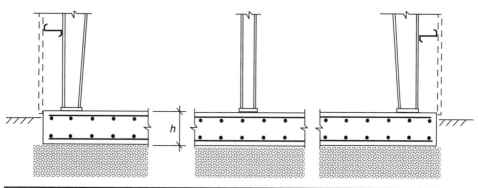

FIGURE 8.16 Basic mat foundation used in metal building systems.

Obviously, the real-life calculations are more complicated than this. For example, should the total (live + dead) loads considered to contribute to the soil pressure under the mat, or just the dead loads? Either way, some pressure fluctuations are unavoidable, a fact that should be considered by the designers of the floating mats.

In a basic version of the mat foundation used in metal building systems the mat is 12 to 24 in. deep and reinforced with two identical layers of deformed bars placed at the top and bottom in two directions (Fig. 8.16).

8.3.2 The Basics of Design

Mat foundations have significant advantages for the use in metal building systems. The foundations are heavy, which helps them overcome the uplift loading on the columns and develop frictional resistance against the soil. The mats are well reinforced, which helps them behave as distributed tie rods. The possibility of having someone cut through a thick mat as through a slab on grade is usually remote. Accordingly, the challenges that require careful consideration in the other types of foundations for pre-engineered buildings can be easily overcome using mats.

The basic behavior of the mat foundation subjected to horizontal and vertical frame reactions is diagrammatically illustrated in Fig. 8.17. For the load combination that produces the maximum downward load in combination with outward horizontal reactions, the latter can be resisted by the mat reinforcement acting as distributed tie rods (Fig. 8.17a). Design for concentrated vertical loads acting near the edge of the mat typically requires computer analysis. The resistance to wind loading is handled in a manner discussed earlier for the slab with haunch, by developing the active and passive soil pressure at the mat edges, as well as friction resistance under the mat (Fig. 8.17b).

Fortunately, there are plenty of available software programs for mat design, such as the popular *spMats* by Structure Point, formerly known as PCA MATS. The published procedures for mat design include ACI 336.2R, *Suggested Analysis and Design Procedures for Combined Footings and Mats*. In addition, formulas in NAVFAC DM-7.2, *Foundations and Earth Structures*, can be used to find the moment and shears in the mats modeled as the beams on elastic foundation.

As discussed in Chap. 3, mat foundations generally have high cost and high reliability, but rather low versatility, because mats cannot be used in buildings with large trenches, pits, and depressions (see Table 3.1).

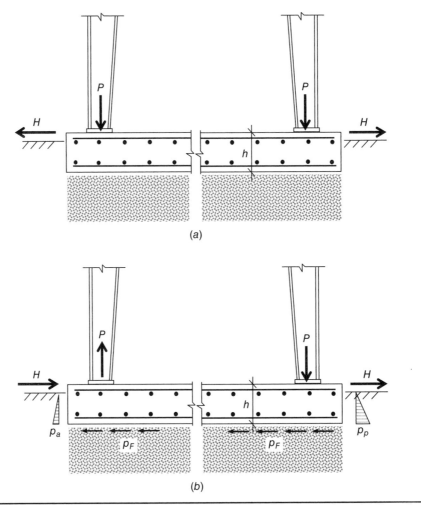

Figure 8.17 Mat foundation subjected to horizontal and vertical frame reactions: (a) At maximum downward load; (b) at wind load acting from left.

8.3.3 Typical Construction in Cold Climates

The basic mat construction shown in Fig. 8.16 allows for a straightforward bar placement and support. However, this configuration can only be used in warm climates, where the minimum foundation depth is not controlled by frost. What about the locations where the frost depth is significant?

Section 1809.5 of the 2009 *International Building Code*® (IBC) requires that shallow building foundations be protected from frost, with some exceptions that usually do not apply to large pre-engineered buildings. The methods of frost protection include:

1. Extending the foundation below the frost line
2. Building the foundation in accordance with ASCE 32
3. Placing the foundation on solid rock

The first method, extending the mat foundation below the frost line, entails the obvious complications. Unless the depth of the frost line is relatively shallow, making the mat as thick as this depth (and probably thicker by about 6 in., to account for raising the floor elevation above the adjacent soil) is often uneconomical (Fig. 8.18a).

Another possibility is to lower the bottom of the mat to the frost line and place a separate slab on grade on top of compacted gravel or crushed stone. While this design saves concrete, it involves the additional efforts of forming and finishing two elements (mat and slab) rather than one. Also, in this scenario the bottom of the column and the

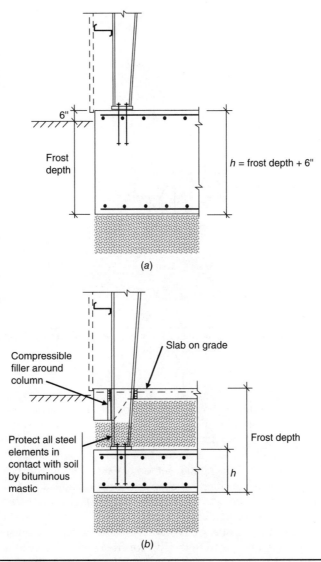

FIGURE 8.18 Constructing mat foundations in the areas subjected to frost: (a) Increased mat thickness; (b) depressed mat in combination with a separate slab on grade.

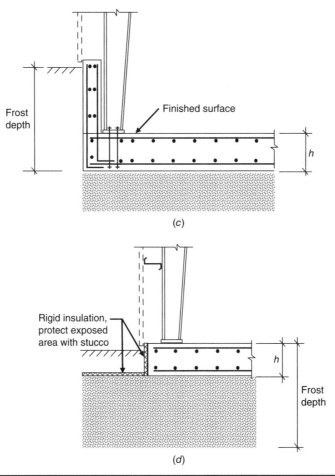

Figure 8.18 (*Continued*) Constructing mat foundations in the areas subjected to frost: (c) lowered floor elevation; (d) frost-protected mat foundation.

column anchors are located well below ground and require corrosion protection. In addition, compressible filler is needed between the column and the slab on grade to prevent column fixity from developing and to protect the concrete from being impacted by the frame movements under fluctuating loading (Fig. 8.18b).

The third approach of using the method or extending the mat below the frost line is to lower the interior floor elevation—the top of the mat—below the adjacent grade (Fig. 8.18c). This approach requires constructing floor ramps for access from the outside. Because the finished floor is now located below grade, this design might not be desirable for office buildings and similar occupancies where moisture-susceptible finishes are used.

The second method given in the IBC for frost protection is to follow the provisions of ASCE/SEI 32-01, *Design and Construction of Frost Protected Shallow Foundations*. Using this document allows the designer to place the bottom of the mat at an elevation above the frost line. A typical frost-protected foundation involves placing the perimeter rigid insulation vertically and horizontally outside the foundation to reduce the heat loss from within the building. The type and thickness of the rigid insulation, as well as the extent of its placement, are determined by analysis. The perimeter vertical rigid insulation must be protected from the elements by covering it with additional materials such as stucco (Fig. 8.18*d*).

One unfortunately common idea for using mat foundations in the areas where the minimum foundation depth is controlled by frost is to place the mat at grade and turn down the edges to the frost line (Fig. 8.19). This approach should not be used in building foundations. Why not? While the perimeter of the mat is supported at the proper depth, the rest of the mat is not, and turning down the edges does not help prevent soil freezing a short distance away. A system like that can be used with frost-protected foundations, but not by itself. Another possible use of the turned-down edges lowered to the frost line exists in the equipment-support foundations, where the weight of the foundation is carried by the structural slab spanning between the deep grade beams at the edges. However, the structural slab in those foundations is not a mat that bears on soil.

Placing the mat on rock, the third IBC method for frost protection, is a relatively rare occurrence. As discussed in Chap. 2, the top of the rock tends to be uneven and sloping, a situation that is not conducive to using a mat of the constant thickness.

8.3.4 Using Anchor Bolts in Mats

Two challenges of using mat foundations involve the anchor bolts. First, the depth of the mat might be insufficient to develop the required length of the bolt embedment.

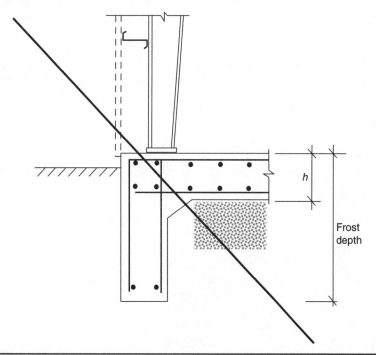

FIGURE 8.19 Simply turning down the edges of the mat below the frost line should *not* be done in lieu of other options.

The depth of a mat foundation is generally established independently from the anchor bolt design, and sometimes the desired length of the bolt embedment cannot be developed in a thin mat. This situation can be resolved in one of three ways:

1. The overall mat thickness can be increased as needed to develop the anchors. This solution might significantly raise the foundation cost, particularly if the difference between the original and the anchor-bolt driven depth is large.
2. The mat can be deepened locally at the column locations. When carefully designed and detailed, this approach could be very cost-effective.
3. The anchor design could be changed. As discussed in Chap. 10, the issues of designing anchor bolts and concrete embedments are quite complex, and there are many available solutions. It might be possible, for example, to enlarge the column base plate, so that more concrete is engaged in uplift resistance, or to provide supplementary tension reinforcement.

The second challenge of using anchor bolts in mats involves their placement. In other foundation types, anchor bolts can usually be suspended from the formwork placed around the foundation or the column pier. This method is difficult to use in large mats, particularly at interior columns. Short of sticking the anchor bolts into fresh concrete from a helicopter—an unwise practice to be sure—how can the anchors be reliably placed in the middle of a sea of concrete?

One approach involves using a so-called mud slab. A mud slab is a thin slab on grade placed under the mat for the purpose of providing a level surface underneath. A mud slab avoids drastic changes in the mat thickness caused by an uneven sub-base, as unfortunately happens quite commonly in slabs on grade. But the mud slab can also be used to facilitate placement of anchor bolts.

Once the mud slab is placed and cured, it is rather straightforward to mark the anchor bolt locations on it and to secure the anchor bolts by attaching them to small steel frames with diagonal braces ("kickers") fastened to the slab. The frames should be sturdy enough to resist the forces of the concrete being placed and consolidated around them. A good introduction to this topic can be found in two articles by Nasvik (2005).

References

2009 International Building Code® (IBC), International Code Council, Country Club Hills, IL, 2009.

ACI 336.2R, Suggested Analysis and Design Procedures for Combined Footings and Mats, American Concrete Institute, Farmington Hills, MI, 2002.

ASCE/SEI 32-01, *Design and Construction of Frost Protected Shallow Foundations*, American Society of Civil Engineers, 2001.

CRSI Design Handbook, Concrete Reinforcing Steel Institute, Schaumburg, IL, 2002.

Nasvik, Joe, "Constructing a Mat Slab," *Concrete Construction*, February 2005 and "Setting Anchor Bolts," *Concrete Construction*, November 2005.

NAVFAC DM-7.2, *Foundations and Earth Structures*, Department of the Navy, Naval Facilities Engineering Command, Alexandria, VA, 1982.

Newman, Alexander, *Metal Building Systems: Design and Specifications*, 2d ed., McGraw-Hill, New York, 2004.

CHAPTER 9
Deep Foundations

9.1 Introduction

The goal of this chapter is to explain how deep foundations can be used for supporting metal building systems. The design of deep foundations is a complex topic well covered in the engineering textbooks and other publications. We simply outline the specific issues related to their use in pre-engineered buildings. There are two main types of deep foundations: Deep piers and piles.

9.2 Deep Piers

9.2.1 The Basics of Design and Construction

Deep piers, also called "caissons" or "drilled shafts," are foundations that bypass the weak soil strata near the surface and bear on the stronger soils below. These elements are similar to the column pedestals discussed in the previous chapters, but extend much deeper. They typically have a circular cross-section.

According to one definition, deep piers have a length-to-thickness ratio of five and larger. These foundations are sometimes constructed using formwork erected in open excavations, but they are most commonly placed in the cylindrical holes drilled in the ground by special rigs.

The deep piers can have either straight shafts or belled bottoms. Straight shafts can be economical for bearing on very firm soils and on rock. The rock-bearing shafts are typically socketed into the rock to provide lateral restraint at the base of the piers. For other soils, belled bottoms provide an increase in the bearing area.

Most caissons are designed as end-bearing columns with continuous lateral support by the soil. Is *any* solid soil considered capable of providing lateral support to deep foundations? The answer is yes, per *2009 International Building Code*® (IBC-09) Section 1810.2.1:

> Any soil other than fluid soil shall be deemed to afford sufficient support to prevent buckling of deep foundation elements . . .

In some cases, a combination of end-bearing and skin-friction resistance of the deep pier can be achieved. The common shaft diameters range from 2 to 6 ft. If used, the bells are typically formed with a minimum 60° angle from the horizontal line. The length of the pier shaft, from its top to the beginning of the flared area at the bell, is generally limited to 30 shaft diameters. The load tables for drilled piers are included in *CRSI Design Handbook* and other references.

Unlike the pile foundations, where the minimum number of piles for lateral stability and for accommodating installation eccentricities is two or three (see Sec. 9.3.2), a single deep pier can be used to support a building column. To qualify for this treatment, IBC-09 Section 1810.2.2 requires that the least horizontal dimension of the pier be at least 2 ft, that lateral support be provided along the length of the pier, and that its height does not exceed 12 times the least horizontal dimension.

Deep piers hold a number of advantages over pile foundations: The drilled shafts are typically less expensive to construct, and they are placed on the soil that can be more thoroughly investigated than the soil receiving the piles. The bearing stratum for the piers is open to observation and is inspected prior to concrete placement.

9.2.2 Resisting Uplift and Lateral Column Reactions with Deep Piers

Resisting wind-induced uplift is quite straightforward with deep piers, as most are heavy enough to provide the necessary "ballast." In addition, the weight of the perimeter grade beams and the soil on top of the belled area can be added to the weight resisting the uplift force. Depending on the soil characteristics, skin friction resistance could also be added.

Developing the pier's resistance to lateral column reactions is more complicated. The passive pressure developed by the bearing of the pier shaft against the soil can develop a certain amount of lateral resistance, similar to a flagpole embedded in the soil. To improve the shaft's bending capacity, a full or a partial reinforcement cage can be provided in its upper part. According to *CRSI Design Handbook*, partial reinforcement should extend the distance of at least three shaft diameters, but not less than 10 ft, from the top of the pier.

Figure 9.1 shows a prefabricated reinforcement cage for a deep pier used for the support of the electric tower. Note that the vertical reinforcing bars become the anchor bolts for the tower superstructure. Note also the longer bar cage in the background of

Figure 9.1 Prefabricated reinforcement cage for a deep pier.

the picture for the pier where the reinforcement cage extends the full height of the shaft. Of course, the reinforcement only increases the flexural capacity of the shaft; it does nothing to increase the passive-pressure resistance of the soil. The capacity of the pier shaft alone to resist lateral loading is limited by its diameter.

The top of an isolated deep pier subjected to a horizontal force has to move laterally to develop the passive-pressure resistance of the soil, similar to the behavior of the cantilevered retaining wall discussed in Chap. 7. As the top of the pier moves laterally, the bottom stays in place, and the pier tilts. A significant amount of such movement might be detrimental to the building. First, it causes the columns of the metal building frame to spread apart, which increases the stresses in the frame. As discussed in Chap. 6, the degree of the potential frame overstress caused by lateral foundation movement is debatable, but it should not be simply ignored altogether. Second, the lateral movement of the pier might cause some distress to the building's nonstructural finishes.

To increase lateral capacity of the drilled piers, one or both of the following two steps can be taken:

1. A continuous concrete grade beam is placed on the perimeter of the building.
2. Soil anchors are used.

The continuous grade beam should extend deep enough into the soil to develop the passive pressure necessary to resist the horizontal column reactions (Fig. 9.2a). The lateral resistance developed by the grade beams could be added to that of the pier shaft,

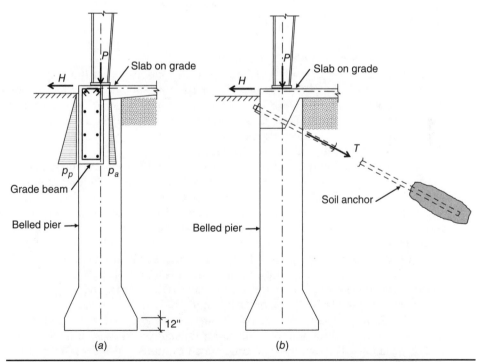

Figure 9.2 Increasing the pier's capacity to resist lateral loads: (a) By using grade beam to develop passive pressure; (b) by using soil anchors.

or the pier resistance could be simply neglected. The grade beam should be reinforced with longitudinal bars and stirrups to develop the bending capacity in the horizontal plane. The longitudinal bars should extend through the pier to develop the beam continuity and avoid torsion. The location of the longitudinal bars needs to be coordinated with the shaft reinforcement to avoid interference.

Soil anchors represent another option that can increase the drilled pier's resistance to horizontal loading. Soil anchors, also known as tiebacks, ground anchors, and earth anchors, are installed in soil or in rock for the purpose of transmitting the applied tensile forces into the ground. The service life of properly designed and installed anchors in permanent installations could reach 75 to 100 years (FHWA, 1999). Tiebacks are widely used in the construction of contemporary excavation-support systems.

Design of soil anchors involves specialized expertise. These structural elements can be designed by specialized consultants or contractor's engineers (particularly in the temporary excavation-support systems), but for permanent foundation systems the engineer of record might directly engage a subconsultant.

There are many types of the available soil-anchor designs, ranging from the mechanically anchored tie-downs to the post-tensioned grouted systems (Fig. 9.2b). The design of soil anchors used in metal building systems should include a prediction of the total system elongation under load. A program to monitor creep deformation of the anchors could be useful (see a description in FHWA, 1976). The information about the design and construction of soil anchors is widely available (FHWA, 1997; FHWA, 1999; ASCE, 1997; Schnabel, 2002; NAVFAC DM-7.2; MIL-HDBK-1007/3).

9.3 Piles

9.3.1 The Basic Options

Pile foundations are commonly used in the same situations where the deep piers might be selected: where the poor soils near the surface are underlain by more competent strata. A variety of pile materials has been used in metal building system foundations. These include steel pipe, steel H sections, cast-in-place concrete, precast concrete, grouted, and wood piles. Piles are typically driven into the ground, with the exception of cast-in-place concrete and grouted varieties.

There are two basic types of piles: end-bearing and friction. As the name suggests, end-bearing piles are similar to building columns, but with continuous lateral support provided by the soil. Steel end-bearing piles sometimes have bearing plates at the bottom.

Friction piles derive their load-carrying capacities from skin friction between their perimeter surfaces and the soil. Cohesive soils are generally able to develop such frictional resistance. According to IBC-09 Section 1810.3.3.1.4, the allowable frictional resistance must not exceed one-sixth of the presumptive soil bearing capacity listed in the IBC-09 Table 1806.2, with the maximum value of 500 pounds per square foot (psf). A higher value may be justified by the appropriate analysis or testing.

Except for the piles with a large self-weight, only friction piles can develop resistance to significant tension forces. Two examples of the piles that might need tension capacities are those designed for wind-uplift resistance and those used as batter piles, as discussed later. For cast-in-place concrete and grouted piles the tension capacity is provided by continuous (or properly spliced) reinforcement placed within the pile. One common solution includes a single reinforcing bar located in the middle of the pile.

As with deep piers, design of pile foundations is a specialized area, and assistance of knowledgeable geotechnical engineers is generally needed. Structural engineers typically focus on the design of the pile caps and the overall layout of the piles. Following the introductory remarks in Sec. 9.1, our discussion is limited to the challenges of using pile foundations in pre-engineered buildings.

9.3.2 The Minimum Number of Piles

As discussed in Chap. 3, some pile designs are more reliable than others. For the lightly loaded metal buildings even a single pile might be able to resist the vertical column reaction. In some projects the designers and contractors alike are seduced by the apparent simplicity of this approach—one column, one pile!—but a system of isolated single piles without grade beams is fraught with potential danger.

The single-pile system requires flawless placement of the piles exactly at the column centerlines, which is difficult to achieve in practice. The almost inevitable eccentricities between the centers of the column and the pile might result in a potential pile overstress and even failure (Fig. 9.3). Therefore, this design cannot be recommended.

What magnitude of the potential eccentricities are we talking about? Obviously, the precision of the pile driving or placement depends on the contractor's expertise, but it is quite common to have as-installed pile locations deviate from 3 to 6 in. from the intended locations. Sometimes, the deviations are much larger, particularly in driven piles that encounter obstacles during placement. According to IBC-09 Section 1810.3.1.3, *Mislocation*, either the deep foundation itself or the superstructure must be designed to resist the effects of at least a 3-in. mislocation. IBC-09 allows a 110% compressive overload of the deep foundation for this condition.

A much more reliable system uses two-pile groups interconnected by grade beams (Fig. 9.4). In this system, the metal building column is located on top of the grade beam, rather than on the pile cap, which allows for the eccentricities in two directions to be accommodated. Obviously, the edge of the pile cap protrudes underground beyond the edge of the building, which is acceptable for most cases. A problem arises only in the rare situation where the building is constructed tight against the property line.

The reinforcement dowels between the grade beam and the pile cap shown in Fig. 9.4 are essential for providing resistance against uplift and horizontal column reactions. The grade beam resists a combination of torsion and flexure caused by the horizontal column reactions and any eccentricity of the vertical reactions. The dowels help resolve the torsional effects within the bearing area of the grade beam on top of the pile cap.

An even more robust system has at least *three* piles under each column. Now, the placement tolerances in two directions can be readily accommodated without an aid of the grade beam (Fig. 9.5). According to IBC-09, a three-pile group where the piles are placed at least 60° apart is considered laterally braced. By contrast, a two-pile group is considered laterally braced only in the plane of the piles; it requires a separate brace in the perpendicular direction. A grade beam of Fig. 9.4 would qualify.

What about supporting a loadbearing wall on piles? While such walls are relatively rare in metal building systems, a loadbearing "hard" wall can occasionally be encountered. There are two options for this design:

1. Place a series of two-pile caps at regular intervals. Depending on its composition, the wall could be designed to span between the pile caps unassisted, or be placed on top of a separate grade beam, such as that of Fig. 9.4.

178 Chapter Nine

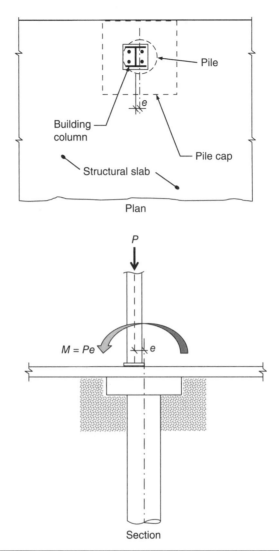

Figure 9.3 Using a single pile under the metal building column results in almost inevitable eccentricity, which can lead to a potential pile overstress.

2. A single line of piles could be placed in a zigzag fashion under the wall. This approach is somewhat controversial. IBC-09 Section 1810.2.2 allows a single line of such supports only "for one- and two-family dwellings and lightweight construction" not more than two stories high above grade, nor 35 ft in height; the pile centers must be located within the wall's footprint.

9.3.3 Using Structural Slab in Combination with Deep Foundations
One issue that repeatedly causes controversy and misunderstanding involves the design of the floor slabs in buildings where deep foundations are used. Is it possible

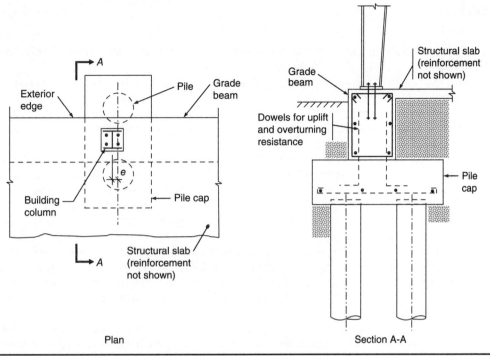

Figure 9.4 Using a series of two-pile groups interconnected by grade beams.

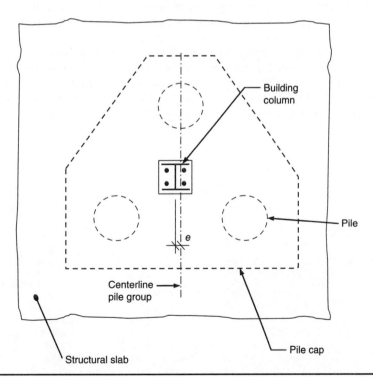

Figure 9.5 The three-pile support accommodates the installation tolerances in two directions without an aid of the grade beam.

179

to use a slab on grade in these circumstances—or must a structural slab be used instead?

The general reason for selecting deep foundations is an insufficient load-carrying capacity of the soil at the site. If the soil is so poor that shallow foundations become uneconomical or even infeasible, would such soil be adequate to support the floor loading? In many cases the answer is no, yet it is unfortunately rather common to see a slab on grade specified in combination with piles or deep piers.

There are certainly situations where deep foundations are used because the soil near the surface is acceptable but rather weak. The vertical column loads are simply too large to be economically supported on it, because the spread footing sizes would become excessive. It would then be cost-effective to extend the deep foundations down to a lower, stronger, stratum in these cases, but there is no problem with a slab on grade used at grade.

However, a different situation arises where the soil near the surface is unsuitable. In this case deep foundations are typically specified to bypass the upper stratum, a common scenario in pre-engineered buildings. In such a case, a slab on grade placed directly on top of the unsuitable soil will likely suffer from performance problems, such as differential settlement and cracking.

A particular challenge arises whenever expansive soils are present at the site. These problem soils are discussed in Sec. 2.2. Quite often expansive soils extend for a limited depth, such as 5 to 8 ft, in which case deep foundations may provide the most cost-effective solution. However, the floor slab must then be of the structural type, that is, spanning between the pile or pier caps, rather than bearing on the ground. Indeed, in these cases a space is often provided between the expansive soil and the bottom of the floor slab to allow for an anticipated soil heave.

Since the column spacing in metal building systems is larger than in many other types of framing—at least in the direction of the frame span—it is rare that the structural slab will be able to span the distance between the columns. In most cases intermediate single-pile supports are provided at regular intervals between the deep foundations at the column locations (Fig. 9.6).

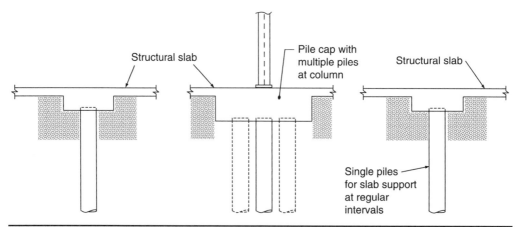

Figure 9.6 Intermediate single-pile supports under the structural slab located at close intervals between the column locations.

One additional advantage of using structural slabs in combination with deep foundations is that the reinforced slabs can often function as seismic ties that interconnect the pile caps, as required by the building codes for certain seismic applications. One alternative is to provide a gridwork of grade beams interconnecting the pile caps.

Obviously, a foundation system that utilizes not only deep foundations, but also structural floor slabs is rather expensive. Occasionally the cost of this foundation system exceeds the cost of the pre-engineered building superstructure. And yet the unsophisticated building owners are quite often preoccupied with getting the best possible prices out of the building manufacturers, while treating the foundations as an afterthought.

9.3.4 Resisting Uplift with Piles

As deep piers, pile foundations used in metal building system are typically designed to resist uplift and horizontal reactions. Unlike the deep piers, however, piles might not be heavy enough to provide the required "ballast." Then either a pile cap of substantial weight is needed, or friction piles. Friction piles can generally be used in cohesive soils, as discussed earlier, and might not always be a viable option in sands and gravels. This leaves the weight of the piles and the pile caps as the main components resisting the uplift loading in cohesionless soils.

Another possibility is to use a contribution of the structural slab, when present (see Sec. 9.3.3). The structural slab can develop a certain amount of resistance in bending, when subjected to wind uplift, similar to a system of tree roots (Fig. 9.7). The rigorous analysis of this behavior requires sophisticated computer software capable of handling the pile-supported slabs subjected to concentrated upward loads.

9.3.5 Resisting Lateral Column Reactions with Piles

In general, pile foundations can resist horizontal column reactions in one of two methods. The first method relies on flexural resistance of the piles designed as cantilevered posts fixed at some point below grade (Fig. 9.8a). The distance to the point of

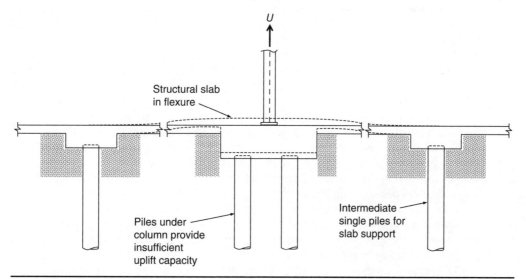

FIGURE 9.7 Using structural slab to develop additional uplift resistance in a pile-supported floor.

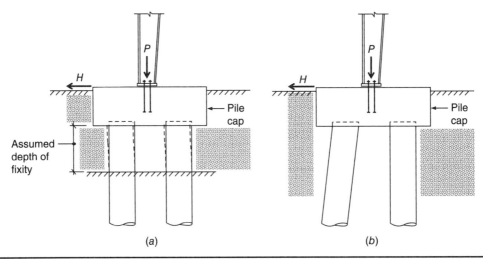

FIGURE 9.8 Resisting horizontal column reactions in pile foundations: (a) By flexural resistance of piles; (b) by using battered piles.

fixity depends on the stiffness of the pile and the soil: The higher those are, the less the required depth of embedment.

While the building codes might not specify the required depth to the assumed point of fixity for piles in competent soils, they do mention the minimum depth of embedment for deep foundations that "stand unbraced in air, water or fluid soils." According to IBC-09 Section 1810.2.1, the point of lateral support is permitted to be taken at a distance 5 ft deep into stiff soils and 10 ft into soft soils. A more sophisticated design model analyzes the piles as the embedded flagpole-type structures.

Either way, determination of the design parameters is generally left to the geotechnical engineers, who might utilize some of the available computer software packages for the analysis of pile bending. A representative value of a pile's resistance to horizontal loading computed by either approach might be 5 to 10 kips.

From the standpoint of considering piles as embedded posts for lateral resistance, there is a definite advantage of using the structural floor slab system supported by intermediate piles at close intervals, as more piles are available for sharing the load. When structural slab is used, all the piles within the tributary area of the frame may be counted upon to share the horizontal loading. The tributary area consists of a strip with the width equal to the bay spacing (such as 25 ft) and the length equal to the width of the building from one side to another. By contrast, a foundation system that combines piles only at the columns with a slab on grade elsewhere will have fewer piles to participate in resisting horizontal column reactions.

One drawback of using this method of resisting horizontal column reactions is that the pile needs to undergo some horizontal displacements at the top to develop the required flexural resistance. As with the other methods of resisting horizontal reactions described in the previous chapters, these displacements might be problematic, and result in damage to brittle finishes.

The second method of resisting horizontal column reactions in pile foundations involves using battered piles, as shown in Fig. 9.8b. The amount of the batter—the pile slope from the plumb line—is determined by analysis. The minimum batter is typically specified as 1:12.

If the behavior of an embedded pile in the first method can be compared to that of a moment-resisting frame with the fixed top and bottom, the battered piles act similarly to a braced-frame system. Their lateral resistance is provided in one of two ways:

1. By developing large axial forces in the battered piles, so that their horizontal components equal or exceed the applied horizontal loading. This approach is more commonly used in heavy industrial facilities rather than in lightweight metal building systems. When a combination of wind and dead loading is checked, the available load in the pile will often be insufficient to counteract the horizontal wind force.
2. By using skin-friction resistance of the piles. In this case, the overturning and sliding resistance is developed by the battered piles acting as tripods. As discussed in the previous section, skin-friction resistance can only be developed in certain soils.

The issues of designing individual piles are adequately addressed in the authoritative publications, including Chapter 18 of the *2009 International Building Code*.

References

2009 International Building Code® *(IBC-09)*, International Code Council, Country Club Hills, IL, 2009.

CRSI Design Handbook, Concrete Reinforcing Steel Institute, Schaumburg, IL, 2002.

FHWA-IF-99-015, Geotechnical Engineering Circular No. 4, Ground Anchors and Anchored Systems, 1999.

FHWA-RD-75-130, Lateral Support Systems and Underpinning, V. III, April 1976.

Guidelines of Engineering Practice for Braced and Tied-back Excavation, ASCE Geotechnical Special Pub. No. 74, 1997.

MIL-HDBK-1007/3, Soil Dynamics and Special Design Aspects, (formerly NAVFAC DM-7.3) 1997.

NAVFAC DM-7.2, Foundations and Earth Structures, 1982.

Schnabel, Harry Jr. and Schnabel, Harry W., *Tiebacks in Foundation Engineering and Construction*, 2d ed., Balkema Publishers, 2002.

CHAPTER 10
Anchors in Metal Building Systems

10.1 General Issues

10.1.1 Terminology and Purpose

In this chapter, we extend the discussion of designing foundations for metal building systems into the interface between the substructure and the superstructure. Anchor bolts serve an important role in any steel-framed building, but this role rises to a critical level in pre-engineered buildings, as explained later.

The term *anchor bolts* is only one of the many names these elements are called. The *2009 International Building Code®* (IBC-09) uses the term *anchors*, as well as *embedded bolts*. The American Concrete Institute (ACI 318) uses the terms *anchors* and *cast-in-anchors* to describe these devices. The American Institute of Steel Construction (AISC) calls them *anchor rods*, to avoid confusion with the structural bolts connecting steel elements together. At least one source—the Occupational Safety and Health Administration (OSHA)—still uses the term *anchor bolts*. Since most design professionals prefer the traditional name *anchor bolts*, we use it here as well, while occasionally using the others.

Anchor bolts serve two purposes:

1. They help stabilize the steel structure during erection.
2. They help transfer horizontal and vertical reactions from the column base plate to the foundation throughout the service life of the building.

To help stabilize the structure during erection, OSHA Safety and Health Standards for the Construction Industry, 29 CFR 1926 Part R, *Safety Standards for Steel Erection* specifies the minimum loading criteria for which the anchor bolts must be designed. Paragraph 1926.755(a)(2) requires that the column anchors must be able to resist the tension caused by the bending moment from a person weighing 300 lb located 18 in. away from the face of the column.

Anchor bolts historically fulfilled the second purpose as well—the transfer of horizontal and vertical reactions from the column base plate to the foundation. However, this practice may be changing. As discussed later, anchor bolts are better suited for the transfer of tension loading, but have very limited capacities for transferring horizontal column reactions. Needless to say, a column simply bearing on concrete can transfer the *downward* forces to the foundation without any anchorage.

10.1.2 The Minimum Number of Anchor Bolts

According to OSHA 29 CFR 1926 Part R, the minimum required number of anchor bolts in a column base plate is four, with an exception for "posts," where only two bolts may be used. OSHA defines a post as an "essentially vertical" member fitting one of the following two criteria:

1. Weighs 300 lbs or less and is axially loaded
2. Is not axially loaded, but is laterally restrained by the above member

Most columns in the primary frames of metal building systems do not qualify as "posts" and therefore require a minimum of four anchor bolts. Some lightweight endwall columns and the framing around doors and windows might so qualify.

We should note that these OSHA regulations are still relatively new, and it is common to see only two anchors at the bases of many existing pre-engineered buildings. The "outside" specifying engineers should check the manufacturers' shop drawings to ensure that a minimum of four anchor bolts are provided at each frame column.

10.2 Anchor Bolts: Construction and Installation

10.2.1 Typical Construction

In the past, an anchor bolt typically consisted of a bent L or J-shaped steel rod with a nut at the threaded top. Some engineers preferred anchor bolts with steel plates attached at the bottoms of two adjacent anchors. These designs, while still used by some, have largely lost their popularity in favor of headed anchors. In fact, a discussion on p. 14-10 of AISC *Steel Construction Manual* (AISC, 2005), states that hooked bolts are appropriate only as the means of temporary column support during erection; they may only be used in columns subjected to axial compression. The L and J bolts should not be used when a calculated tension exists.

A headed anchor bolt consists of either a fully treaded rod or a rod with treaded ends, with a nut at the bottom and a nut at the top (Fig. 10.1). The nut is prevented from slipping off at the bottom by damaging the threads or by tack welding. A thick plate washer typically supplements the nut at the top, to keep it from pulling through the oversized hole in the base plate.

Anchor rod material typically conforms to ASTM Standard F1554, "Standard Specification for Anchor Bolts, Steel, 36, 55, and 105 Yield Strength" (ASTM, 2007). This standard has superseded the previously popular ASTM 307, but the most common material for the rods still has the same yield strength of 36 ksi. Using the high-strength steel Grades 55 and 105 is generally unnecessary, since the steel strength rarely governs anchor bolt design, as explained later.

In fact, high-strength steel materials might have a problem with weldability: Special carbon-content requirements must be met in order for them to be weldable. The different

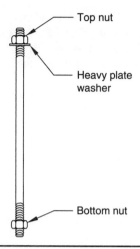

FIGURE 10.1 Headed anchor bolt.

grades can be easily identified in the field by their color coding. The rods made of Grade 36 steel are marked blue, those of Grade 55, yellow, and of Grade 105, red.

The high-strength structural bolts conforming to ASTM 325 or ASTM 490 should not be used as anchor rods. Not only those high-strength bolts are unnecessarily strong to be used as anchor rods, but also their maximum length is only 8 in., much less than is typically needed. Another difference between the high-strength structural bolts and the mild-steel anchor rods is that the latter are usually not pretensioned.

The nuts used for anchor bolts should conform to ASTM Standard F1554, which lists a variety of options. The strength of the nuts and their threads normally exceeds the strength of the matching rod material. The washers are typically made of the heavy stock conforming to ASTM Standard A36 (ASTM, 2008).

10.2.2 Field Installation

Anchor bolts are typically used in groups of four, which is the minimum number required by OSHA (see Sec. 10.1.2). They are set in concrete with an aid of temporary plywood templates, which help maintain the appropriate spacing between the bolts in the cluster. The templates should be firmly attached to the concrete formwork to prevent the anchors from being displaced during concreting operations. Of course, the whole template could still be placed in the wrong location, which is why we recommend that a registered surveyor or engineer supervises this work. Setting the templates should not be delegated to construction laborers.

We also strongly recommend surveying the final anchor bolt locations prior to fabrication of the steel frames. AISC 303-05 *Code of Standard Practice for Steel Buildings and Bridges* (COSP) Paragraph 7.5.4 specifically requires such a survey for structural-steel buildings. But do these requirements apply to metal building systems?

According to AISC 303-05 Section 2.1, the definition of structural steel includes built-up beams and columns made of plates—the typical composition of primary frames in pre-engineered buildings. Also, the anchor rods that receive structural steel are included in the definition of structural steel.

Another authoritative publication that addresses the erection of the primary frames in pre-engineered metal buildings is MBMA *Metal Building System Manual* (MBMA, 2006). Although the manual does not get as specific as AISC COSP, its Guide Specifications direct the erector of the framing to verify that the embedded anchors are in correct position.

Despite these requirements, the problems with incorrect placement of anchor bolts are all too common. As discussed in the following sections, these problems go well beyond minor fit-up challenges. They bring into question the whole idea of relying on anchor bolts for transferring horizontal column reactions from the bases of the metal building frames to the foundations. Instead, more and more engineers turn toward other concrete embedments for this purpose.

10.2.3 Placement Tolerances vs. Oversized Holes in Column Base Plates

Why are anchor bolts losing popularity for transferring horizontal forces? To a large degree, the imperfect installation techniques are to blame. As just discussed, even when plywood templates are used, the whole group of anchors could be placed in a wrong location. The incorrect anchor placement means that the frames or the secondary members would not fit properly without replacement or some alteration.

To be sure, the design and construction professionals fully realize that erecting a steel frame is not the same as crafting a Swiss watch, and some deviations from the theoretical locations are inevitable. The question is only how much deviation is acceptable.

According to AISC 303-05 Section 7.5, the maximum dimensional variation between the centers of adjacent groups of anchor rods is ¼ in. The tolerance for setting the group of anchor rods from its theoretical position is also ¼ in. These permissible tolerances can be accommodated within the oversized holes in the column base plates. These oversized holes have been historically used to facilitate column placement on the mislocated bolts.

The accuracy of anchor-bolt placement has never been ideal—and seems to be further slipping lately. To compensate, the maximum sizes of the holes in the base-plates have been getting even bigger, as evidenced by the changes in AISC *Steel Construction Manual* from the 9th (1989) to the 13th (2005) edition. For example, the maximum size of the hole for a ¾-in. anchor rod has increased from 1-1/16 in. in the 9th edition to 1-5/16 in. in the 13th edition (see Table 14-2 in the 13th edition of AISC manual).

Regrettably, even with the larger hole sizes the reports of misplaced anchor rods are heard routinely. In this situation the column cannot be placed in the intended position without some field rework. The latter might include enlarging the base-plate holes in the field or even removing and replacing the base plate with a larger one.

Unfortunately, the base-plate holes enlarged in the field are not always made as accurately as the holes made in the shop. The author was involved in the investigation of a construction accident where the field holes were made so sloppily that they contributed to a structural failure that caused a loss of life. In the accident, the nut of an expansion bolt and a thin plate washer pulled right through the field-burned hole the bolt was supposed to restrain.

We should note that even under the best of circumstances the corners of the standard hexagonal nut used with a ¾-in. anchor rod barely extend beyond the edges of the oversized hole (Fig. 10.2). This makes using thick plate washers critical.

Another authoritative source of information on the placement tolerances of anchor bolts is ACI *Specifications for Tolerances for Concrete Construction and Materials and Commentary*.

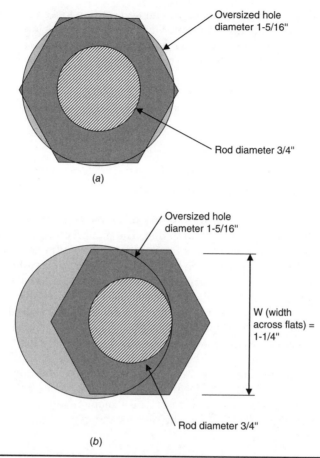

FIGURE 10.2 Corners of the typical nut of a ¾-in. anchor rod barely extend beyond the oversize hole in the column base plate, making a thick plate washer critical: (a) Rod perfectly centered on the oversize hole; (b) rod bears against an edge of the oversize hole.

In both 2006 and 2010 editions (ACI 117-06 and ACI 117-10), the tolerances for centerline of *individual anchor bolts* from the specified location are:

- ¼ in. for ¾ and 7/8 in. bolts
- 3/8 in. for 1, 1¼, 1½ in. bolts
- ½ in. for 1¾, 2 and 2½ in. bolts

Therefore, for the common anchor bolt diameters of ¾ and 7/8 in. the tolerances are the same as in AISC 303, but the ACI tolerances exceed those of AISC for the anchors of larger diameters. Thus a problem could arise when, say, a 2-in. bolt is mislocated within the ACI tolerances. Also, please note that the maximum diameter of the anchor rods listed in AISC Table 14-2 is 1¼ in.

To further confuse the matters, both ACI 117-06 and ACI 117-10 also establish the tolerances for horizontal and vertical deviation for the centerline of an *assembly* of embedded items from the specified location. According to ACI 117-10 Section 2.3.2, this tolerance is 1 in. Some might interpret this section, for which no commentary is provided, as applying to the base plates with embedded anchors, even though the tolerances for individual anchor bolts are much tighter.

10.2.4 Using Anchor Bolts for Column Leveling

AISC *Steel Construction Manual* (AISC, 2005) discusses three methods of column alignment that can be used for structural-steel columns with shop-attached base plates:

1. Leveling nuts beneath the base plates
2. Shim stack placed under the base plates
3. Steel leveling plates set on a pad of grout

All three methods assume that a pad of nonshrink grout exists between the column base plate and the supporting foundation. The main reason for using the grout is the tolerances for as-built elevations of the foundations from the intended line. According to ACI 117-10 Section 3.3, deviation from elevation for the top surfaces of foundations is (+½ in.) and (–2 in.), meaning that the top of the foundation is considered acceptable if it is no higher than ½ in. above and no lower than 2 in. below the intended elevation.

Whichever method of column alignment is used, the grout fills the distance between the as-built top of concrete and the base plate (Fig. 10.3a). In addition, grout helps to even out the sometimes rough surface of concrete.

However, grouting under the column base plates is typically avoided in metal building systems (see Figs. 1.6, 1.7, 3.8, 10.3b, and 10.4). According to MBMA Common Industry Practices (MBMA, 2006), grouting is excluded from the scope of work of the metal building manufacturer, and many typical manufacturers' details do not show any grout underneath the base plates. The owners and general contractors usually do not

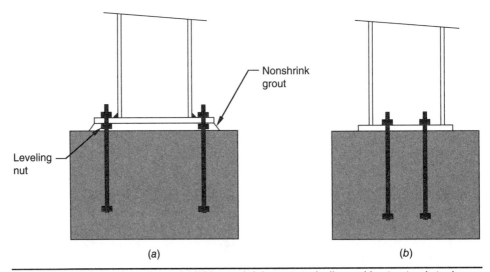

Figure 10.3 Column base plates: (a) With nonshrink grout, typically used in structural-steel buildings; (b) without grout, typically used in pre-engineered buildings.

Figure 10.4 Typical base plate of a pre-engineered building column with flush girts.

see a need to add something that is not required by the manufacturer. (This issue is further discussed in Chap. 3.)

The absence of grout does not allow adjusting the elevations of the column base plates, leading to potential misalignment of the building framing. We hope that this practice is going to change in the near future. In the interim, we strongly recommend that the specifiers require a grout pad under the metal building columns in their projects.

In the relatively infrequent (for now) situations where grout *is* used, it probably is placed after the steel frame is erected, by somebody other than the steel erector. Thus the leveling-nut approach (AISC manual method 1) is arguably the most practical for metal building systems: It allows the steel erector to place the column on top of the properly aligned top nuts, where the column should be stable for a short period of time until the grout is provided.

Curiously, the question of placing grout must be also debated among the designers of steel light poles. In some installations grout exists under the base plates; in others, it doesn't. The anchor bolt diameters are the same in both cases.

10.2.5 Should Anchor Bolts Be Used to Transfer Horizontal Column Reactions?

As discussed in more detail later, the existence of oversized holes in column base plates makes transfer of horizontal column reactions problematic. How so? An anchor that does not touch the edge of the plate in the direction of the horizontal load has to deform to bear against it. If the deformation is substantial, the anchor could fracture well before this happens.

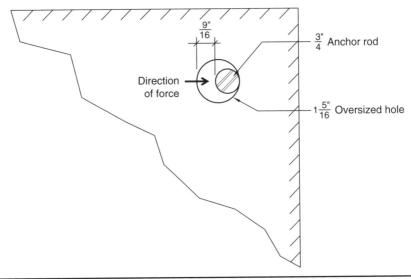

Figure 10.5 A ¾-in. anchor rod that touches one edge of the 1-5/16-in. hole has to deform 9/16 in. under horizontal loading to reach the opposite edge.

Consider the worst-case scenario, when a common ¾-in.-diameter anchor touches one edge of the 1 5/16-in. hole and the horizontal load acts in the opposite direction (see Figs. 10.5 and 10.2b). The anchor would have to deform perpendicular to its length an incredible 9/16 in. to bear against the opposite edge of the hole! Even in an "neutral" case, where the anchor happens to be located exactly in the middle of the oversized hole (Fig. 10.2a), the amount of shear deformation in any direction would be 9/32 in.—still a very substantial amount. A deformation of this magnitude might damage the concrete near the surface or cause yielding or fracture of the rod. This is one of the reasons the American Institute of Steel Construction discourages the use of anchor rods for transfer of shear loading (the horizontal column reactions), although the anchors may still resist vertical upward reactions (uplift).

The engineers have come up with two ways to sidestep the issue of large anchor-bolt movements within the oversized hole. These are:

1. Welding a heavy plate washer to the top of the base plate (Fig. 10.6). The washer has a regular-sized hole—1/16 in. larger than the nominal bolt diameter (see ACI 351.2R-10). Doing so does not eliminate bending in the anchor, but it changes the rod to a fixed-top, fixed-bottom element controlled by its flexural capacity rather than by the need to bear against the edge of the plate.

2. Fill the oversized holes with weld material, shims, or nonshrink grout after the anchors are placed but before the top nuts are attached. This approach, while feasible in theory, might be difficult to achieve on the construction site in a reliable manner. In particular, grouting small uneven spaces seem problematic.

The issue of designing anchor bolts for shear is addressed in Sec. 10.5.

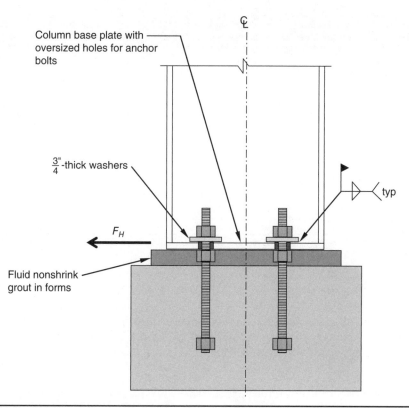

FIGURE 10.6 Welding a heavy plate washer to the top of the base plate eliminates the need for anchor rod bearing against the plate.

10.3 Design of Anchor Bolts: General Provisions

10.3.1 Provisions of the International Building Code

Some provisions for designing embedded anchors in concrete exist in the International Building Code. IBC-09 contains two sections dealing with the matter: Section 1911, Anchorage to Concrete—Allowable Stress Design (ASD) and Section 1912, Anchorage to Concrete—Strength Design.

Section 1911, Anchorage to Concrete—Allowable Stress Design, applies only to headed bolts and headed stud anchors in normal-weight concrete. It does not apply to post-installed anchors or to the anchors designed for earthquake loading. (For those, IBC-09 Section 1912 may be used.)

The IBC provisions assume that the anchors comply with ASTM A307 or an approved equivalent. As stated earlier, today's preferred material for anchor bolts is ASTM Standard F1554, which includes Grade 36. The anchors conforming to ASTM F1554 Grade 36 are considered equivalent to those provided under ASTM A307.

Bolt Diameter, in.	Min. Embedment in.	Min. Edge Distance, in.	Min. Spacing, in.	Tension (P_t), lb	Shear (V_t), lb
¾	5	4.5	9	2250	3560
		7.5	9	2950* 3200†	4300* 4400†
1	7	6	12	3250* 3650†	4500* 5300†
1¼	9	7.5	15	4000	5800

Source: IBC-09 Table 1911.2.
Notes:
*For f'_c = 3000 psi
†For f'_c = 4000 psi

TABLE 10.1 Representative Allowable Service Loads on Embedded Bolts in Tension (P_t) and Shear (V_t) from IBC-09 Table 1911.2

IBC-09 includes Table 1911.2, Allowable Service Loads on Embedded Bolts, which provides the allowable tension (P_t) and shear (V_t) capacities for anchor bolts, as a function of the bolt diameter, minimum embedment depth, spacing, edge distance, and 28-day concrete compressive strength f'_c. Table 10.1 summarizes the allowable values for three representative bolt diameters. The tabulated values are permitted be modified as follows:

- Reduce by 50% if the edge distance and spacing are reduced by 50% (use linear interpolation if reduction less than 50%)
- Can increase by one-third if allowed by IBC-09 Section 1605.3.2 for wind loading
- Can increase P_t (but not V_t) by 100% when special inspection is provided

IBC-09 provides the following formula for combined tension and shear loading:

$$(P_s/P_t)^{5/3} + (V_s/V_t)^{5/3} \leq 1$$

where P_s and V_s = applied tension and shear
P_t and V_t = allowable tension and shear

Section 1912, Anchorage to Concrete—Strength Design, specifies that headed bolts, headed studs, J or L bolts embedded in concrete, as well as post-installed expansion and undercut anchors be designed per ACI 318-08 Appendix D, Anchoring to Concrete (hereinafter called Appendix D). The section further states that strength design of anchors that are not within the scope of Appendix D "shall be in accordance with an *approved* procedure." Some provisions of Appendix D are modified by IBC Sections 1908.1.9 and 1908.1.10.

The anchor rods used in metal building systems are generally spaced close together (a common layout is a grid of 4 × 4 in.), and their edge distances are relatively small. The simple provisions of IBC-09 Section 1911 are of little help for the small values of anchor spacing and edge distances. The following design example demonstrates this situation.

Design Example 10.1: Attempted Design of Anchor Bolts for Metal Building System per IBC-09 Section 1911

Problem: Design the anchor bolts for a small metal building system with the anchor spacing of 4 in. and concrete $f'_c = 3000$ psi, using IBC-09 ASD basic load combinations.
The column reactions are:

Horizontal: dead 1.9 kips, roof live 10.2 kips (both acting outward)

Vertical: dead 4.6 kips, roof live 20.6 kips (both acting downward)

Wind reactions on the left-side column, wind from left to right: Horizontal: 13.8 kips (inward); vertical: 13.7 kips (uplift)

Solution Combine loads per IBC-09 ASD basic load combinations

Case 1: $D + L_r$:

$$H_1 = 1.9 + 10.2 = 12.1 \text{ (kips) outward}$$
$$V_1 = 4.6 + 20.6 = 25.2 \text{ (kips) down}$$

Case 2: 0.6D + W to right:

$$H_2 = 13.8 - 0.6 \times 1.9 = 12.66 \text{ (kips) inward}$$
$$V_2 = 13.7 - 0.6 \times 4.6 = 10.94 \text{ (kips) upward}$$

The directions of column reactions are shown in Fig. 10.7. By observation, Case 2 provides the critical combination for combined shear and tension loading.

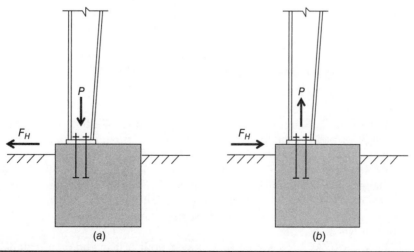

FIGURE 10.7 Load combinations for design of anchor bolts: (a) Case 1, dead + collateral + snow (or roof live load); (b) Case 2, dead + wind + lateral earth pressure.

Using four anchors, the load on each anchor is

$$V_s = 12.66/4 = 3.165 \text{ kips} = 3165 \text{ lb}$$
$$P_s = 10.94/4 = 2.735 \text{ kips} = 2735 \text{ lb}$$

Using the largest bolts from Table 10.1, try 1¼-in. bolts with 7½-in. edge distance and 15-in. spacing.

$$P_t = 4000 \text{ lb and } V_t = 5800 \text{ lb}$$

Combined loading per IBC formula:

$$(2735/4000)^{5/3} + (3165/5800)^{5/3} = 0.531 + 0.364 = 0.895 < 1.0$$

This seems OK, but this check assumes a 15-in. anchor spacing, not 4-in. actual spacing. Even if we try to reduce the minimum spacing by one-half to 7.5 in. the shear capacity becomes

$$V_t = 0.5 \times 5800 \text{ lb} = 2900 \text{ lb}$$

The tension capacity is also decreased by one-half, but assuming special inspection is present, the allowable capacity may be doubled, so the total tension capacity, and the ratio of actual to allowable tension values stays the same. The combined ratio becomes

$$0.907 + 0.364 = 1.27 > 1.0 \quad \text{NG}$$

Conclusion The largest anchors from the IBC table do not work for the relatively minor loads in the example. Therefore, for practical applications involving anchor bolts in metal building systems, ACI 318 Appendix D must be used. ▲

10.3.2 ACI 318-08 Appendix D

ACI 318 Appendix D, Anchoring to Concrete, is today's primary authoritative document dealing with the design of anchors in concrete. As just discussed, the International Building Code refers to Appendix D for strength design of embedded anchors, for design of post-installed anchors and for the anchors subjected to earthquake loading.

The first version of Appendix D appeared in the 2002 edition of ACI 318. Its provisions have been changing in every edition of ACI 318 even since, sometimes drastically. Our objective is to provide a concise explanation of the provisions in the Appendix, while keeping in mind that many of these provisions are likely to change in the subsequent editions of ACI 318.

The scope of ACI 318 Appendix D covers both cast-in and post-installed anchors; the latter must be qualified by a testing program outlined in ACI 355.2, *Qualification of Post-Installed Mechanical Anchors in Concrete & Commentary*. Appendix D does not cover adhesive/grouted anchors, through-bolts, specialty inserts, and multiple anchors connected to a single steel plate at embedded end. It also does not apply to powder-actuated anchors and the anchors used mostly for high-cycle fatigue or impact loading.

The general methodology of Appendix D is based on the Concrete Capacity Design (CCD) method, although other models in agreement with test results may be used. The specific design approach depends on several factors, such as the type of applied loading (including earthquake loading triggers a particular set of provisions), rigidity of the base plate, and the embedment length of the anchor.

For example, according to Appendix D Section D.3.3, the following applies to the anchors subjected to earthquake loading in the structures assigned to seismic design categories C through F:

- Values of ϕN_n (pullout strength) and ϕV_n (shear strength) associated with concrete failure modes are multiplied by 0.75, assuming the concrete is cracked, unless demonstrated otherwise.
- Anchor strength must be governed by a ductile steel element, not concrete embedment, unless
  - The attachment is designed to yield at a force $\leq \phi N_n$ or 0.75 ϕV_n governed by concrete failure modes.
  - The design strength of nonductile anchors is taken as $(0.4)(0.75)\phi N_n$ or $(0.4)(0.75)\phi V_n$ (with additional provisions for stud bearing walls that do not apply to the present discussion).

According to Appendix D Section D.4, the following limit states are considered for strength design of anchors:

- Steel strength of anchor in tension and in shear
- Concrete breakout strength of anchor in tension and shear
- Pullout strength of anchor in tension
- Concrete side-face blowout strength in tension
- Concrete pryout strength in shear
- Splitting failure of concrete (this can be eliminated by providing the required edge distances, spacing, and concrete thickness)

Many engineers complain that the design provisions of ACI 318 Appendix D are unnecessarily complex. They point out that the anchor-bolt designs that have been used for decades in metal building systems can no longer be supported by the calculations that follow the requirements in the Appendix. Despite the fact that the continual changes to Appendix D seem to attempt to make it more practical, the general approach of this document is simply incompatible with the traditional anchor-bolt design and construction practices. Yet as long as Appendix D exists within ACI 318 and is referenced by the building codes, it remains the law of the land, and it is these design and construction practices that must be changed to comply. This topic is explored in the sections that follow.

10.4 Design of Anchor Bolts for Tension per ACI 318-08 Appendix D

10.4.1 Tensile Strength of Anchor Bolt vs. Tensile Strength of Concrete for a Single Anchor

While various design solutions exist for resisting horizontal column reactions, anchor bolts are typically relied upon to resist uplift. What is the mechanism for developing this resistance? A headed anchor bolt derives its holding capacity from bearing of the bottom nut against the concrete. Two failure modes could control the design:

1. The tensile capacity of the anchor's shank
2. The tensile capacity of the concrete

The first failure mode is preferred in today's design approach: The anchor should fail in a ductile fashion by yielding of the shank before the concrete breaks. This requirement is given in ACI 318-08 Paragraph 15.8.3.3, independent of the provisions in Appendix D:

> Anchor bolts and mechanical connections shall be designed to reach their design strength before anchorage failure or failure of surrounding concrete.

This provision is easier to meet when the metal building column is placed directly on top of a large concrete footing. It may be difficult to meet when the column bears on top of a concrete pedestal with a modest footprint. This is one of the reasons why the anchors made of high-strength steel are generally unnecessary in metal building foundations.

Consider a single headed anchor embedded in a large concrete footing away from the edges and subjected to a pullout force. When concrete holding the anchor finally fails, it will develop the so-called concrete breakout prism, which is really a cone-shaped failure surface. In previous practice the angle of the cone was assumed to be 45°, but in today's codes the slope is shallower: One part vertical to 1.5 parts horizontal, or about 35° (Fig. 10.8a). The larger the effective depth of embedment h_{ef}, the larger the concrete tensile capacity.

For simplicity of calculations, the projected area of the breakout prism may be taken as a square with the sides equal to $3h_{ef}$ (Fig. 10.8b). According to Appendix D Section D.4.2.2, provisions for concrete breakout strength apply for anchors with diameters not more than 2 in. and the embedment length not more than 25 in.

10.4.2 Tensile Strength of an Anchor Group

As already discussed, the minimum number of anchor bolts in metal building systems is four. How is the applied tension force, such as wind uplift, distributed among the bolts in a group?

If the force is applied concentrically to the center of the group—the typical case—the force is distributed equally among the anchors. One caveat applies to the base plate: It should be thick enough to prevent yielding under load or to allow prying action to develop.

Sometimes the tension force is applied eccentrically to the centroid of the group. This occurs when the anchors are placed asymmetrically, or then a fixity moment exists.

Anchors in Metal Building Systems

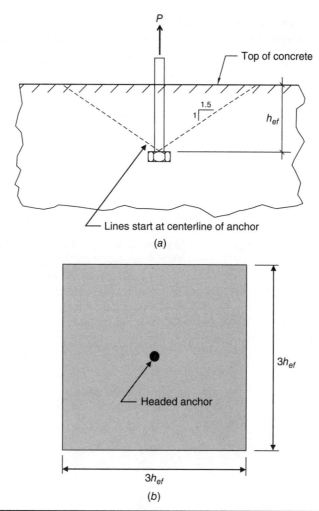

FIGURE 10.8 Concrete breakout prism for a single headed anchor bolt: (a) Section; (b) assumed projected area.

In these cases the force is distributed among the anchors on the basis of elastic analysis, although plastic analysis can be used when the requirement for anchor ductility applies. The plastic analysis follows the basic procedure of the reinforced-concrete design, with the applied moment resisted by a couple with the tension and compression components.

The tension component of the couple is obviously located at the pair of the anchor bolts on the tension side, but the exact location of the compression component might be difficult to find. According to the guidance in *PCA Notes*, its location depends on whether web stiffeners exist at the column base. When web stiffeners are present, the compression force may be assumed to be at the outer edge of the base plate. When no web stiffeners are present, as is typical in metal building system columns, the compression force may be assumed to be at the leading edge of the column flange.

10.4.3 Tensile Strength of Steel Anchors

The nominal tensile strength of anchors can be found from the following equation:

$$N_{sa} = nA_{se,N} f_{uta}$$

where N_{sa} = nominal strength of anchor in tension, governed by steel capacity
n = number of anchors in the group
$A_{se,N}$ = effective area of an anchor in tension, in.²
f_{uta} = specified tensile strength of anchor material

Essentially, this equation defines the breaking strength of the rod (Fig. 10.9a). In the equation, f_{uta} shall not be taken larger than the smaller of $1.9f_{ya}$ or 125,000 psi, where f_{ya} is the specified yield strength of anchor steel. This check does not control for the typical anchor steel with the specified yield strength of 36,000 psi. For the rods made of steel conforming to ASTM F1554 Gr. 36, steel f_{uta} = 58,000 psi.

The effective area of an anchor in tension $A_{se,N}$ can be found in Table 10.2. The table also provides the bearing area for a heavy-hex bolt head or nut, A_{brg}. To convert the nominal strength of anchors, as determined earlier, into the design strength, the nominal strength must be multiplied by a strength reduction factor ϕ. The strength reduction factor is taken as follows:

$$\phi = 0.75 \text{ in ductile steel elements}$$

$$\phi = 0.65 \text{ in brittle steel elements}$$

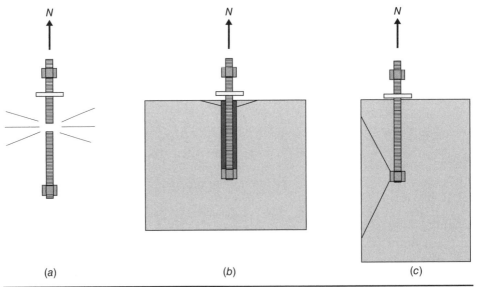

FIGURE 10.9 Limit states for a headed anchor bolt in tension: (a) Steel strength; (b) pullout; (c) side-face blowout.

Bolt Diameter, in.	$A_{se,N}$ (same as $A_{se,V}$)	A_{brg}
3/4	0.334	0.911
7/8	0.462	1.188
1	0.606	1.501
1-1/8	0.763	1.851
1 1/4	0.969	2.237

Source: PCA Notes on ACI 318-08.

TABLE 10.2 Values of $A_{se,N}$ (same as $A_{se,V}$) and A_{brg} for Commonly Used Anchor Diameters

An anchor rod of ASTM F1554 Gr. 36 steel is considered a ductile element. As an example, the design ultimate tensile strength of four anchor bolts with ¾-in. diameter is

$$N_u = 0.75(4)(0.334)(58) = 58.1 \text{ kips.}$$

ACI 318-08 Appendix D specifies various values for ϕ for other limit states. For cast-in anchors governed by concrete breakout, side-face blowout, pullout, or pryout:

$$\phi = 0.75 \text{ if Condition A applies}$$

$$\phi = 0.70 \text{ if Condition B applies}$$

Condition A exists when supplementary reinforcement (described next) is present, except for pullout and pryout. Condition B exists when no supplementary reinforcement is provided and for pullout and pryout limit states. (See also ACI 318 Appendix D Section D.4.4 for ϕ factors for post-installed anchors and Section D.4.5 for use with the load factors of ACI 318 Appendix C.)

10.4.4 Pullout Strength of Anchor in Tension

As the name of this limit state suggests, a cast-in headed or hooked anchor can fail by pulling out of concrete (Fig. 10.9b). ACI 318 Appendix D provides the following formula to determine the nominal pullout strength of a single anchor in tension:

$$N_{pn} = \psi_{c,P} N_p$$

where N_{pn} = nominal pullout strength
 $\psi_{c,P}$ = 1.4 for uncracked concrete (if shown by analysis); 1.0 for cracked concrete
 $N_p = 8 A_{brg} f'_c$ for headed bolt or stud
 A_{brg} = net bearing area of the heavy-hex bolt head or nut, in.^{2}

The values of A_{brg} for typical anchor diameters can be found in Table 10.2. Since pullout strength is a function of concrete crushing under anchor head or nut, it does not depend on the embedment length. For typical applications it also does not depend on the number of anchors, as each anchor can develop it pullout capacity independently. As stated in the previous section, $\phi = 0.70$ for a limit state of pullout.

Appendix D provides another formula for hooked anchors in pullout, such as J and L bolts. As explained in Sec. 10.2.1, L and J bolts should not be used when a calculated tension exists. Although Appendix D does not specifically state this, AISC *Steel Construction Manual* (AISC, 2005) does.

10.4.5 Concrete Side-Face Blowout Strength of Headed Anchors in Tension

A *single* headed anchor in tension located close to the edge of concrete can fail in a limit state of side-face blowout (Fig. 10.9c). A small edge distance in combination with large embedment depth increases the chance that this type of failure will occur. Appendix D Section D.5.4 provides the following formula to determine the nominal side-face blowout strength of a single anchor in tension when its effective embedment depth h_{ef} exceeds 2.5 times the minimum edge distance c_{a1} ($h_{ef} > 2.5c_{a1}$):

$$N_{sb} = 160c_{a1}(A_{brg})^{1/2}\lambda(f'_c)^{1/2}$$

where c_{a1} = minimum edge distance
A_{brg} = net bearing area of the heavy-hex bolt head or nut, in.^{2}
$\lambda = 1.0$ for normal weight concrete

When the edge distance c_{a2}, measured in the direction perpendicular to the minimum edge distance c_{a1}, is less than three times c_{a1} ($c_{a2} < 3c_{a1}$), the value of N_{sb} should be multiplied by the factor $(1 + c_{a2}/c_{a1})/4$, where $1 \leq c_{a2}/c_{a1} \leq 3$.

The ϕ factors are discussed in Sec. 10.4.4.

For a *group* of headed anchors in tension located close to the edge of concrete, the following design procedure applies.

When the effective embedment depth of the anchors h_{ef} exceeds 2.5 times the minimum edge distance $c_{a,1}$ ($h_{ef} > 2.5c_{a1}$) and the anchor spacing s is less than $6c_{a,1}$ ($s < 6c_{a,1}$), the maximum nominal side-face blowout strength for those anchors susceptible to this limit state (that is, only those where $h_{ef} > 2.5c_{a1}$) is

$$N_{sbg} = [1 + s/(6c_{a1})]N_{sb}$$

where s = distance between outer anchors along the edge
N_{sb} = as determined earlier without modification for c_{a2}

Once determined, N_{sbg} is compared to the proportion of the tension force applied to those anchors susceptible to this limit state (that is, only those where $h_{ef} > 2.5c_{a1}$).

10.4.6 Concrete Breakout Strength of Anchors in Tension

As described in Section 10.4.1, ACI 318 Appendix D assumes that a single isolated cast-in headed anchor located far away from the edge will develop a concrete breakout prism shown in Fig. 10.8a. (The large edge distance precludes the design capacity from being controlled by side-face blowout.) The projected area of the breakout prism is taken as a square with the sides equal to $3h_{ef}$ (Fig. 10.8b). As with other limit states considered in Appendix D, provisions for concrete breakout strength apply for anchors with diameters not more than 2 in. and the embedment length not more than 25 in.

In typical metal building foundations the anchor bolts are spaced closely and are not able to develop the full concrete breakout prism. Instead, a truncated prism made of overlapping concrete cones is developed. The resulting combined concrete failure surface may be larger than the surface of the concrete breakout prism for a single anchor but smaller than a sum of four such surfaces.

The primary variables for finding the combined projected area of the breakout prism are the anchor's effective depth of embedment, spacing, and the edge distance. The general definition of the anchor edge distances and spacing in metal building systems is illustrated in Fig. 10.10.

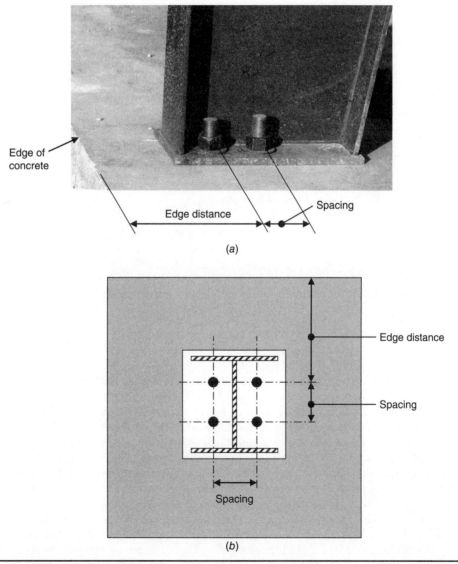

FIGURE 10.10 Anchor edge distances and spacing in metal building systems: (*a*) Picture; (*b*) plan.

The design requirements for this limit state are rather complex. The reader is encouraged to review the full text of ACI 318 Appendix D, including the Commentary, before embarking on the design, as some provisions could be interpreted in different ways.

For a single anchor, the nominal concrete breakout strength N_{cb} is determined as

$$N_{cb} = (A_{Nc}/A_{Nco})\psi_{ed,N}\psi_{c,N}\psi_{cp,N}N_b$$

For a group of n anchors, the nominal concrete breakout strength N_{cbg} is determined as

$$N_{cbg} = (A_{Nc}/A_{Nco})\psi_{ec,N}\psi_{ed,N}\psi_{c,N}\psi_{cp,N}N_b$$

In these two formulas, A_{Nco} is projected area of one anchor with an edge distance at least equal to $1.5h_{ef}$ ($c_{a1} \geq 1.5h_{ef}$):

$$A_{Nco} = 9h_{ef}^2$$

A_{Nc} is the projected area of the group, which should not exceed the value nA_{Nco}:

$$A_{Nc} \leq nA_{Nco}$$

N_b = basic concrete breakout strength for a single anchor located away from other anchors and edges in cracked concrete:

$$N_b = k_c\lambda(f'_c)^{1/2}h_{ef}^{1.5}$$

where $k_c = 24$ for cast-in anchors

Alternatively, for cast-in headed bolts and studs with 11 in. $\leq h_{ef} \leq$ 25 in., maximum value of N_b is

$$N_b = 16\lambda(f'_c)^{1/2}h_{ef}^{5/3}$$

The foregoing provisions of Appendix D apply when the anchor bolts have the smallest edge distance c_{a1} being at least equal to $1.5h_{ef}$ (Fig. 10.11a). This situation typically occurs in the columns placed directly on large foundations. It can also occur when the column is located in the center of a large pedestal, as may be possible in the buildings with bypass girts. In the metal buildings with flush or semiflush girts, the columns are typically placed relatively close to the exterior edge of concrete, so that $c_{a1} < 1.5 h_{ef}$ (Fig. 10.11b).

The calculations of the projected area of the group A_{Nc} depend on the number of concrete edges where the edge distances are less than $1.5h_{ef}$ (that is, $c_a < 1.5h_{ef}$). When two edges are critical, the projected area of the breakout prism A_{Nc} is shown in Fig. 10.12a. When three edges are critical, the projected area A_{Nc} is shown in Fig. 10.12b.

One of the occasionally overlooked provisions of Appendix D concerns the situation illustrated in Fig. 10.12b, where the edge distance on three or more edges is less than $1.5h_{ef}$. According to Appendix D Section D.5.2.3, when that happens, the value of

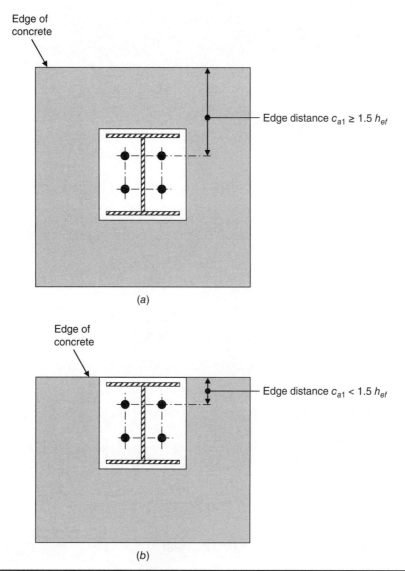

FIGURE 10.11 Relative locations of the column on the foundation or pedestal: (a) Column in the center, building with bypass girts; (b) column near the edge, building with flush girts.

h_{ef} used in several equations should be "the greater of $c_{a,max}/1.5$ and one-third of the maximum spacing between anchors within the group." Since in metal building systems the spacing between anchors within the group is relatively close, the edge-distance criterion typically controls the value of h'_{ef} (the adjusted value of h_{ef}):

$$h'_{ef} = c_{a,max}/1.5$$

Obviously, this greatly affects the concrete breakout capacities of the anchors in tension. The application of this concept is illustrated in Design Example 10.2.

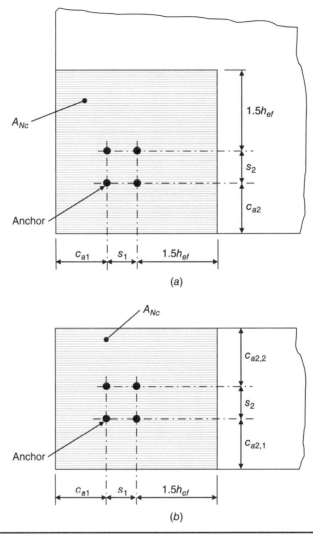

FIGURE 10.12 The number of concrete edges where $c_{a,1} < 1.5h_{ef}$: (a) Two edges; (b) three edges.

Design Example 10.2: Determination of h'_{ef} and A_{Nc} for Anchors Close to the Edge on Three Sides

Problem Find h'_{ef} and A_{Nc} for four anchors shown in Fig. 10.12b,

where $c_{a1} = 6$ in.
$c_{a2,1} = 8$ in.
$c_{a2,2} = 12$ in.
$s_1 = s_2 = 4$ in.

Solution

$c_{a,max} = 12$ in. and $c_{a,max}/1.5 = 8$ in. Controls for determination of h'_{ef}
$s_{max}/3 = 4$ in./$3 = 1.33$ in.
Use $h'_{ef} = 8$ in.
$1.5\, h'_{ef} = 12$ in.
$A_{Nc} = (6 + 4 + 12)(8 + 4 + 12) = 528$ in.^{2} ▲

The various modification factors ψ (psi) apply when certain conditions exist. These factors are:

- $\psi_{ec,N}$ = modification factor for anchor groups loaded eccentrically (see ACI Equation D-9)
- $\psi_{ed,N}$ = modification factor for edge effects for single anchor or groups in tension

    If $c_{a,max} \geq 1.5 h_{ef}$ $\psi_{ed,N} = 1.0$
    If $c_{a,max} < 1.5 h_{ef}$ $\psi_{ed,N} = 0.7 + 0.3(c_{a,min})/(1.5 h_{ef})$

    Note that h_{ef} here is reduced to h'_{ef} if three edge distances are critical

- $\psi_{c,N}$ = modification factor for an (upward) adjustment for uncracked concrete

    $\psi_{c,N} = 1.25$ for cast-in anchors in uncracked concrete

- $\psi_{cp,N}$ = modification factor for tensile strength of post-installed anchors in uncracked concrete without supplementary reinforcement

    $\psi_{cp,N} = 1.0$ for cast-in anchors

10.4.7 Using Anchor Reinforcement for Tension

What if the concrete breakout strength for anchors in tension computed as discussed earlier is insufficient to resist the applied uplift loading? This is a rather common scenario for metal building columns bearing on isolated concrete pedestals of moderate sizes. When this occurs, additional tension capacity must be introduced. This can be done by using anchor reinforcement for tension.

One type of such anchor reinforcement is shown in Fig. 10.13. In this design, a pair of inverted U-shaped dowels is placed symmetrically on each side of the anchor group. The distance between the anchors and the dowels should be as small as practical and should not exceed $0.5 h_{ef}$ (otherwise, it does not qualify, according to Appendix D Commentary).

Note that Figure RD.5.2.9 in Appendix D shows this distance both in the plane of the dowels and perpendicular to that plane, which could be interpreted by some that the actual distance measured as a square root of sum of squares is larger. A better interpretation seems to limit the actual distance (measured as square root of sum of squares) to $0.5 h_{ef}$.

The dowels extend a distance that equals or exceeds the development length of the dowel bars L_d, measured from the point of the dowel intersection with the breakout prism of the anchors. As stated in the previous paragraph, this point is determined as a square root of sum of squares of the distances between the dowels and anchor centerlines in two directions. Essentially, the force is transferred from the anchor bolts to the dowels and then to the concrete below the breakout prism.

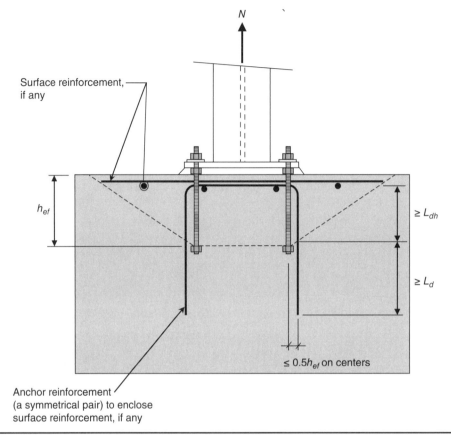

FIGURE 10.13 Anchor reinforcement for tension consisting of inverted U-shaped dowels. (After *ACI 318-08*.)

The dowels are also developed *above* the assumed surface of the breakout prism by a distance that equals or exceeds the development length of the hook for the dowel bars L_{dh}. This distance is measured from the point of the dowel intersection with the breakout prism of the anchors to the top of the dowels, determined again as a square root of sum of squares of the distances between the dowels and anchor centerlines in two directions.

According to ACI 318 Appendix D Commentary Section RD.5.2.9, the research on which these provisions are based was limited to the dowel sizes with the maximum size of No. 5 bar. The Commentary also states that it is beneficial for the anchor reinforcement to enclose the surface reinforcement, as shown in Fig. 10.13. (We should note that such surface reinforcement is not typically present in metal building foundations.)

In another design, the anchor bolts are "spliced" in a similar manner with the vertical pier reinforcement, rather than with the U-shaped dowels (Fig. 10.14). As in the previous design, the inclined surface of the breakout prism generated at the top face of each bottom nut intersects the adjacent vertical bar at a certain point. The vertical bar should extend above and below the point of intersection at least the distance of its development length L_d.

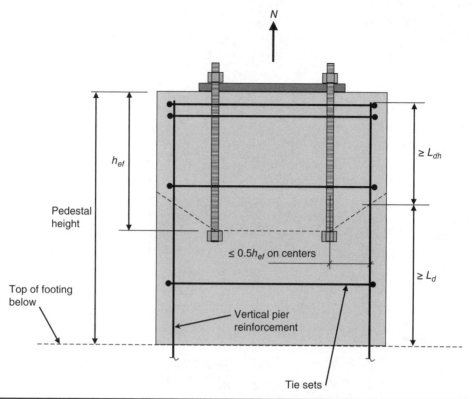

Figure 10.14 Using vertical pedestal bars as anchor reinforcement for tension.

This design could be useful when the pier cross-section is small, or the anchors are spaced wider than the common 4 in. Otherwise, it may be difficult to comply with the provision that the distance between the anchor and pier bar centerlines does not exceed $0.5h_{ef}$. Also, the typical sizes of the vertical pier reinforcement exceed No. 5. However, since this statement is included in the Commentary, rather than in the text of the Appendix D itself, it is not clear whether engaging the vertical pier reinforcement that is located farther away than the distance $0.5h_{ef}$ from the anchors and/or has a size larger than No. 5, is prohibited. This design has been successfully used for many years, and further research might validate its continuing use.

Naturally, the load path for tension transfer should continue through the foundation. The designer should ensure that the vertical bars are hooked into the footing, that the embedment length of the hooks L_{dh} is sufficient to meet the ACI 318 provisions, and that the overall weight of the foundation is sufficient to resist the applied uplift force.

When anchor reinforcement is specified, its strength can be used instead of the concrete breakout strength. The Commentary recommends using the strength reduction factor ϕ of 0.75 for designing anchor reinforcement.

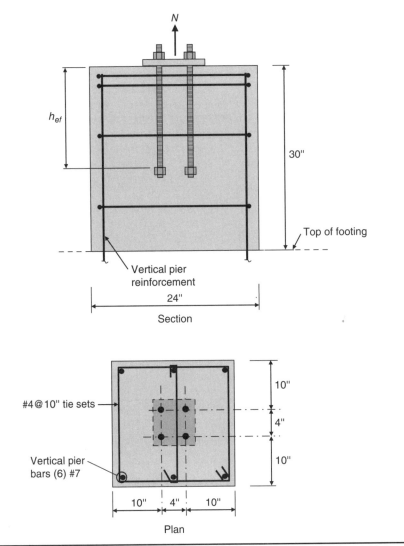

FIGURE 10.15 Concrete pedestal and anchors used for Design Examples 10.3 and 10.4.

Design Example 10.3: Design of Anchor Bolts for Tension, Moderate Tension Load*

Problem Design four headed anchors to support $N_u = 12,000$ lb at the base of metal building frame in Seismic Design Category B placed in the middle of the column pier. The normal-weight concrete has $f'_c = 3000$ psi. The column pier is 30 in. deep, 24 × 24 in. in plan, reinforced as shown in Fig. 10.15. The anchor bolts are spaced 4 in. on centers. The shear loading is considered separately in Design Example 10.6.

---

*Following the procedure of Example 34.6 of PCA Notes on ACI 318-08.

Solution Try (4) ¾-in.-diameter ASTM F1554 Gr. 36 anchors 18 in. long

$h_{ef} = 18$ in.

$c_{a1} = 10$ in., $c_{a2} = 10$ in.

$s = 4$ in.

$h_a = 30$ in. (concrete thickness)

Steel Strength of Anchor in Tension ($\phi = 0.75$)

$\phi N_{sa} = 0.75(4) A_{se,N} f_{uta}$

For ¾-in.-diameter anchors (see Table 10.2) $A_{se,N} = 0.334$, $A_{brg} = 0.911$ (in.²).

$f_{uta} = 58,000$ psi

$\phi N_{sa} = 0.75 \times 4 \times 0.334 \times 58,000 = 58,116\#$.

Concrete Breakout Strength for Group

$\phi = 0.70$, assume no supplementary reinforcement:

$\phi N_{cbg} = \phi (A_{Nc}/A_{Nco}) \psi_{ec,N} \psi_{ed,N} \psi_{c,N} \psi_{cp,N} N_b$

$c_{a,\max} = 10$ in. $< 1.5 h_{ef} = 27$ in.

Since the edge distance on three or more sides is less than $1.5h_{ef}$, reduce h_{ef} per Appendix D Section D.5.2.3:

$h'_{ef} = c_{a,\max}/1.5$ or $h'_{ef} = s/3$ (4 in./3 = 1.33 in. does not control)

Use $h'_{ef} = 10$ in./1.5 = 6.67 in. and $1.5 h'_{ef} = 10$ in.

$A_{Nc} = 24$ in. $\times 24$ in. $= 576$ in.² (area of the whole pedestal).

A_{Nco} (projected area of one anchor) $= 9 h'^2_{ef} = 9(6.67)^2 = 400$ in.²

Check

$A_{Nc} = 576 \leq (4)400 = 1600$ (in.²)   OK

Determine modification factors.

$\psi_{ec,N} = 1.0$ (no eccentricity)

Find $\psi_{ed,N}$ (modification factor for edge distance):
Since

$c_{a,\max} < 1.5 h_{ef}$

$\psi_{ed,N} = 0.7 + 0.3 (c_{a,\min}) / (1.5 h'_{ef})$

$\psi_{ed,N} = 0.7 + 0.3(10)/(1.5 \times 6.67) = 1.0$

Find $\psi_{c,N}$ (adjustment for uncracked concrete)

$\psi_{c,N} = 1.25$ for cast-in anchors in uncracked pier (with multiple ties, this is a reasonable assumption)

$\psi_{cp,N} = 1.0$ for cast-in anchors

Find N_b (basic concrete breakout strength for a single anchor located away from other anchors and edges)

$$N_b = k_c \lambda (f'_c)^{\frac{1}{2}} h_{ef}^{1.5}$$

where $k_c = 24$ for cast-in anchors
$N_b = 24 \times 1 \times (3000)^{\frac{1}{2}} 6.67^{1.5} = 22,644\#$
$\phi N_{cbg} = 0.7[(576)/(400)](1)(1)(1.25)(1)22,644 = 28,532\#$

Pullout Strength of Anchors

$$\phi N_{pn} = n\phi \psi_{c,P} N_p$$

where $n = 4$
$\phi = 0.7$ for pullout
$\psi_{c,P} = 1.4$ for uncracked concrete pier
$N_p = 8A_{brg} f'_c$ for headed bolt or stud
A_{brg} = net bearing area of the head = 0.911 (in.²)
$\phi N_{pn} = 4(0.7)(1.4)(8)(0.911)3000 = 4(21,427) = 85,707\#$

Concrete Side-face Blowout Strength

Check if $h_{ef} > 2.5 c_{a1}$

$$h_{ef} = 18 \text{ in.} < 2.5 \times 10 = 25 \text{ in.}$$

Side-face blowout limit state is not applicable.

Summary of Tension Capacities

Steel strength of anchor 58,116#
Concrete breakout 28,531#   Controls
Pullout strength   85,707#
Concrete side-face blowout strength N/A

∴ Use tension capacity $\phi N_n = 28,532\# > 12,000\#$ OK ▲

Design Example 10.4: Design of Anchor Bolts for Tension, High-Tension Load

Problem Design four headed anchors in Design Example 10.3 to support a higher factored uplift load $N_u = 30,000\#$. All other parameters stay the same as in Design Example 10.3. Use the minimum pier reinforcement consisting of six No. 7 vertical bars and No. 4 ties at 10 in. on centers.

Solution Using the summary of tension capacities at the end of Design Example 10.3, only the concrete breakout strength in tension is insufficient:

$$\phi N_{cbg} = 28,532\# < 30,000\#$$

Try supplemental anchor reinforcement for tension.

First, look at a possibility of using vertical pier bars as supplemental reinforcement. Two issues discussed earlier in Sec. 10.4.7 are present:

1. The size of vertical pier bars is No. 7, which exceeds the maximum No. 5 bars recommended by the Commentary.
2. The distance between the vertical pier bars, assuming 2-in. cover and $h_{ef} = 18$ in., is $(10 - 2 - 7/16)1.414 = 10.7$ in. $> 0.5h_{ef} = 9$ in.   NG

This approach would violate both limitations on the supplemental anchor reinforcement for tension stated in ACI 318 Appendix D Commentary Section RD.5.2.9. Also, the development length of No. 7 pier bars, whether straight of hooked, becomes difficult to achieve in the 30-in.-deep pier. Therefore, consider a design that uses inverted U-shaped dowels.

Try supplemental tension reinforcement with a pair of hooked No. 4 bars located 2 in. from the centers of anchors (the minimum practical placement), measured both in the plane of the dowels and perpendicular to that plane (see Fig. 10.16).

Assuming again $h_{ef} = 18$ in.

$$2 \text{ in.} < 0.5h_{ef} = 9 \text{ in.}   \text{OK}$$

L_{dh} for #4 bars = $15.34d_b = 7.7$ in. ≈ 8 in. (see footnote No. 4 for Table 6.1)
Extend the dowels below and above the intersection with the assumed breakout failure plane by at least 8 in.

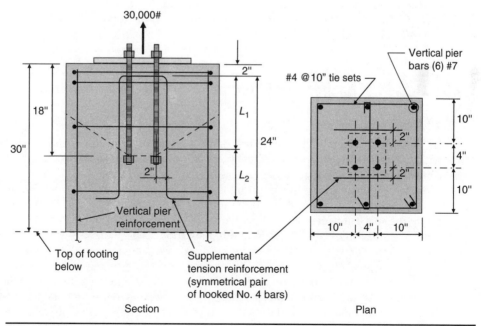

Figure 10.16 Supplemental tension reinforcement for Design Example 10.4.

From Fig. 10.16, the available length above the assumed breakout failure plane, with 2-in. cover to the top of U-shaped dowel and 1:1.5 slope of the breakout plane, is

$$L_1 = 18 - 2 - 2(1.41)/1.5 = 14.1 \text{ in.} > L_{dh} = 8 \text{ in.} \quad \text{OK}$$

Here, the term $2(1.41)/1.5$ is the length of the hypotenuse of a triangle with 2-in. sides divided by the slope of the breakout plane, to establish the vertical dimension where the dowel intersects the breakout plane.

The minimum required length below the assumed breakout failure plane is

$$L_2 = L_{dh} = 8 \text{ in.}$$

The minimum required total length of the U-shaped dowels is

$$14.1 + 8 = 22.1 \text{ in.}$$

Use dowels 24 in. tall to allow for tolerances and placement irregularities. These dowels will fit within the 30-in.-deep pier. (OK)

Compute the capacity of supplemental tension reinforcement of four No. 4 bars per ACI Section D.5.2.9, using strength reduction factor 0.75:

$$A_s = 4 \times 0.2 = 0.8 \text{ in.}^2$$
$$\phi N_n = 0.75(0.8)(60,000 \text{ psi}) = 36,000\# > 30,000\# \quad \text{OK}$$

∴ The supplementary reinforcement is adequate. Note that two No. 4 tie sets are required within the 5-in. distance measured from the top of the pedestal, per ACI 318-08 Section 7.10.5.6. ▲

10.5 Design of Anchors for Shear per ACI 318-08 Appendix D

10.5.1 Introduction

As discussed in Sec. 10.2.5, there are practical reasons why the anchor bolts placed within oversized holes in column base plates should not be used for resisting horizontal column reactions. If they are nevertheless used for this purpose, certain steps can be taken to facilitate the load transfer from the anchors to the base plate. These steps might involve welding heavy washers to the top of the base plate (see Fig. 10.6) or perhaps filling the space between the anchor shaft and the edges of the oversized hole, with the caveats expressed in Sec. 10.2.5. In another solution, cast-in anchors are welded to the bottom of an embedded plate, and the column base plate is welded to the embedded plate during steel erection.

Our goal in this section is to outline the process of designing embedded anchors for horizontal column reactions, which result in shear stresses in the anchors. We postpone the discussion about the mechanics of load transfer from the base plate to the anchors until Chap. 11.

Anchors in Metal Building Systems

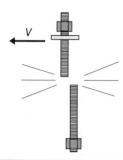

FIGURE 10.17 Anchor bolt failure in shear.

As with the provisions regulating the capacities of embedded anchors for tension, shear provisions of Appendix D Section 6 consider a number of limit states:

1. Steel strength of anchors in shear (Fig. 10.17)
2. Concrete breakout strength in shear
3. Concrete pryout strength in shear

10.5.2 Steel Strength of Anchors in Shear

The nominal steel strength in shear of headed stud anchors and cast-in headed and hooked bolts is computed differently. As opposed to a headed anchor bolt, a headed stud anchor is welded to a plate or another shape before its embedment in concrete (Fig. 10.18).

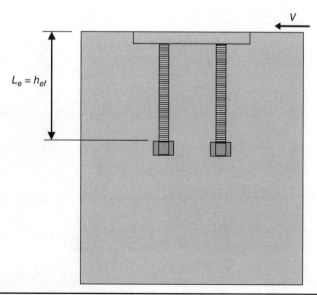

FIGURE 10.18 Cast-in headed studs.

For cast-in *headed stud anchors*, the nominal steel strength in shear V_{sa} is given by the following formula:

$$V_{sa} = nA_{se,V} f_{uta}$$

where n = number of anchors in group
$A_{se,V}$ = effective area of anchor in shear, in.² $A_{se,V} = A_{se,N}$ (see Table 10.2)
f_{uta} = same as for tension (see Sec. 10.4.3). For rods made of steel conforming to ASTM F1554 Gr. 36 f_{uta} = 58,000 psi

For *cast-in headed and hooked bolts* (and some post-installed anchors):

$$V_{sa} = n0.6 A_{se,V} f_{uta}$$

These formulas apply when the embedded plate or the base plate bears directly on concrete. If grout pads are used, V_{sa} is reduced by multiplying the previous formulas by a factor of 0.8.

10.5.3 Concrete Breakout Strength in Shear: General

The concept of concrete breakout strength of anchors in tension was discussed in Secs. 10.4.1 and 10.4.6. As described there, a single cast-in headed anchor located far away from the edge of the foundation will develop a concrete breakout prism shown in Fig. 10.8. However, the close edge distances and anchor spacing typically found in metal building systems might not allow for the development of a full concrete breakout prism. The typical anchor edge distances and spacing are illustrated in Fig. 10.10.

As will become clear soon, the anchor bolts located in this fashion will have a very small shear capacity, which is typically controlled by concrete breakout. In a classic concrete breakout a piece of concrete, roughly in the shape of an inverted half-pyramid, splits away from the rest at the anchor bolt locations (Fig. 10.19).

Figure 10.19 Concrete breakout.

Anchors in Metal Building Systems

The subject of concrete breakout strength in shear is described in ACI 318 Appendix D Section D.6.2. It provides the design equations for the nominal concrete breakout strength for a single anchor (V_{cb}) and for a group of anchors (V_{cbg}) for the four cases listed as follows.

1. Shear perpendicular to edge for single anchor (this often controls for anchors close to the edge):

$$V_{cb} = (A_{Vc}/A_{Vco})\psi_{ed,V}\psi_{c,V}\psi_{cp,V}V_b$$

2. Shear perpendicular to edge for a group of anchors:

$$V_{cbg} = (A_{Vc}/A_{Vco})\psi_{ec,V}\psi_{ed,V}\psi_{c,V}\psi_{h,V}V_b$$

3. Shear parallel to edge: It is permitted to take twice the values determined earlier for the case of shear assumed to act perpendicular to edge and $\psi_{ed,V} = 1.0$.

4. For corner anchors, use smaller of V_{cb} or V_{cbg} computed for both directions separately.

In the previous equations:

V_b is basic concrete breakout strength determined as discussed later.

Modification factors $\psi_{ec,V}$, $\psi_{ed,V}$, $\psi_{c,V}$, $\psi_{h,V}$ are determined as discussed later.

A_{Vco} is the projected area for one anchor in a deep member, with edge distance $\geq 1.5c_{a,1}$ in the direction perpendicular to the shear force (Fig. 10.20).

$$A_{Vco} = 4.5(c_{a,1})^2$$

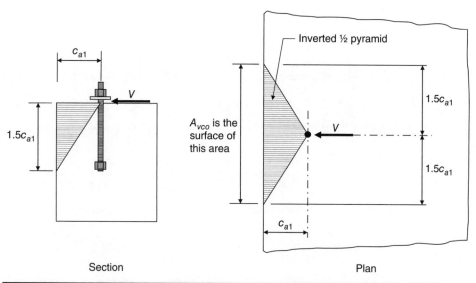

Figure 10.20 A_{Vco}, the projected area for one anchor in a deep member with edge distance $\geq 1.5c_{a,1}$ in the direction perpendicular to the shear force.

A_{Vc} is the total projected area of the failure surface on the side of concrete for a group of anchors.

$$A_{Vc} \leq nA_{Vco} \text{ or } n4.5(c_{a,1})^2$$

n is the number of anchors.

In metal building systems, four anchors typically support the building column. Two of them are closer to the edge of concrete than the other two (Fig. 10.11b). How should the horizontal loading be apportioned in this situation? In some older codes, such *1999 BOCA National Building Code*, the anchor capacity was controlled by the anchors closest to the edge, and the capacity of the group was taken as their capacity times the number of anchors. By contrast, ACI 318 Appendix D provides two possible approaches.

In Case 1 for calculation of A_{Vc}, the horizontal force is divided equally between the anchors closest and farthest from the edge (Fig. 10.21). Obviously, the capacity of the closest anchor controls.

In Case 2 for calculation of A_{Vc}, the horizontal force is resisted solely by the anchor located farthest from the edge. The closest anchor is neglected (Fig. 10.22).

Both cases should be checked, and the one producing the lowest capacity should be used in the design. When the anchors consist of welded studs, only Case 2 needs consideration, as stated in Appendix D Commentary Figure RD.6.2.1(b). The reason: If the anchor closest to the edge starts failing, the load shifts to the back anchor.

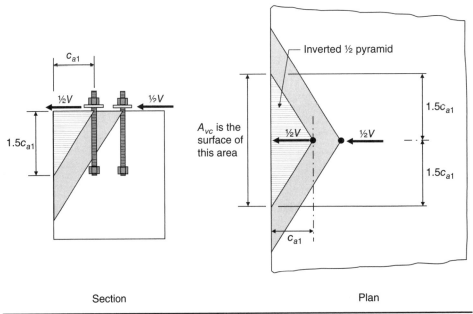

FIGURE 10.21 Case 1 for calculation of A_{Vc}: The horizontal force is divided equally between the anchors closest and farthest from the edge.

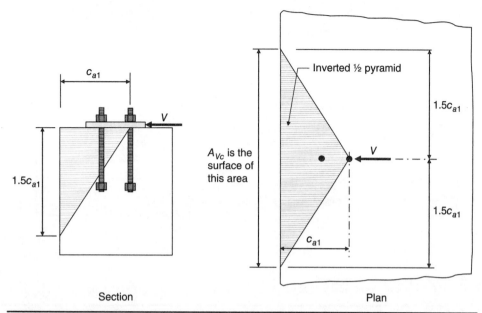

FIGURE 10.22 Case 2 for calculation of A_{Vc}: The horizontal force is resisted solely by the anchor located farthest from the edge; the closest anchor is neglected.

For an anchor located near a corner of the foundation, even the inverted half-pyramid shown in Figs. 10.20, 10.21, and 10.22 cannot be developed. In this case, A_{Vc} is the projected surface area of a truncated inverted half-pyramid (Fig. 10.23).

For anchors located in thin concrete of the thickness h_a (where $h_a < 1.5c_{a1}$), A_{Vc} is modified by taking the height of the inverted half pyramid equal to h_a rather than to $1.5c_{a1}$ (Fig. 10.24).

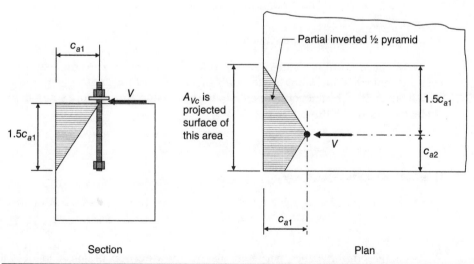

FIGURE 10.23 Finding A_{Vc} for an anchor located near a corner of the foundation.

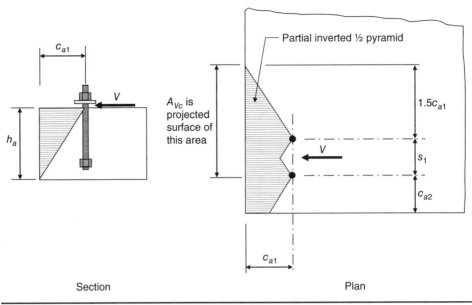

FIGURE 10.24 Finding A_{vc} for anchors located in thin concrete where $h_a < 1.5c_{a,1}$.

10.5.4 Basic Concrete Breakout Strength in Shear V_b

The basic concrete breakout strength in shear for one anchor in cracked concrete (V_b) is defined by the following Appendix D formula:

$$V_b = 7(L_e/d_a)^{0.2}(d_a)^{1/2} \lambda (f'_c)^{1/2} (c_{a1})^{1.5}$$

where L_e = load-bearing length of anchor for shear
$\qquad L_e = h_{ef}$ (the embedment depth) for headed studs (Fig. 10.25a).
Also, $\quad L_e < 8d_a$
$\qquad d_a$ = shaft diameter of the anchor

For cast-in headed studs, headed bolts and for hooked bolts "continuously welded to steel attachments" [including plate] with the minimum thickness t_{min} of 3/8 in. and at least one-half of the anchor diameter d_a, the basic concrete breakout strength in shear can be increased as follows (see Fig. 10.25b):

$$V_b = 8(L_e/d_a)^{0.2}(d_a)^{1/2} \lambda (f'_c)^{1/2} (c_{a1})^{1.5}$$

This formula applies only when:

- For anchor groups, the strength of the farthest from the edge row of anchors is used.
- The anchor spacing $s \geq 2.5$ in.
- If the edge distance perpendicular to the direction of the load $c_{a2} \leq 1.5h_{ef}$, reinforcement is provided at the corners.

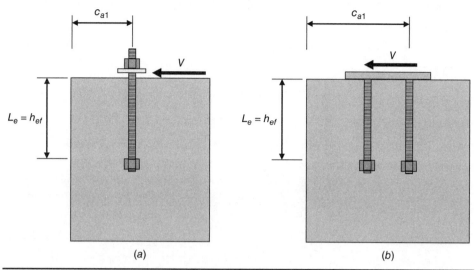

FIGURE 10.25 Determination of L_e, load-bearing length of anchor for shear: (a) Single headed anchor; (b) headed studs welded to a common plate.

10.5.5 Concrete Breakout Strength in Shear for Anchors Close to Edge on Three or More Sides

As discussed in Sec. 10.4.6 and illustrated in Design Example 10.2, concrete breakout strength in tension is substantially reduced the when the edge distance on three or more edges is less than $1.5h_{ef}$. A similar situation occurs with concrete breakout strength in shear.

Appendix D Section D.6.2.4 states: "Where anchors are influenced by three or more edges," the value of c_{a1} used in determination of A_{Vco}, V_b and the modification factors $\psi_{ec,V}$, $\psi_{ed,V}$, $\psi_{c,V}$, and $\psi_{h,V}$ (Appendix D Equations D-23 through D-29) should not exceed the larger of:

$c_{a2}/1.5$ in either direction (Appendix D Figure RD.6.2.4 states that $c_{a2,\,max}$ be used for this purpose)

$h_a/1.5$

$s/3$ (one-third of the maximum spacing between anchors in the group)

As ACI 318 Commentary Section RD.6.2.4 explains, this can be visualized by moving the actual distance c_{a1} closer to the edge to achieve a distance c'_{a1}. Since in metal building systems the spacing between anchors within the group is relatively close and the foundation thickness h_a is typically sufficient, this often results in using the adjusted edge distance in the direction of the load being

$$c'_{a1} = c_{a2}/1.5 \text{ in either direction}$$

For small foundation thicknesses, h_a could control c'_{a1}.

The design procedure for navigating these provisions is illustrated in Design Example 10.5.

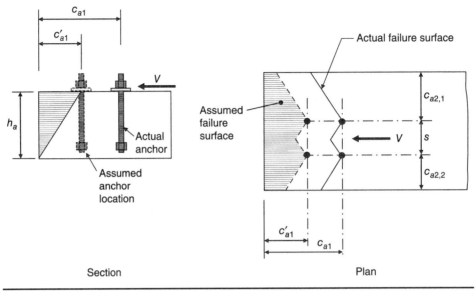

FIGURE 10.26 Determination of A_{Vc} for Design Example 10.5.

Design Example 10.5: Determination of A_{Vc} When Anchors Are Close to Three Edges

Problem Determine A_{Vc} for the concrete foundation with four anchor bolts located close to three edges, using the following dimensions:

$c_{a1} = 11$ in., $c_{a2,1} = 6$ in., $c_{a2,2} = 5$ in., $s = 4$ in., $h_a = 12$ in. (see Fig. 10.26).

Solution

$$c_{a2,\,max} = 6 \text{ in.}$$

c'_{a1} is the larger of :

$c_{a2}/1.5 = 6$ in./$1.5 = 4$ in.
$h_a/1.5 = 12$ in./$1.5 = 8$ in. Controls
$s/3 = 4$ in./$3 = 1.33$ in.

Use $c'_{a1} = 8$ in.

$$A_{vc} = (6 + 4 + 5)(1.5 \times 8) = 180 \text{ in.}^2 \ \blacktriangle$$

10.5.6 Concrete Breakout Strength in Shear: Modification Factors

There are four modification factors (ψ) used to determine concrete breakout strength in shear.

- Factor for anchor groups loaded eccentrically in shear $\psi_{ec,V}$

  This factor is used when the horizontal force acting in the plane of the top of concrete is applied eccentrically to the group. The maximum value of $\psi_{ec,V}$ is 1.0. The formula is given in ACI 318-08 Eq. D-26:

$$\psi_{ec,V} = \cfrac{1}{\left(1 + \cfrac{2e'_v}{3c_{a1}}\right)}$$

The definition of the force eccentricity c'_v is shown in Fig. 10.27a. The factor $\psi_{ec,V}$ generally does not apply when the anchor layout is symmetrical in reference to the column centerline, as is typically the case in metal building systems.

- The edge effect factor $\psi_{ed,V}$ applies as follows (see Fig. 10.27b):
  - When edge distance $c_{a2} \geq 1.5c_{a1}$, $\psi_{ed,V} = 1.0$
  - When edge distance $c_{a2} < 1.5c_{a1}$, $\psi_{ed,V} = 0.7 + 0.3(c_{a2})/(1.5c_{a1})$

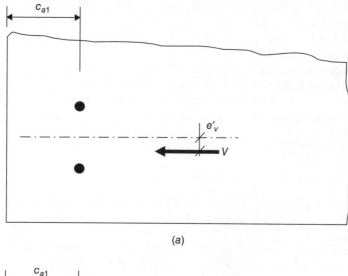

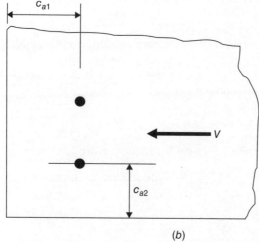

Figure 10.27 The terms for concrete modification factors (ψ): (a) Factor for anchor groups loaded eccentrically in shear $\psi_{ec,V}$; (b) edge effect factor $\psi_{ed,V}$.

- Note that the edge distance in the direction of the force c_{a1} used in these equations should be adjusted as stated in Sec. 10.5.5, if that section applies.
- Cracked concrete factor $\psi_{c,V}$ applies as follows:
  - If no cracking at service loads can be shown by analysis, $\psi_{c,V} = 1.4$.
  - Otherwise, $\psi_{c,V}$ varies from 1.0 to 1.4, as described in ACI 318 Section D6.2.7.
- Shallow concrete factor $\psi_{h,V}$ applies when the foundation thickness h_a is small ($h_a < 1.5c_{a1}$):

$$\psi_{h,V} = (1.5c_{a1}/h_a)^{1/2}$$

$\psi_{h,V}$ should be ≥ 1.0

10.5.7 Using Anchor Reinforcement for Concrete Breakout Strength in Shear

When one attempts to use the methodology of ACI 318 Appendix D for establishing shear capacities of embedded anchors, a realization quickly sets in that the common anchor edge distances and spacing used in metal building systems do not work even for moderate horizontal column reactions.

In many practical situations when the horizontal column reaction acts toward the "free" (discontinuous) edge of the concrete, the formulas in Appendix D predict insufficient capacity and likely failure of the anchor bolts. In most cases the controlling mode of failure is concrete breakout. In these cases, one of two design solutions are pursued:

1. Using concrete embedments rather than anchor bolts to transfer horizontal reactions into the foundation
2. Using anchor reinforcement for shear

In this section the second solution is discussed. Anchor reinforcement typically consists of hairpin bars wrapped around one of the anchors and extending into the concrete away from the free edge, as required to achieve the development length L_d. Alternatively, anchor reinforcement could be developed on each side of the breakout prism.

When anchor reinforcement is employed, its strength may be used in lieu of the concrete breakout strength for the determination of ϕV_n, where the strength reduction factor ϕ for anchor reinforcement is 0.75.

To be effective, the hairpins or other anchor reinforcement should be in contact with the anchor, typically the outer anchor nearest the free edge. The hairpins should be placed as close as possible to the surface of concrete, which in practice means 1.5 to 2 in below the top of the foundation, unless the base plate is recessed.

As the ACI 318 Commentary Section RD.6.2.9 recommends, the hairpin size should not exceed No. 6 bars. The research on which the relevant provisions of Appendix D are based used the maximum size of bars similar to No. 5 bars. Also, IBC-09 Section 1908.1.10 *requires* anchor reinforcement for anchors with diameters exceeding 2 in. that resist shear loading.

The hairpins could be of two basic configurations: straight (Fig. 10.28) and flared (Fig. 10.29). The straight hairpins that enclose the anchor are designed as follows:

$$\phi V_n = 0.75(A_s)60{,}000 \text{ psi}$$

where A_s = total area of hairpin bar (two legs)

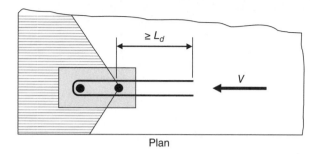

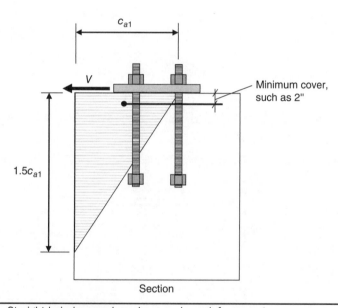

FIGURE 10.28 Straight hairpins used as shear anchor reinforcement.

The development lengths for straight bars (L_d) and for hooked bars (L_{dh}) are determined from standard ACI formulas. The results are summarized as follows for reference, using the "top" bar adjustments from *PCA Notes* Table 4-2.

No. 3 Hairpin Bars

$$A_s = 0.22 \text{ in.}^2; \ \phi V_n = 0.75(0.22)60{,}000 = 9900\#$$

- For $f'_c = 3000$ L_d for top bars = $44 d_b \times 1.3 = 44(3/8)(1.3) = 21.5$ in.
  $L_{dh} = 8.2$ in.
- For $f'_c = 4000$ L_d for top bars = $38 d_b \times 1.3 = 38(3/8)(1.3) = 18.5$ in.
  $L_{dh} = 7.1$ in.

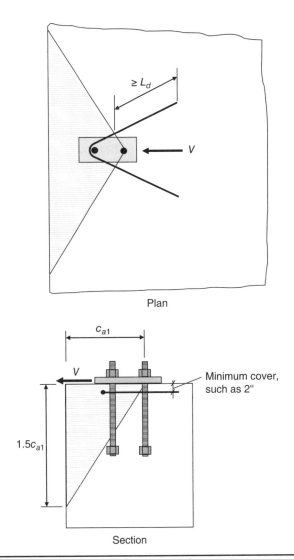

FIGURE 10.29 Flared hairpins used as shear anchor reinforcement.

No. 4 Hairpin Bars

$$A_s = 0.40 \text{ in.}^2; \phi V_n = 0.75(0.40)60{,}000 = 18{,}000\#$$

- For $f'_c = 3000$ L_d for top bars $= 44d_b \times 1.3 = 44(½)(1.3) = 28.6$ in.
  $L_{dh} = 11.0$ in.
- For $f'_c = 4000$ L_d for top bars $= 38d_b \times 1.3 = 38(½)(1.3) = 24.7$ in.
  $L_{dh} = 9.5$ in.

10.5.8 Using a Combination of Edge Reinforcement and Anchor Reinforcement for Concrete Breakout Strength in Shear

Another design for supplementary reinforcement used to increase anchor capacity in shear relies on a combination of edge and anchor reinforcement. In one version of this design a reinforcement bar is placed along the concrete edge toward which the load is acting, and anchor reinforcement bars enclose the edge bar in the breakout zone (Fig. 10.30).

In another version, anchor reinforcement simply consists of stirrups and ties. Only the bars located no more than the smaller of $0.5c_{a1}$ or $0.3c_{a2}$ from the centerline of the anchors are counted.

10.5.9 Concrete Pryout Strength in Shear

Yet another limit state that ACI 318-08 Appendix D requires to be checked for anchors subjected to shear loading is pryout strength. Concrete pryout is illustrated in Fig. 10.31.

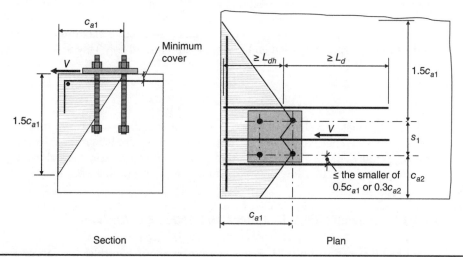

Figure 10.30 A combination of edge and anchor reinforcement for shear resistance.

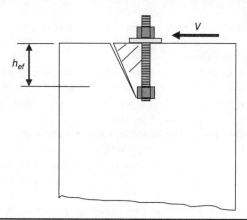

Figure 10.31 Concrete pryout.

The following provisions of ACI 318-08 Section D.6.3 apply to the nominal pryout strength of a single anchor (V_{cp}) or a group (V_{cpg}):

- For a single anchor $V_{cp} = k_{cp} N_{cb}$
- For a group $V_{cpg} = k_{cp} N_{cbg}$

where $k_{cp} = 1.0$ for $h_{ef} < 2.5$ in.
$k_{cp} = 2.0$ for $h_{ef} \geq 2.5$ in.

The latter ($k_{cp} = 2.0$) typically applies in real-life designs found in metal building foundations, as the anchor-bolt embedment is almost always more than 2.5 in. The values N_{cb} and N_{cbg} determined as for concrete breakout strength for tension (see Sec. 10.4.6).

10.5.10 Combined Tension and Shear

In most practical metal building applications, anchor bolts must resist both tension and shear. According to ACI 318-08 Section D.7, interaction of tension and shear loading on the anchors can be handled in two ways, depending on the relative values of both. When tension and shear forces exceed 20% of their respective maximum allowable values determined separately (that is, $V_{ua} > 0.2\phi V_n$ and $N_{ua} > 0.2\phi N_n$) the following equation should be used:

$$N_{ua}/\phi N_n + V_{ua}/\phi V_n \leq 1.2$$

When the factored shear force $V_{ua} \leq 0.2\phi V_n$, the full tension strength ϕN_n may be used in a tension check alone ($\phi N_n \geq N_{ua}$), neglecting shear. When the factored tension force $N_{ua} \leq 0.2\phi N_n$, the full shear strength ϕV_n may be used in shear check alone ($\phi V_n \geq V_{ua}$), neglecting tension. Other interaction formulas are allowed, provided they are corroborated by tests.

10.5.11 Minimum Edge Distances and Spacing of Anchors

As discussed earlier, the anchor edge distance greatly influences its tension and shear capacities. When the limit state of concrete breakout controls, the larger the edge distance, the larger the tension and shear capacity of the anchor. To increase the edge distance, using a certain girt orientation in the metal building superstructure can help. Recall the discussion in Chap. 1 about bypass and flush girts: A bypass girt is placed wholly outside of the column, while the outer flange of a flush girt is approximately even with the exterior face of the column.

The wall girts of the bypass inset result in the columns—and their anchor bolts—located farther away from the edge of the concrete than the columns with flush girts. From the standpoint of the anchor-bolt design, bypass girts provide larger anchor edge distances and are preferred over flush girts.

ACI 318-08 Appendix D Section D.8 requires that certain minimum edge distances and spacing be provided, unless supplementary reinforcement is used to prevent splitting of concrete at the free edge. The minimum edge distances had been recommended by the manufacturers of post-installed anchors for many years. Those edge distances were determined by testing, and Appendix D continues the tradition by allowing smaller edge distances to be used in post-installed anchors when they are tested in accordance with ACI 355.2.

Since testing is rarely performed for cast-in anchors used in metal building systems, the minimum values of edge distances and spacing given in ACI 318-08 Appendix D Section D.8 are worth becoming familiar with. They are, as a function of the bolt diameter d_a:

- Spacing s: $4d_a$ for untorqued cast-in anchors and $6d_a$ for torqued cast-in and post-installed anchors. It follows that the common 4-in. spacing of untorqued anchor bolts in metal building systems satisfied this provision only as long as $d_a \leq 1$ in; supplementary reinforcement is required for larger-diameter anchors.
- Minimum edge distance for untorqued cast-in anchors is the same as the minimum cover for steel reinforcement; it is $6d_a$ for torqued cast-in anchors. According to ACI 318 Section 7.7, the minimum cover for concrete exposed to earth in forms is 2 in. (unless governed by fire or corrosion protection requirements). Additional provisions for post-installed anchors of various types are given in Section D.8.

Design Example 10.6: Design of Anchor Bolts for Tension and Shear

Problem Using the data in Design Example 10.3, add shear loading $V_u = 6000$ lb acting from right to left in Fig. 10.15. Assume the anchors are welded to base plate.

Solution From Design Example 10.3, the controlling tension capacity was

$$\phi N_n = 28,532\text{\#} \text{ (controlled by concrete breakout).}$$

Design for Shear

Steel Strength ($\phi = 0.65$)

$\phi V_{sa} = \phi n 0.6 A_{se,V} f_{uta}$
$A_{se,V} = 0.334$ (in.²)
$f_{uta} = 58,000$ psi
$\phi V_{sa} = 0.65(4)(0.6)(0.334)58,000 = 30,220\text{\#}$

Concrete Breakout Strength For anchors welded to plate use Case 2 (only the back anchors resist V), therefore $c_{a,1} = 14$ in. (the edge distance in the direction of applied force V). See Figs. 10.22 and 10.32.
Check if ACI 318 Section D.6.2.4 controls:

$$c_{a2}/1.5 \text{ in either direction} = 10 \text{ in.}/1.5 = 6.67 \text{ in.}$$
$$h_a/1.5 = 30 \text{ in.}/1.5 = 20 \text{ in.} \quad \text{Controls}$$
$$s/3 = 4 \text{ in.}/3 = 1.33 \text{ in.}$$

Since 20 in. > 14 in., Section D.6.2.4 does not control and $c_{a,1} = 14$ in. Find A_{Vco} (projected area for 1 anchor in deep member).

$$A_{Vco} = 4.5(c_{a1})^2 = 4.5(14)^2 = 882 \text{ in.}^2$$

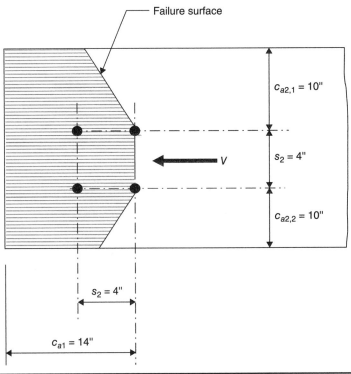

FIGURE 10.32 Determination of edge distance $c_{a,1}$ for Design Example 10.6.

Find A_{Vc} (projected area for group, taken as the total width of the pedestal times $1.5c_{a,1}$)

$$A_{Vc} = (24)(1.5 \times 14) = 504 \text{ in.}^2$$

Check $A_{Vc} \leq nA_{Vco}$ 504 < 2(882) OK

Find group design shear strength $V_{cbg} = (A_{Vc}/A_{Vco})\psi_{ec,V}\psi_{ed,V}\psi_{c,V}\psi_{h,V}V_b$

$\phi = 0.7$ (no supplementary reinforcement)

$\psi_{ec,V} = 1.0$ (no eccentricity)

$\psi_{c,V} = 1.4$ (no cracking in pier by analysis)

$\psi_{h,V} = 1.0$ (since $h_a = 30$ in. $> 1.5c_{a,1} = 21$ in.)

$\psi_{ed,V}$ for $c_{a2} < 1.5c_{a1}$ 10 in. < 21 in.

$\psi_{ed,V} = 0.7 + 0.3(c_{a2})/(1.5c_{a1}) = 0.7 + 0.3(10)/21 = 0.84$

The single anchor shear strength V_b (assume corner reinforcement is provided in the form of pedestal ties) is

$$V_b = 8(L_e/d_a)^{0.2}(d_a)^{1/2}\lambda\,(f'_c)^{1/2}(c_{a1})^{1.5}$$

Per ACI 318 Section D.6.2.2:

$L_e = h_{ef}$ for headed studs, but not more than $8d_a = 6$ in.

∴ Use $L_e = 6$ in.

$V_b = 8(6/0.75)^{0.2} (0.75)^{½} (1.0)(3000)^{½} (14)^{1.5} = 30{,}129\#$

$\phi V_{cbg} = \phi(A_{Vc}/A_{Vco}) \Psi_{ec,V} \Psi_{ed,V} \Psi_{c,V} \Psi_{h,V} V_b$
$= 0.7(504/882)(1)(0.84)(1.4)(1)(30{,}129) = 14{,}173\#$

Pryout Strength ϕV_{cp} probably does not govern for this configuration, but check anyway
Group $\phi V_{cpg} = \phi k_{cp} N_{cbg}$

$\phi = 0.7$ and $k_{cp} = 2.0$ for $h_{ef} \geq 2.5$ in.

From concrete breakout strength for tension in Design Example 10.3:

$\phi N_{cbg} = 0.7[(576)/(2916)](1)(0.811)(1.25)(1)100{,}388 = 14{,}072\#$
and $N_{cbg} = 14{,}072/\phi = 14{,}072/0.7 = 20{,}103\#$
$\phi V_{cpg} = 0.7(2.0)(20{,}103) = 28{,}144\#$

Summary of Shear Capacities
- Steel strength $\phi V_{sa} = 30{,}220\#$
- Concrete breakout $\phi V_{cbg} = 14{,}173\#$ Controls
- Pryout strength $\phi V_{cpg} = 28{,}144\#$

∴ Use $\phi V_n = 14{,}173\#$

Tension and Shear Interaction

Applied loads: $N_{ua} = 12{,}000\#$ and $V_{ua} = 6000\#$

Capacities: $\phi N_n = 28{,}532\#$ and $\phi V_n = 14{,}173\#$

By observation, neither $V_{ua} \leq 0.2\phi V_n$ nor $N_{ua} \leq 0.2\phi N_n$ apply, so must use interaction equation;

$$N_{ua}/\phi N_n + V_{ua}/\phi V_n \leq 1.2$$
$$12{,}000/28{,}532 + 6000/14{,}173\# = 0.84 < 1.2 \quad \text{OK}$$

Check Minimum Anchor Edge Distances and Spacing

- Minimum spacing, s: $4d_a$ (3 in.) for untorqued cast-in anchors and $6d_a$ (4.5 in.) for torqued cast-in anchors
  - Actual 4 in. OK for untorqued anchors, NG for torqued
  - The cast-in anchors welded to plate are not torqued (OK)
- Minimum edge distance for untorqued cast-in anchors: Same as cover for rebars (2 in.)

Actual 10 in.   (OK)

Conclusion Four ¾-in.-diameter-18-in.-long anchors welded to common plate are OK.

A Note of Caution The design load in shear in this example was very light. For higher loads, supplemental reinforcement for shear will probably be needed, as illustrated in Design Example 10.7. ▲

Design Example 10.7: Design of Anchor Bolts for Tension and Shear Using Supplemental Reinforcement for Shear

This design example illustrates the use of supplemental reinforcement for shear.

Problem Using the data from Design Example 10.1, design the anchor bolts for a small metal building system. The column reactions are:

Horizontal: dead 1.9 kips, roof live 10.2 kips (both acting outward)

Vertical: dead 4.6 kips, roof live 20.6 kips (both acting downward)

Wind reactions on the left-side column, wind from left to right:

Horizontal: 13.8 kips (inward); vertical: 13.7 kips (uplift)

The directions of column reactions are shown in Fig. 10.7.

Solution As discussed in Design Example 10.1, Case 2, $0.6D + W$ *to right* controls. Converting the loads to USD format, the load combination in Case 2 becomes:

$$0.9D + 1.6W \text{ (wind acting from left to right)}$$

Tension $N_{ua} = 0.9 \times 4.6 - 1.6 \times 13.7 = -17.78$ (kips) acting upward

Shear $V_{ua} = 0.9 \times 1.9 - 1.6 \times 13.8 = -20.37$ (kips) acting inward

Ultimate Design Capacities Tension (concrete breakout controls, as computed in Design Example 10.3):

$$28.531 \text{ kips} > 17.78 \text{ kips} \quad \text{OK}$$

Shear (concrete breakout controls if no supplemental reinforcement is provided, from Design Example 10.6):

$$14.173 \text{ kips} < 20.37 \text{ kips} \quad \text{NG}$$

Find shear capacity with supplemental reinforcement. Capacity of supplemental shear reinforcement of four No. 4 hairpins per ACI Section D.5.2.9, using strength reduction factor 0.75:

$$A_s = 4 \times 0.2 = 0.8 \text{ in.}^2$$

$$\phi N_n = 0.75(0.8)(60{,}000 \text{ psi}) = 36{,}000\# = 36 \text{ kips} > 20.37 \text{ kips} \quad \text{OK}$$

Shear and tension interaction with supplemental reinforcement for shear:

$$N_{ua}/\phi N_n + V_{ua}/\phi V_n \leq 1.2$$

$$17.78/28.531 + 20.37/36 = 1.19 < 1.2 \quad \text{OK}$$

The details of the pedestal, including the supplemental shear reinforcement, are shown in Fig. 10.33. Note that two tie sets are provided at the top of the pedestal per ACI 318-08 Section 7.10.5.6. ▲

10.5.12 Concluding Remarks

As can be seen from Design Example 10.6, a substantial 24 × 24-in. column pedestal designed under the provisions of ACI 318-08 Appendix D could barely resist a relatively minor factored shear loading of 6000 lb. Recall that the column base plate was centered on the pedestal, which required using bypass girts.

The designs with larger horizontal column reactions or those with flush girts would likely require using supplemental reinforcement for shear, such as that shown in Figs. 10.28, 10.29, 10.30, and 10.33. For high uplift loading, supplemental reinforcement for tension, such as that designed in Design Example 10.4, might also be needed. Since the horizontal column reactions act in two directions—and when a wall cross-bracing is present, in the perpendicular direction as well—the many overlapping hairpins, dowels, and ties might cause serious congestion at the top of the pedestal and thus become impractical.

Historically, supplemental reinforcement for tension and shear has not been provided in the foundations used for metal building systems. Implementing the design provisions of ACI 318-08 Appendix D requires a total change in the design and construction practices of metal building foundations. The anchor-bolt designs used in pre-engineered buildings for decades will no longer work.

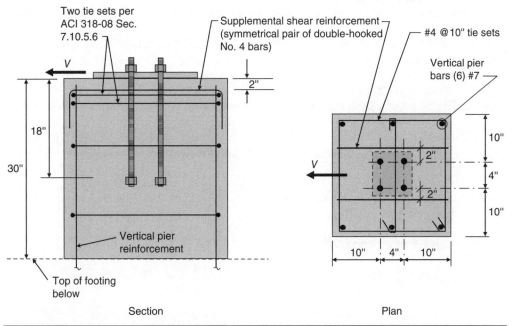

Figure 10.33 Details of the pedestal, including the supplemental shear reinforcement, for Design Example 10.7.

We hope the foregoing discussion sufficiently demonstrates the challenges and opportunities in the design of anchor bolts following today's codes. Regardless of the future code changes, it is clear that the traditional designs of column base plates used in metal building systems will need to evolve. The changes should affect the anchor spacing, grouting practices under the base plates—and probably the whole idea of using anchor bolts for shear transfer. Other concrete embedments might offer cost-effective alternatives for resisting high horizontal forces.

References

2009 International Building Code®, International Code Council, Country Club Hills, IL, 2009.

ACI 117-06, *Specifications for Tolerances for Concrete Construction and Materials and Commentary*, American Concrete Institute, Farmington Hills, MI, 2006.

ACI 117-10, *Specifications for Tolerances for Concrete Construction and Materials and Commentary*, American Concrete Institute, Farmington Hills, MI, 2010.

ACI 318-08, *Building Code Requirements for Structural Concrete and Commentary*, American Concrete Institute, Farmington Hills, MI.

ACI 351.2R-10, *Report on Foundations for Static Equipment*, American Concrete Institute, Farmington Hills, MI.

ACI 355.2-07, *Qualification of Post-Installed Mechanical Anchors in Concrete & Commentary*, American Concrete Institute, Farmington Hills, MI.

AISC 303-05, *Code of Standard Practice for Steel Buildings and Bridges,* American Institute of Steel Construction, Chicago, IL, 2005.

AISC 360-05, *Specification for Structural Steel Buildings,* American Institute of Steel Construction, Chicago, IL, 2005.

AISC *Design Guide 1: Base Plate and Anchor Rod Design*, American Institute of Steel Construction, Chicago, IL, 2006.

AISC *Steel Construction Manual*, 9th ed., American Institute of Steel Construction, Chicago, IL, 1989.

AISC *Steel Construction Manual*, 13th ed., American Institute of Steel Construction, Chicago, IL, 2005.

ASTM Standard A36 /A36M, "Standard Specification for Carbon Structural Steel," ASTM International, West Conshohocken, PA, 2008.

ASTM A307, "Standard Specification for Carbon Steel Bolts and Studs, 60,000 psi Tensile Strength," ASTM International, West Conshohocken, PA, 2010.

ASTM Standard F1554, "Standard Specification for Anchor Bolts, Steel, 36, 55, and 105 Yield Strength," ASTM International, West Conshohocken, PA, 2007.

Metal Building System Manual, Metal Building Manufacturers Association (MBMA), Cleveland, OH, 2006.

PCA Notes on ACI 318-08, Building Code Requirements for Structural Concrete, Portland Cement Association, 2008.

CHAPTER 11

Concrete Embedments in Metal Building Systems

11.1 The Role of Concrete Embedments

11.1.1 Prior Practices vs. Today's Code Requirements

In this chapter we continue the discussion in Chap. 10, extending it to concrete embedments. As explained in Chap. 10, the design provisions of ACI 318 Appendix D require a complete overhaul of the design and construction practices related to column anchorage in metal building systems. The changes affect the manner in which the columns are attached to the foundations, the presence of grout under the base plates, and the minimum sizes and reinforcement of column pedestals.

While in the past a couple of large-diameter anchor bolts were considered sufficient to resist typical horizontal and upward-acting vertical reactions, this is not the case today. The common design of the decades past is shown in Fig. 11.1: A large column bearing on a small pedestal, with two anchor bolts and no grout. Note that the column even overhangs the edge of the pedestal at left, a result of imprecise concrete placement practices discussed in Chap. 10. Today, this design would likely be considered totally inadequate.

Under the provisions of ACI 318 Appendix D, even a large column pedestal reinforced with vertical bars and ties could have a rather modest capacity to resist horizontal forces. As demonstrated in Design Example 10.6, four anchors placed in a substantial 24 × 24-in. pedestal could barely resist a relatively small factored horizontal column reaction of 6000 lb. The limit state of concrete breakout controlled the capacity. To attain even that small capacity, the column base plate had to be centered on the pedestal, which required using bypass girts, and the anchors had to be welded to the underside of the base plate. Essentially, the anchors were welded studs rather than true anchor bolts placed in oversized holes in the column base plate. No grout was used, as that would have further diminished the shear capacity of the steel anchors.

11.1.2 Two Options for Resisting High Horizontal Column Reactions

But what to do if much higher values of the horizontal column reactions must be resisted? Two basic options are available for this common situation:

1. Anchor bolts with supplemental reinforcement for shear
2. Concrete embedments

236 Chapter Eleven

Figure 11.1 Common design of the decades past: a large column bearing on a small pedestal with two anchor bolts and no grout.

As demonstrated in Chap. 10 design examples, following the design provisions of ACI 318 Appendix D typically produces very modest shear resistance capacities of anchor bolts. These capacities are often controlled by the limit state of concrete breakout. As a result, supplemental reinforcement for shear is often needed, which complicates the design and construction of the foundation. When the horizontal reactions exist in two orthogonal directions, as happens at the wall bracing locations, the overlapping supplemental reinforcement bars tend to become unwieldy. This option is further discussed in Sec. 11.2.

The second option—concrete embedments discussed in this chapter—might prove more cost effective for resisting high horizontal column reactions than multiple layers of supplemental reinforcement for shear. The embedments discussed in Sec. 11.3 resist horizontal forces by engaging concrete in direct bearing.

11.1.3 Transfer of Uplift Forces to Foundations: No Alternative to Anchor Bolts?

Before discussing the inner workings of concrete embedments, we should reiterate that the embedments discussed in Sec. 11.3 are intended for transfer of *horizontal* column reactions from columns to the foundations. But what about transfer of *uplift* reactions? Here, the anchor bolts and similar embedded anchors reign supreme.

Theoretically, it is possible to develop an embedment design that does not rely on classic anchor bolts. For example, the cap plate design of Sec. 11.6.1 could be modified to resist the uplift forces by supplementing the cap plate with horizontal rods or anchors embedded or drilled through the column pedestal. The rods or anchors would be placed in shear under the uplift loading on the column.

In another possible solution, vertical pier reinforcement could be extended above the top of concrete, where it would become column anchors. This design is commonly used for anchoring electric towers and similar equipment (see Fig. 9.1).

Both these design solutions are rarely, if ever, used in metal building foundations because of the construction complications involved. Regular anchor bolts, perhaps with supplemental reinforcement for tension discussed in Chap. 10, remain the most practical option for resisting uplift loading in foundations for metal building systems.

11.2 Using Anchor Bolts to Transfer Horizontal Column Reactions to Foundations

11.2.1 Some Problems with Shear Resistance of Anchor Bolts

Summarizing the discussion in Chap. 10, one of the main problems with using anchor bolts for transferring horizontal column reactions from columns to the foundations is the presence of the oversized holes in the column base plates. An anchor bolt placed in the oversized hole cannot possibly be in contact with all the edges of the plate at the same time. The column and the base plate would have to move in the direction of the horizontal force to bear against the anchor bolt.

As explained in Chap. 10, the most common number of anchor bolts used in metal building system columns is four. As discussed there, designers should consider the placement tolerances for embedded items—the main reason for having oversized holes in base plates. Because the tolerances apply not only to the dimensional but also to the rotational position of anchor-bolt groups, the four anchors usually end up being placed in the slightly different positions within their oversized holes in the base plate.

To engage all four anchor bolts, the base plate would have to not only move laterally under load, but also to slightly rotate at the base. This is difficult to imagine, given the fact that the base plate is welded to the column! More likely, some of the anchor bolts would be slightly bent by the plate under horizontal loading until the load is shared by all four.

Naturally, to move any distance at all, the column base would have to overcome the friction force resulting from the load the column carries. We should note that such frictional resistance represents the real-life primary mechanism of transferring horizontal forces between the column and the foundation. Still, the building codes generally do not allow relying on friction alone for the transfer of lateral forces.

As discussed in Chap. 6, lateral movement (that is, spreading apart) of the foundations supporting pre-engineered rigid frames is usually undesirable. It might cause damage to brittle finishes and overstress the frames. The amount of the frame movement required to engage the anchor bolts in oversized holes could be quite large. As shown in Fig. 10.5 for a common ¾-in.-diameter anchor bolt placed in a typical oversized hole in the base plate, the horizontal movement could reach as much as 9/16 in. Figure 11.2 illustrates this plate movement in a larger scale (in the cross-section).

The authors of the American Institute of Steel Construction (AISC) *Steel Design Guide 1*, who address this issue as it pertains to structural-steel buildings, recognize the problem and recommend a "cautious approach" to using anchor rods for shear transfer, such as relying only on two anchor rods rather than on all four. Elsewhere, the American Institute of Steel Construction states: "AISC recommends avoiding resisting shear by anchor rods if possible" (Zoruba, 2006). Finally, the Commentary on AISC Specification observes:

> Shear at the base of a column is seldom resisted by bearing of the column base plate against the anchor rods.

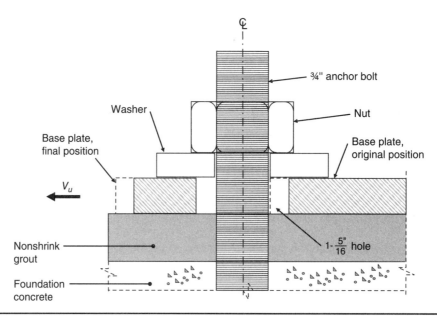

FIGURE 11.2 The amount of column movement needed to bring the edge of an oversized hole in the base plate in contact with the anchor bolt placed in the middle of the hole.

Another problem with using anchor bolts for transferring horizontal forces is the presence of grout underneath the base plate. Depending on the design assumptions about grout performance in this situation, grout could severely reduce the flexural capacity of the anchor bolts—or reduce it only a little or not at all, as explained later.

11.2.2 Possible Solutions to Enable Resistance of Anchor Bolts to Horizontal Forces

Two design solutions could be used to minimize large movements of anchor bolts within the oversized holes:

1. Welding a heavy plate washer to the top of the base plate (see Fig. 10.6). The washer has a regular-size hole—1/16-in. larger than the nominal anchor bolt diameter (ACI 351.2R-10). The anchor functions as a cantilevered column in flexure, fixed at the base and the top; it needs not achieve full bearing against the edge of the hole if it develops the required bending capacity. Obviously, a sufficient clearance must exist between the anchor bolts and the column flanges to allow the washer to be installed and for the welds to be made.

2. Fill the oversized holes with weld material, shims, or nonshrink grout after the bearing plate is in place. No bending of anchor bolts would then take place. This approach, popular among some engineers, might be difficult to achieve on the construction site in a reliable manner. In particular, grouting small uneven spaces around the anchor bolts seems problematic.

The following discussion explains how to analyze anchor bolts in such circumstances. When a heavy plate washer is welded to the top of the base plate, the load path for the

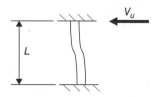

FIGURE 11.3 Anchor rod in reverse-curvature bending.

horizontal force travels through the washer. The anchor rod is then subjected to flexure within the oversized hole in the base plate. The rod can be analyzed as a mini-column fixed at the base and at the top and acting in reverse-curvature bending (Fig. 11.3).

AISC *Steel Design Guide 1* recommends taking the lever arm of the anchor rod in reverse-curvature bending as one-half of the distance between the "center of bearing of the plate washer" and the top of grout. From a practical standpoint, the midthickness of the plate washer can be taken as the point where the rod is fixed at the top and where the horizontal force is applied to the anchor rod. But where is the assumed point of fixity at the bottom?

Taking this point at the top surface of the grout pad is only one of three possible assumptions. It is illustrated as the distance L_1 in Fig. 11.4. Under this assumption, the grout acts as a perfect support for the rod and undergoes no crushing under load.

In a more conservative assumption, some local grout crushing takes place, lowering the bottom point of fixity by some amount (such as ¼–½ in.) and increasing the assumed rod length to the distance L_2 in Fig. 11.4. In the most conservative assumption, the grout

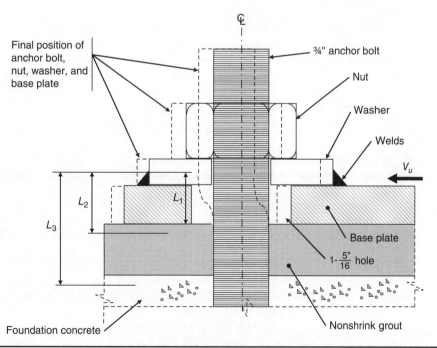

FIGURE 11.4 Bending of an anchor rod within the oversize hole of the base plate. (The deflected shape of the rod and the amount of the base plate movement are exaggerated for clarity.)

is considered crushed or otherwise totally ineffective, and the bottom point of fixity is lowered still farther: to the top surface of concrete (distance L_3 in Fig. 11.4).

The last assumption is necessary when the quality of grout placement—or even its presence—cannot be assured, as many a light pole can attest. If the designer wishes to be even more conservative, the distance L_3 could be taken to some depth *below* the surface of concrete, to allow for some local concrete crushing under load. Gomez et al. (2009) provide additional discussion on this topic.

11.2.3 Design of Anchor Bolts for Bending

The Commentary on 2005 AISC Specification (AISC 360-05) Section J9 states that anchor rods should be designed as threaded parts per AISC 360-05 Table J3.2. In addition, AISC Manual Tables 7-1 and 7-2 list the available shear and tension strengths of bolts. For the anchor material conforming to ASTM A307 (the closest designation in the tables for the typical steel used in anchor rods), the information given in these sources is listed as follows.

- Nominal shear stresses F_{nv} for threaded parts, when threads are not excluded from shear planes: $F_{nv} = 0.40F_u$ or $0.4 \times 58 = 23.2$ ksi. (For solid rods $F_{nv} = 24$ ksi. For allowable stress design (ASD), using the safety factor $\Omega = 2$ gives $F_v = 12$ ksi. From Table 7-1 the available ASD shear strength for ¾-in. bolts with gross area $A = 0.442$ in.², $V_{ASD} = 5.3$ kips.)

- Nominal tension stresses for threaded parts, when threads are not excluded from shear planes: $F_{nt} = 0.75F_u$ or $0.75 \times 58 = 43.5$ ksi. (For solid rods $F_{nt} = 45$ ksi. From Table 7-2 the available ASD tensile strength for ¾-in. bolts is $T_{ASD} = 9.94$ kips.)

These stresses are already reduced by the effect of threads for tension (0.75) and for shear (0.8), so that they may be used with the nominal rod areas. (For example, for ¾-in. bolts the nominal area $A_g = 0.442$ in.².) We should note that ACI 318 Appendix D uses reduced "tensile stress area" for calculations, as described in Chap. 10. Table 10.2 lists the effective area of ¾-in. anchor rod for shear ($A_{se,V}$) and tension ($A_{se,N}$) as 0.334 in.². The ratio of $A_{se,V}$ to A_g is indeed the same as just stated:

$$0.334 / 0.442 = 0.75$$

The AISC Specification, Commentary, and Manual provide no explicit equations for designing anchor rods for bending. The Manual includes a design example for the solid round rod acting as a simple-span beam (Example F.13, Round Bar in Bending). The Manual's design example uses the gross area of the rod, since no threads are present.

But for the purposes of using AISC equations for threaded rod flexure, the *net* rod diameter should be considered, with the depth of the threads subtracted. The data for such net (reduced) diameters of the threaded rods are not easy to find, as the threads could be of different types and there are tolerances for making them.

The simplest approach is to take the effective area of the rod for shear ($A_{se,V}$) and tension ($A_{se,N}$) from Table 10.2 and to arrive at the reduced diameter d_{red} using the formulas:

$$A_{se} = \pi d_{red}^2 / 4, \text{ from where}$$
$$d_{red} = (4A_{se} / \pi)^{1/2}$$

For example, for a ¾-in. threaded rod:

$$d_{red} = (4 \times 0.334/3.1415)^{1/2} = 0.652 \text{ (in.)}$$

This value falls between the minor diameter of 0.6733 in. for fine-thread series and 0.6273 in. for coarse-thread series listed in *Marks' Standard Handbook for Mechanical Engineers* (indeed, the average of these two numbers is 0.650 in.). Once the reduced diameter is determined, the section modulus, plastic modulus, and the moment of inertia can be computed.

The following design example illustrates one possible procedure for flexural design of anchor rods. As the design example demonstrates, the maximum horizontal force the bolt can resist, as controlled by flexure, is rather modest. Other solutions must be sought for larger horizontal forces.

Design Example 11.1: Anchor Bolt in Flexure

Problem Find the design bending capacity of a ¾-in. Grade 36 anchor bolt used in a metal building system foundation and the maximum horizontal force the anchor can resist, as controlled by its flexure. The column base plate is ¾-in. thick, and the horizontal forces are transmitted through a 3/8-in.-thick washer welded to the top of the plate (Fig. 11.4). The base plate is placed on top of a 1-in.-thick grout bed.

Solution For a ¾-in. rod, using $d = d_{red} = 0.652$ in.:

$$S_x = \pi d^3/32 = \pi(0.652)^3/32 = 0.0272 \text{ (in.}^3)$$

(for comparison, using the nominal diameter of ¾ in. gives $S_x = 0.0414$ in.^{3})

$$Z_x = d^3/6 = (0.652)^3/6 = 0.0462 \text{ (in.}^3)$$

(for comparison, using the nominal diameter of ¾ in. gives $Z_x = 0.0703$ in.^{3})

$$I_x = \pi d^4/64 = \pi(0.652)^4/64 = 0.0089 \text{ (in.}^4)$$

(for comparison, using the nominal diameter of ¾ in. gives $I_x = 0.0155$ in.^{4})

Nominal flexural strength of the rod M_n can now be determined. From flexural yielding limit state:

$$M_n = M_p = F_y Z_x \leq 1.6 M_y$$
$$F_y Z_x = 36 \text{ ksi}(0.0462 \text{ in.}^3) = 1.663 \text{ kip-in.}$$
$$1.6 M_y = 1.6 F_y S_x = 1.6(36 \text{ ksi})(0.0272 \text{ in.}^3) = 1.567 \text{ kip-in.} < F_y Z_x$$
$$\therefore \text{ Use } M_n = 1.567 \text{ kip-in.}$$

Using the ASD Format

$$\Omega_b = 1.67$$
$$M_{all} = M_n/\Omega_b = 1.567/1.67 = 0.938 \text{ (kip-in.)}$$

The maximum horizontal force H_{all} the bolts can resist, as controlled by bolt flexure is

$$H_{all} = 2M_{all}/L$$

The force H_{all} depends on the assumed distance L (see Fig. 11.4). For this example use $L = L_1$ (the distance from the mid-thickness of the plate washer to the top of grout):

$$L_1 = \tfrac{3}{4} + \tfrac{1}{2}(3/8) = 0.9375 \text{ (in.)}$$
$$H_{all} = 2M_{all}/L_1 = 2(0.938)/0.9375 = 2.00 \text{ (kips)}$$

Using the LRFD Format

$$\phi_b = 0.90$$
$$\phi_b M_n = 0.9(1.567) = 1.410 \text{ (kip-in.)}$$

Using $L_1 = 0.9375$ in.,

$$H_u = 2\phi_b M_n/L_1 = 2(0.9)(1.410)/0.9375 = 2.71 \text{ (kips)}$$

For a group of four anchor bolts, $H_{uo} = 2.71 \times 4 = 10.83$ kips.
Finally, find the maximum deflection at the top of the bolt from $H_{all} = 2.00$ kips.

$$\Delta_{max} = PL^3/(12EI) = 2.00(0.9375/2)^3/[12(29,000)0.0089] = 0.0001 \text{ (in.)}$$

The deflection is almost negligible.

Note The design bending capacity of a ¾-in. Grade 36 threaded rod used in this example, and the maximum horizontal force the rod could resist as controlled by flexure, were rather modest. For a group of four ¾-in. anchor bolts, $H_{uo} = 10.83$ kips, which is much smaller than the shear capacity determined in Design Example 10.6. We should note that the steel capacity in shear should be reduced by a factor of 0.8 when a grout pad is used (see ACI 318 Section D.6.1.3); this would presumably apply to the rod's bending capacity in this design example but not in Design Example 10.6. The maximum horizontal force would be even smaller if the larger distances L_2 and L_3 were used, as illustrated in Fig. 11.4. ▲

11.3 Concrete Embedments for the Transfer of Horizontal Column Reactions to Foundations: An Overview

If anchor rods have limited capacities to transfer horizontal column reactions—and perhaps should not even be used for this purpose—what *should* be used? A more reliable solution is to employ the concrete embedments described in this section. They are:

- Shear lugs
- The Newman lug
- Recessed column base
- Other (cap plate, embedded plate with studs, and so on)

The unifying feature of all these embedments is that they transfer horizontal forces from columns to the foundations by bearing of steel bars or plates against the concrete. The anchor rods are still present, but only for stability during erection and for uplift resistance;

they are not designed for shear loading. As demonstrated in the sections that follow, the embedments are much better suited for resisting horizontal forces than anchor bolts.

Unlike anchor bolts, these concrete embedments are not specifically addressed in the International Building Code and in ACI 318 Appendix D. Fortunately, there are other sources of information, including Appendix D of ACI 349-06 *Code Requirements for Nuclear Safety-Related Concrete Structures and Commentary*. The embedment provisions of ACI 349 are based on the paper by Rotz and Reifschneider (1989). Another useful publication is AISC *Steel Design Guide 1: Base Plate and Anchor Rod Design*.

11.4 Shear Lugs and the Newman Lug

11.4.1 Construction of Shear Lugs

The shear lug, also known as the shear key, is a steel plate welded to the bottom of the column base plate and embedded in a slot cut into the top of the foundation. The space around the shear lug is filled with fluid nonshrink grout after the column is installed. Generally, a grout pad exists under the column base plate. The shear lug is usually supplemented by anchor bolts to ensure column stability during erection (Fig. 11.5).

The oversized holes in the column base plate help accommodate the anchor-bolt placement tolerances. The column base plate can be aligned vertically with the help of leveling nuts or shim packs. When leveling plates are used, thick washers are needed above them, to prevent the nuts from protruding inside the oversized holes in the base plates. Separate leveling plates placed on top of a bed of nonshrink grout do not work well with shear lugs, since grouting of the slot in concrete cannot be done after the bed of grout is placed above it. AISC *Steel Design Guide 1* recommends that the thickness of the column base plate equal the thickness of the shear lug.

Shear lugs have been used in heavily loaded structural steel buildings for decades, particularly on the West Coast of the United States. The use of shear lugs in metal building

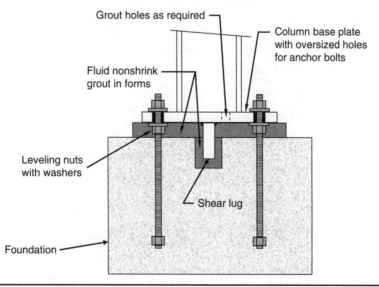

FIGURE 11.5 Shear lug used in combination with anchor bolts.

systems has been relatively infrequent. Some of the reasons these embedments have not been more popular in pre-engineered buildings are:

- The need to form a precisely placed pocket (slot) in concrete
- The need to use nonshrink grout under the column base plate and inside the pocket
- The need to change the typical close spacing of the anchor bolts, as demonstrated later

Nevertheless, if the traditional anchor bolts are incapable of resisting large horizontal column reactions, there is little choice but to change the common practices used by metal building manufacturers and contractors. The manufacturers need to change their typical base-plate designs and the anchor-bolt layout details, and the contractors need to get used to grouting the column bases and forming pockets in concrete.

In some cases, using two or more shear lugs might be advantageous, as discussed later. In other cases, shear lugs could exist in two orthogonal directions, which requires a cross-shaped slot in concrete (Fig. 11.6). One example where this might be needed involves a cross-bracing bent or a portal frame placed in the sidewall perpendicular to the primary frame, as illustrated in Chap. 1.

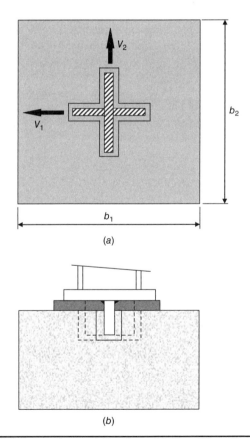

Figure 11.6 Shear lugs in two orthogonal directions: (a) Plan; (b) section.

11.4.2 Minimum Anchor Bolt Spacing and Column Sizes Used with Shear Lugs

The slot in concrete made to receive the shear key must be wide enough to accomplish two objectives:

1. To provide sufficient space for grout around the shear key
2. To allow for less-than-precise placement of the slot in concrete

To accomplish the first objective, the minimum clearance around the shear key to allow for proper grouting needs to be determined. This information is not widely available, but two sources provide a clue. First, the manufacturers of nonshrink grout recommend the minimum grout thickness of 1 in. under the base plate (BASF, 2007). It might be easier to place fluid grout in a shallow slot than under a wide base plate, and a smaller clearance for filling the slot seems appropriate. Second, the minimum clearance for fine grout around reinforcing steel used in masonry structures is ¼ in. (TMS 602-11). Therefore, the minimum clearance around the shear key could be taken as ¼ in. and perhaps larger.

To accomplish the second objective, the discussion on concrete tolerances needs to be recalled. According to ACI 117-06 and ACI 117-10 Section 2.3.2, the tolerance for the centerline of embedded items from specified location is 1 in. Since a slot in concrete is generally made by casting a removable embedded item, a 1-in. tolerance applies to the location of the slot.

Combining the tolerances for making the slot (1 in.) and the minimum grout cover (¼ in.) gives the *minimum* grouted space around the shear key of 1¼ in., as shown in Fig. 11.7. A larger clearance might be appropriate in some circumstances involving congested construction, very large base plates, or unusual shear-lug configurations.

Knowing the minimum required grout clearance allows the designer to establish the minimum spacing of the anchor bolts around the concrete slot for the shear key. For example, for a 1½-in.-thick shear key, the distance from its center to the edge of the slot is

$$1.5/2 + 1.25 = 2 \text{ (in.)}$$

Assuming the minimum distance from the center of the anchor bolt to the edge of the slot of 1½ in., the minimum spacing of the anchor bolts around the concrete slot

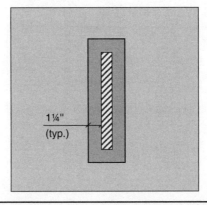

FIGURE 11.7 Suggested clearance around the shear lug in the slot formed in concrete.

becomes 7 in. (see Fig. 11.8). Needless to say, this is larger than the spacing of 4 in. commonly used in metal building systems (see a discussion and illustrations in Chap. 10).

If the anchor bolts are placed between the column flanges, the minimum column depth that works with shear keys can be determined. Assuming the flange thickness of ½ in. and the clear distance from the inside surface of the flange to the center of the anchor of 2 in., the minimum column depth that may be used with shear lugs is

$$7 + 2 + ½ + 2 + ½ = 12 \text{ (in.)}$$

What about the minimum depth of the column base plate in this design? Assuming the metal building column is connected to the base plate by means of ¼-in. fillet welds

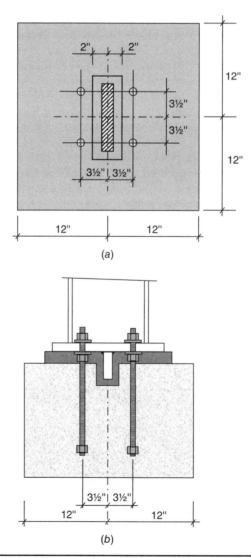

FIGURE 11.8 The minimum suggested spacing of anchor bolts around a shear key.

placed on the outside surfaces, a minimum "shelf" clearance of ½ in. is required at each weld (see AISC Manual 13th ed. Fig. 8.11). Therefore, the minimum depth of the column base plate used with shear lugs is

$$\tfrac{1}{2} + 12 + \tfrac{1}{2} = 13 \text{ (in.)}$$

These minimum sizes of the column and the base plate are wider than is standard for many metal building manufacturers at this time—one more item to keep in mind when specifying shear lugs.

11.4.3 Design of Shear Lugs: General Procedure

The discussion in this section is based on the provisions of ACI 349-06 Appendix D and the information in the AISC Design Guide 1. The Commentary on ACI 349-06 Section RD.11 states that the shear strength of the shear lug is the sum of its bearing strength in concrete and the additional strength from the effects of confinement by anchors. Another response mode is that of shear-friction.

The bearing mode develops first and produces an initial fracture plane in concrete in front of the shear lug. The shear-friction mode follows, resulting in the final fracture plane.

When two or more shear lugs are used, the initial failure plane develops between the shear lugs, followed by the shear-friction mode (Fig. 11.9). The final fracture plane in Fig. 11.9 assumes that the shear lug is placed in a substantial concrete foundation.

The assumed fracture surface for the shear lugs placed into the column pedestals of moderate sizes is different: It approximates the failure surface that develops during the concrete breakout limit state (see discussion in Chap. 10). The configuration of the failure surface is illustrated in Fig. 11.10, after a design example in AISC Design Guide 1 and general requirements in ACI 349-06 Section D.11.2.

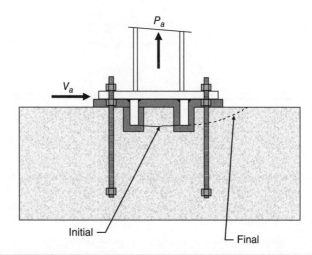

FIGURE 11.9 Initial and final fracture planes that develop at shear lugs placed in a large concrete foundation (after ACI 349-06).

248 Chapter Eleven

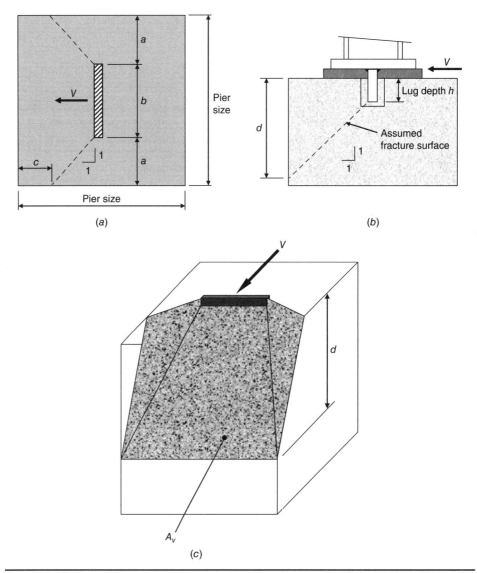

FIGURE 11.10 Determination of the assumed fracture surface for the shear lug placed in column pedestal: (a) Plan; (b) section; (c) isometric. (After AISC *Steel Design Guide 1*, 2d ed.)

The process of designing the shear lug in a column pedestal includes the following steps:

1. Determine the size of the shear lug, controlled by the concrete bearing capacity against the lug.
2. Check the shear capacity of concrete, considering the failure surface in Fig. 11.10.
3. Design the steel section of the shear lug and the welds connecting it to the base plate.

For the extremely high horizontal forces, shapes other than solid bars or plates can be used. For example, wide-flange and tubular shapes have been used as shear lugs. Weisensel (2010) illustrates the use of a heavy wide-flange section in one such situation.

11.4.4 Determination of Bearing Strength

According to ACI 349-06 Section D.4.6.2, the ultimate design bearing stress for concrete or grout placed against shear lugs is

$$\phi f_{ubrg} = \phi 1.3 f'_c$$

The embedded area of lug A_L for a shear lug with the length b and depth h is

$$A_L = b \times h$$

The bearing capacity of the shear lug is

$$\phi P_{ubrg} = \phi 1.3 f'_c A_L$$

Using capacity reduction factor for bearing $\varphi = 0.65$ (per ACI 349-06 Section D.4.4 and ACI 318-08 Section 9.3.2.3), the bearing capacity of the shear lug is

$$\phi P_{ubrg} \approx 0.85 f'_c A_L$$

According to ACI 349-06 Section D.4.6.2, f'_c is the smaller of the 28-day compressive strength of concrete or grout.

ACI 318-08 Section 10.14 provides a different formula for concrete bearing strength for a bearing area A_1:

$$\phi P_{u,brg} = 0.65(0.85) f'_c A_1$$

Some confusion exists about a possible increase of the bearing strength following the provisions of this section, as they apply to shear lugs. When the supporting surface is wider *on all sides* than the loaded area, resulting in the increased loaded area A_2 at the base of the truncated pyramid, the design bearing strength on the loaded area may be increased by a ratio of

$$(A_2/A_1)^{1/2} \text{ but } \leq 2.0$$

With a typical shear lug, only three supporting surfaces are wider than the loaded area, so this increase in bearing strength does not apply, resulting in the following equation for bearing capacity of the lug (taking $A_1 = A_L$):

$$\phi P_{u,brg} = 0.65(0.85) f'_c A_L = 0.55 f'_c A_L$$

This formula will be used in the design examples that follow.

11.4.5 Determination of Concrete Shear Strength

Concrete shear strength in front of the lug is found following the provisions of ACI 349-06 Section D.11.2, Shear toward Free Edge. For each shear lug loaded toward a free edge without supplemental reinforcement:

$$V_u = 4 \phi \sqrt{f'_c} A_v$$

Here, A_v is the effective stress area defined by a 45° plane projecting from the bearing edges to the free surface, excluding the bearing area of the shear lug (Fig. 11.10).

Per ACI 349-06 Section D.4.4 $\phi = 0.8$ for shear lugs, when shear is acting toward free edge. (We should note that AISC Design Guide No. 1, 2d ed., uses $\phi = 0.75$ in their Example 4.9.)

ACI 349-06 allows for an increased shear resistance of the lugs caused by the effects of confinement by anchor bolts. According to ACI 349-06 Section RD11.1, the increase in shear strength resulting from the effects of confinement can be taken as $\phi K_c (N_y - P_a)$,

where $\phi = 0.85$ for shear lugs
N_y = anchor yield strength of the tension anchors
$N_y = nA_{se}f_y$
P_a = factored external axial load on the anchorage (P_a is positive for tension, negative for compression)
$K_c = 1.8$ for a single shear lug located at least the distance h from the front edge of the base plate

The contribution of anchor bolts to shear resistance of the lug is often neglected, since the shear capacity of a reasonably-sized pier is often sufficient, as illustrated in a design example that follows.

11.4.6 The Newman Lug

In heavy industrial construction, massive column base plates with shear lugs are commonly fabricated separately from the columns. This allows setting and aligning the base plate with relative ease and then placing fluid nonshrink grout underneath the base plate and into the concrete slot containing the shear key. Grout holes in the base plates may be used if needed. After the grout has gained the required strength, the column is welded to the base plate.

In metal building construction, the base plates are typically shop-welded to the primary-frame columns and installed with them. The columns are carefully placed so that the holes in the base plates go over the anchor bolts. A shear key protruding from the bottom of the column would complicate the erection. A simple and practical solution has been developed by the author to simplify the erection process.

The Newman lug consists of a separate plate with shop-welded shear lug (Fig. 11.11). It is placed separately from the column, as in heavy industrial construction. Once the assembly is aligned and grouted, the metal building column is welded to the top of the embedded plate. The design of the shear lug itself proceeds no differently than for a shear lug welded to the base plate.

The main advantage of the Newman lug is that it simplifies column erection by eliminating shop-welded shear lug at the bottom; this design also preserves the oversized holes in the column base plates. The column is placed over the anchor bolts as usual. The welding that would be made between the heavy washers placed around the anchor bolts and the top of the base plate is now made between the base plate and the embedded plate (Fig. 11.12). The welds are designed to transfer horizontal column reactions only; the anchor bolts resist uplift. The thickness of the column base plate need not depend on the thickness of the shear lug. Design Example 11.2 illustrates the process.

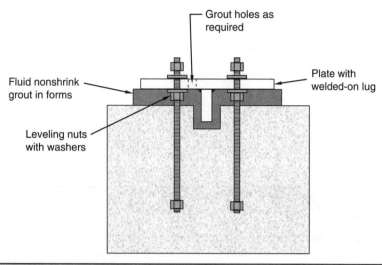

Figure 11.11 The Newman lug.

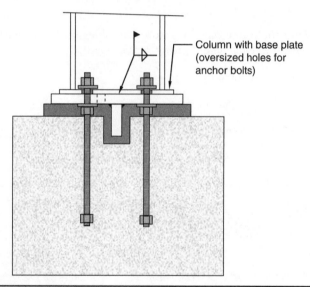

Figure 11.12 Column placed on top of the Newman lug.

Design Example 11.2: Shear Lug*

Problem Design shear lug to support dead and snow loading on a single-span rigid frame used in a metal building system. The frames have an 80-ft span and 18-ft

---

*Design example follows the general procedure of AISC Design Guide No. 1, *Base Plate and Anchor Rod Design*, 2d ed.

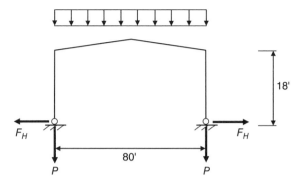

FIGURE 11.13 Rigid frame used in Design Example 11.2.

eave height; they are spaced 25 ft on centers (Fig. 11.13). The roof snow load is 30 pounds per square foot (psf). Use the frame reactions from the tables in the Appendix. (This is the same frame used in Design Example 7.1.) Concrete $f'_c = 4000$ psi and steel $F_y = 36,000$ psi. Foundation pier is 24 × 24 in. and 30 in. deep, with minimum reinforcement. Assume grout thickness $G = 1.5$ in.

Solution Find the frame reactions from the tables in the Appendix for a frame with 80-ft span and 18-ft eave height, using service loads.

Horizontal reactions:

$$\text{Dead} = 2.9 \text{ kips, snow} = 21.8 \text{ kips}$$

Vertical reactions (not needed for this example):

$$\text{Dead} = 4.8 \text{ kips, snow} = 30.9 \text{ kips}$$

Combine the loads using IBC basic load combination for factored loading $1.2D + 1.6S$:

$$F_{H,u} = 1.2(2.9) + 1.6(21.8) = 38.36 \text{ (kips)}$$

The minimum required area of the shear lug $A_{L,\min}$ is controlled by bearing on concrete. Using ACI 318 values of factored bearing stress:

$$\phi P_{ubrg} = 0.65(0.85)f'_c A_{L,\min} = 38.36 \text{ (kips)}$$

$$0.65(0.85)(4.0)A_{L,\min} = 38.36 \text{ (kips)}$$

$$A_{L,\min} = 17.36 \text{ in.}^2$$

Try lug 8.75 × 2 in. deep.

$$A_L = 8.75 \times 2 = 17.5 > 17.36 \text{ in.}^2 \quad \text{OK}$$

Find effective stress area A_v, as controlled by shear capacity of concrete V_u. Assume the lug is 1.375 in. thick. See Figs. 11.10 and 11.14 for abbreviations.

$$c = 12 - 1.375/2 = 11.312 \text{ (in.)}$$
$$d \text{ (depth of stress area)} = 2 + 11.312 = 13.312 \text{ (in.)}$$

Concrete Embedments in Metal Building Systems 253

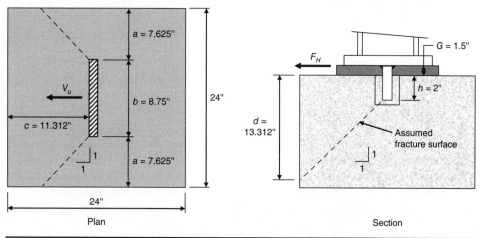

FIGURE 11.14 Concrete dimensions used in Design Example 11.2.

$A_v = 24 \times 13.312 - 2 \times 8.75 = 302$ (in.²)

$V_u = 4\phi f'_c A_v = [4(0.8)(4000)^{1/2} 302]/1000 = 61.12$ (kips) > 38.36 kips   OK

∴ No need to rely on additional shear capacity from confinement provided by anchor bolts.

Find the required lug thickness using $G = 1.5$ in. Design the lug as a cantilever extending from the underside of the base plate. Use AISC specification provisions for the design of steel elements. Fig. 11.15 shows the location of the design horizontal force for the design of the shear lug.

$M_{u,L} = V_u(G + h/2) = 38.36(1.5 + 2/2) = 95.9$ (kip-in.)

$Z_{L,rq} = b(t_{L,rq})^2/4$

$\phi = 0.9$ for bending of steel.

$M_{u,L} = \phi F_y Z_{L,rq}$

$95.9 = 0.9(36)8.75(t_{L,rq})^2/4 = 70.88(t_{L,rq})^2$

$t_{L,rq} = 1.16$ in.

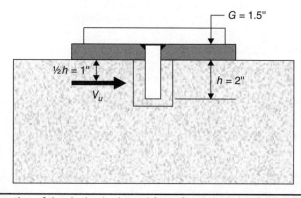

FIGURE 11.15 Location of the design horizontal force for shear lug in Design Example 11.2.

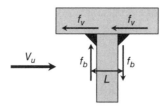

Figure 11.16 Forces on shear-lug welds in Design Example 11.2.

A 1.25-in.-thick lug could be used. The author's preference is to increase the lug thickness by 1/8 in. to allow for the possible effects of corrosion, since some grout cracking and accumulation of moisture at the bottom of the slot could occur. The cost differential between small lengths of 1.25- and 1.375-in. plates is negligible. Use 1.375-in.-thick shear lug.

Design fillet welds connecting shear lug to plate, each 8.75 in. long.

Per AISC 360-05 Table J2.2, minimum size of the fillet weld connecting two 1-3/8-in. plates is 5/16 in., so use two 5/16-in. fillet welds. The welds are under both shear (f_v) and bending (f_b) forces (see Fig. 11.16).

Spacing between the weld centers is

$$L = 1.375 + (5/16)(1/3)2 = 1.583 \text{ (in.)}$$
$$f_b = (M_{u,L})/L = 95.9/(1.583 \times 8.75) = 6.92 \text{ kip-in.}$$
$$f_v = (V_u)/(2 \times 8.75 \text{ in.}) = 38.36/(2 \times 8.75) = 2.19 \text{ kip-in.}$$

Combined force on welds (f_o) is

$$f_o = (f_b^2 + f_v^2)^{0.5} = 7.26 \text{ kip-in.}$$

The design strength of a 5/16-in. weld loaded perpendicular to its length is 10.5 kips/in., following the provisions of AISC 360-05 Section J2.2.4.

$$10.5 > 7.26 \text{ (kip-in.)} \quad \text{OK}$$

Can use two 5/16-in. welds 8.75 in. long.

Conclusion Shear lug 8.75 in. long, 3.5 in. deep and 1.375 in. thick is adequate. As noted earlier, the thickness of the plate to which the shear lug is welded is typically equal to the lug's thickness. ▲

11.5 Recessed Column Base

11.5.1 Construction

The idea behind this concrete embedment design is essentially the same as for the shear lug: an embedded shape transfers horizontal forces into the foundation by bearing against concrete. Instead of the shear lug, the embedded shape consists of the column flange in combination with the column base plate (Fig. 11.17). In a version of this system, only the column base plate is embedded.

Instead of a narrow slot in concrete needed in the shear-lug design, the recessed column base requires a relatively large pocket—large enough to embed the whole column

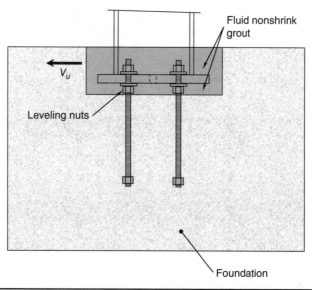

FIGURE 11.17 Recessed column base.

base and allow the room for grout. Accordingly, this embedment should be used in large foundations. As in the shear-lug solution, leveling nuts may be used to vertically align the base plate. One advantage of this system for metal buildings: It works with closely spaced anchor bolts, something not possible with shear lugs (see Sec. 11.4.2).

One conceptual disadvantage of this solution is the fact that the column base embedded in concrete could develop a certain degree of fixity—an undesirable situation, as discussed throughout this book. Another potential problem is a slight rocking movement of the column base caused by the frame's flexural deformations under changing loading levels. Over time, this movement could damage the grout surrounding the column base and render the system ineffective. Another cause for concern is the same as for shear lugs: the need to protect the embedded area from corrosion.

11.5.2 Design

The basic principles applicable to the design of shear lugs apply to the recessed column bases. As with shear lugs, the depth of embedment is determined first, as governed by concrete bearing capacity against the embedded shapes. The concrete shear resistance is checked next, following ACI 349-06 Section D.11.

The bearing strength of concrete is determined per ACI 318-08 Section 10.14. As discussed in Sec. 11.4.4, the following formula applies for the bearing area of the embedded column flange and base plate A_{brg}:

$$\phi P_{u,brg} = 0.65(0.85)f'_c A_{brg} = 0.55 f'_c A_{brg}$$

If needed, the bearing resistance of the column flange and the bearing plate can be computed separately (see Fig. 11.18):

$$V_u < R_{fl} + R_{pl}$$

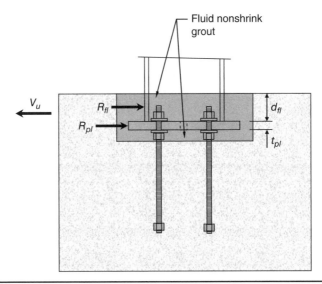

FIGURE 11.18 Bearing resistance of the recessed column flange and the base plate.

The concrete shear strength for recessed shapes is determined by following the provisions of ACI 349-06 Section D.11.2, Shear toward Free Edge. The formula and the design procedure is the same as for shear lugs (see Sec. 11.4.5):

$$V_u = 4\phi f'_c A_v$$

A_v is the effective stress area defined by a 45° plane projecting from the bearing edges to the free surface, excluding the bearing area of the embedded area. Per ACI 349-06 Section D.4.4, $\phi = 0.8$ for "embedded plates and shear lugs, shear toward free edge."

As with shear lugs, additional shear capacity from confinement provided by anchor bolts could be used if needed.

Design Example 11.3: Recessed Column Base

Problem Find the depth of the recessed column base and check the foundation's shear capacity for the metal building system column used in Design Example 11.2, except use a 60-psf roof snow load. The column has 8-in.-wide flanges; the base plate is 9 × 9 in., ½ in. thick. The strength of both concrete and grout $f'_c = 4000$ psi. The column is bearing on the 48 × 48 in. square footing that is 36 in. deep.

Solution Determine frame reactions first. Since the frame reactions in the Appendix do not include an entry for 60-psf roof snow, use twice the snow-load reactions from the table for 30-psf snow loading (the frame reactions for dead load do not change).

The horizontal reactions at service loads are

Dead = 2.9 kips, snow = 21.8 kips × 2 = 43.6 kips

Combine the loads using IBC basic load combination for factored loading 1.2D + 1.6S:

$$F_{H,u} = 1.2(2.9) + 1.6(43.6) = 73.24 \text{ (kips)}$$

The required design bearing strength of the base plate and the embedded column flange is

$$\phi P_{u,brg} = 0.55 f'_c A_{brg} = 0.55(4) A_{brg} = 73.24 \text{ (kips)}$$
$$A_{brg} = 73.24/[0.55(4)] = 33.29 \text{ (in.}^2)$$
$$A_{pl} = 9 \times \tfrac{1}{2} = 4.5 \text{ (in.}^2)$$
$$\text{Minimum } A_{fl} = 33.29 - 4.5 = 28.79 \text{ (in.}^2)$$
$$\text{Minimum } d_{fl} = 28.79/8 = 3.6 \text{ (in.}^2)$$

Use 4-in. flange embedment.

Find effective concrete shear stress area A_v, as controlled by shear capacity of concrete V_u. For simplicity of calculations assume the plate and flange of same width, 8 in. (a slightly conservative assumption). The total depth of embedment is then:

$$4 + \tfrac{1}{2} = 4.5 \text{ (in.)}$$

The effective concrete shear stress area A_v and shear capacity of concrete V_u are determined as follows (see Fig. 11.19):

$$c = 48/2 - 8/2 = 20 \text{ (in.)}$$
$$d = 4.5 + 20 = 24.5 \text{ (in.)}$$
$$A_v = 48 \times 24.5 - 4.5 \times 8 = 1140 \text{ (in.}^2)$$
$$V_u = 4\phi f'_c A_v = [4(0.8)(4000)^{1/2} 1140]/1000 = 230.72 \text{ (kips)} > 73.24 \text{ kips} \quad \text{OK}$$

∴ No need to rely on additional shear capacity from confinement provided by anchor bolts.

Conclusion The total required depth of the recessed column flange and base plate is 4.5 in. The concrete shear capacity is adequate. ▲

11.6 Other Embedments

There are a number of other concrete embedment designs that could be used to transfer of horizontal forces from the metal building columns into the foundations. A few of them are briefly described next.

11.6.1 Cap Plate

In this design solution, nothing is embedded in concrete save for the anchor bolts. Instead, a steel cap is placed over the column pedestal. The cap consists of the top plate with oversized holes for the anchor bolts and the edge plates shop welded on all four sides of the pedestal.

To allow for field tolerances in the pedestal construction, the edge plates are positioned some distance away from the theoretical locations of the vertical pedestal surfaces. The clearance depends on the as-built dimensions of the pedestals. If the plates are fabricated after the concrete pedestals are placed and measured, the vertical edge gaps might be about ¼ in.; if the plate fabrication precedes the concrete placement or if the measurements are not available, larger gaps might be needed.

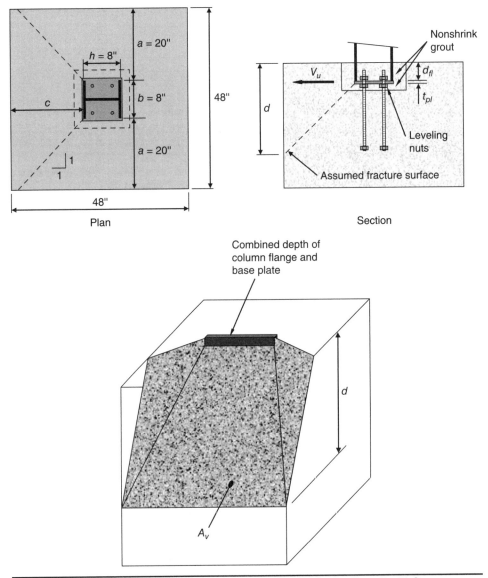

Figure 11.19 Dimensions of the assumed fracture surface in Design Example 11.3.

(Refer to ACI 117-10 for concrete tolerances that apply to various types of cast-in-place concrete foundations.)

After the cap is placed on top of the pier over the anchor bolts, the resulting gaps at the edges are filled with field-welded shims for a tight fit on all sides. To provide vertical adjustability, shim packs can be placed on top of the pedestal and the horizontal void under the cap plate filled with fluid nonshrink grout. The base plates should have grout holes for grout placement. After the cap plate is aligned and secured, the column is placed on top of it and welded to it (Fig. 11.20).

Concrete Embedments in Metal Building Systems 259

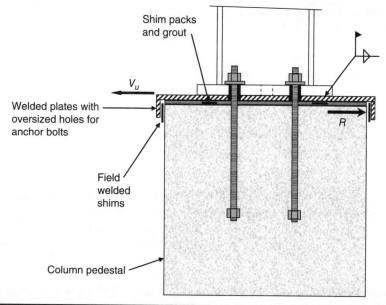

FIGURE 11.20 The cap plate.

The advantages of this design are:

1. The simplicity of constructing the pedestal, as no pockets or slots are required.
2. The design does not depend on the shear capacity of concrete, as in shear lugs and recessed plates. Instead, the whole pedestal section is engaged by the edge plates.
3. A close spacing of anchor bolts, common in metal building systems, can be preserved. The anchor bolts are needed only for tension transfer.

The challenges of using this design are:

1. The corrosion protection of the cap plate and the shims is needed, which may be difficult to achieve in practice. The cap plate and the shims could be galvanized, but welding at the base plate and at the shims would damage the galvanized finish. Plus, a contact between galvanized and plain steel should generally be avoided to prevent corrosion between dissimilar materials. Of course, if the whole primary frame is galvanized and the appropriate touch-up paint compound provided for the welds, the problem is minimized. Alternatively, the cap plate can be coated by one of the high-performance compounds discussed in Chapter 3 of Newman (2001), but then great care must be used to touch up the welds as well.
2. The sides of the entire column pedestal must be exposed for the cap plate to work. The simplest way to achieve this purpose is to elevate the pedestal slightly above the adjacent slab on grade and foundation walls.

The design of the cap plate is rather straightforward. The depth of the edge plates is determined by the bearing capacity of concrete, as with the shear key and the recessed base systems. The thickness of the edge plates can be found from its required flexural capacity, similar to establishing the thickness of the shear key. The thickness of the cap plate should be generally equal to that of the edge plates. Since the welding is done in the shop, complete- or partial-penetration welds can be used for a clean design of the edges.

11.6.2 Embedded Plate with Welded-On Studs

In this design a plate with welded-on studs is embedded into the top of the concrete foundation. Steel shims are welded to the top of the plate if needed for vertical alignment, and then the column base plate is welded to the shims (Fig. 11.21). The welded studs are used to resist both horizontal and vertical column reactions, following the provisions of ACI 318 Appendix D.

The main advantage of this method lies in the fact that the welded studs are not acting as traditional anchor bolts placed within oversized holes in the base plate, thus the issues of anchor bolt bending are sidestepped. The anchors must still be designed in accordance with the provisions of ACI 318 Appendix D, as discussed in Chap. 10, which means that supplemental reinforcement for shear and/or tension might be required. When such reinforcement is needed, the shear-lug or the recessed-plate solutions could be explored instead.

The issues of corrosion protection for the embedded plate and the shims should be considered, as discussed in the previous section. Another issue that needs consideration involves ensuring temporary column stability during erection, since no anchor bolts are used in this design.

In a more elaborate version of this system, both the anchor bolts and the welded-on studs are provided (Fig. 11.22). The anchor bolts are used for field erection and to resist

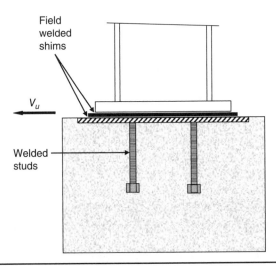

Figure 11.21 Embedded plate with welded-on studs.

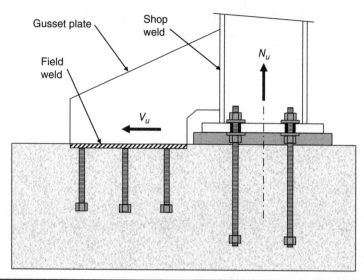

FIGURE 11.22 A combination of anchor bolts and embedded plate with welded-on studs.

applied tension, while the welded-on studs are designed to resist horizontal forces. Obviously, a large foundation is needed to provide the required design capacities for both groups of anchors. This version is more common in structural-steel buildings than in metal building systems, but it offers one more possibility in the designer's palette of choices.

References

2009 International Building Code® (IBC), International Code Council, Country Club Hills, IL, 2009.

ACI 117-06, *Specifications for Tolerances for Concrete Construction and Materials and Commentary*, American Concrete Institute, Farmington Hills, MI, 2006.

ACI 117-10, *Specifications for Tolerances for Concrete Construction and Materials and Commentary*, American Concrete Institute, Farmington Hills, MI, 2010.

ACI 318-08, *Building Code Requirements for Structural Concrete and Commentary*, American Concrete Institute, Farmington Hills, MI, 2010.

ACI 349-06, *Code Requirements for Nuclear Safety–Related Concrete Structures and Commentary*, American Concrete Institute, Farmington Hills, MI.

ACI 351.2R-10, *Report on Foundations for Static Equipment*, American Concrete Institute, Farmington Hills, MI.

AISC 360-05, *Specification for Structural Steel Buildings*, American Institute of Steel Construction, Chicago, IL, 2005.

AISC Commentary on the Specification for Structural Steel Buildings, American Institute of Steel Construction, Chicago, IL, 2005.

AISC *Steel Construction Manual*, 13th ed., American Institute of Steel Construction, Chicago, IL, 2005.

AISC *Steel Design Guide 1: Base Plate and Anchor Rod Design*, 2d ed., American Institute of Steel Construction, Chicago, IL, 2010.

Gomez, Ivan, et al., "Shear Transfer in Exposed Column Base Plates," report presented to the American Institute of Steel Construction, March 2009.

Metal Building System Manual, Metal Building Manufacturers Association (MBMA), Cleveland, OH, 2006.

Newman, Alexander, *Structural Renovation of Buildings: Methods, Details, and Design Examples*, McGraw-Hill, New York, 2001.

Rotz, J. V., and Reifschneider, M., "Combined Axial and Shear Load Capacity of Embedments in Concrete," 10th International Conference, Structural Mechanics in Reactor Technology, Anaheim, CA, August 1989.

TMS 602-11/ACI 530.1-11/ASCE 6-11, *Specification for Masonry Structures*, Masonry Standards Joint Committee, 2011.

Zoruba, Sergio, Answer in Steel Interchange, *Modern Steel Construction*, April 2006.

APPENDIX
Frame Reaction Tables

As discussed in Chap. 3, it is sometimes necessary to estimate the value of column reactions before they are available from the metal building system manufacturer. This is a common situation in the competitive bidding process, where both the manufacturer and the general contractor are selected at the same time. In that situation, the foundations and the anchor rods must usually be designed before the manufacturer is selected—and the final column reaction values are available—so that a set of contract documents can be produced for all bidders.

To aid the specifiers in this process, the following frame reaction tables are reprinted with permission from the *Project and Engineering Manual* by Nucor Building Systems. The specifiers are cautioned that these reactions should be considered as preliminary, because other manufacturers will likely have somewhat different numbers, even for the identical frame configurations. Also, the governing code provisions will likely be different by the time you use these tables. For these reasons and to avoid the unpleasant situation when the final reactions come in much higher than those assumed in the foundation design, it is prudent to increase the tabulated reactions by a substantial factor.

The following notes are reproduced from the Nucor manual:

> The purpose for this section is to provide the column base reactions for a large variety of standard frame geometries for use in foundation design. The defaults used for the variables for the Column Base Reaction Tables are as listed on the next page. Each of the listed variables can be interpolated for non-standard values. Column base reactions for buildings not fitting these specific parameters may be interpolated so long as they do not exceed the limits for interpolation.
>
> If the structure under consideration exceeds the limits defined in the table on the next page, please contact Nucor Building Systems for the column base reactions and anchor bolt configurations.

Variable	Standard Value Used	Limits for Interpolation
Bay Spacing	25 ft.	≤ 35 ft.
Span Width (All spans of a multi-span condition must be equal)	Varies	40 ft ≤ span ≤ 60 ft, multi-spans
		30 ft ≤ span ≤ 120 ft, clear-span gable
		30 ft ≤ span ≤ 100 ft, clear-span single-slope
Eave Height	Varies (12', 18', 24', or 30')	12 ft ≤ height ≤ 30 ft
Slope	1:12 (good up to 2:12)	1/4:12 ≤ slope ≤ 2:12
Live Load	Varies (20, 30, or 40 psf)	≤ 40 psf
Building Code	ASCE 7-95	ANY CURRENT MODEL CODE (ASCE 7-95, BOCA, UBC, SBC)
Wind Speed	80 mph	≤ 110 mph
Exposure Factor	C	B or C

Base reactions have been tabulated for a range of possible endwall column heights and spacings. It is recommended that the tallest corner and interior columns be used when extracting information from the table in order to cover the worst-case condition. These values can also be interpolated for geometries or gravity loadings not present in the tables. However, no extrapolation beyond the bounds is allowed.

Reactions have been tabulated for the effects of longitudinal wind loads transferred to the foundation by 'X' bracing. These can be added to the frame column base reactions at affected column locations. The number of braced bays required by the eave height and building width has also been included.

Additional reactions have also been included for a standard closed fascia which has a 4' offset and is 7' tall (with 3' extension above the eave) and for a standard 5' and 10' canopy. As with the bracing loads, these reactions are intended to be added to the column base reactions of columns where fascia or canopy members attach.

Interpolation is also possible for different wind loads due to building codes, basic wind speeds and exposure factors. Load factors have been developed for the frame column base reactions by which each of the X and Y reactions for the Wind Load (WL) case are to be multiplied to adjust for the specific case under consideration. A column is also included under each exposure factor which gives a single factor by which the endwall column base reactions, the additional longitudinal wind bracing reactions and the additional reactions from a fascia or canopy are to be multiplied.

Introduction Frame Column Base Reactions

Code		Wind Speed (mph)	Exposure B						Exposure C					
			Main Frames					Other *	Main Frames					Other *
			Low Eave Ext. Col.		High Eave Ext. Col.		Int. Col.		Low Eave Ext. Col.		High Eave Ext. Col.		Int. Col.	
			X	Y	X	Y	Y	X or Y	X	Y	X	Y	Y	X or Y
96	BOCA	70	0.43	0.47	0.47	0.53	0.49	0.50	0.74	0.74	0.78	0.80	0.75	0.77
		80	0.56	0.62	0.60	0.66	0.64	0.61	1.00	1.00	1.00	1.00	1.00	1.00
		90	0.74	0.74	0.78	0.80	0.75	0.78	1.25	1.24	1.28	1.29	1.22	1.27
		100	1.00	1.00	1.00	1.00	1.00	0.96	1.61	1.58	1.53	1.54	1.50	1.56
		110	1.25	1.24	1.28	1.29	1.22	1.16	2.00	1.96	1.79	1.78	1.75	1.89
97	UBC	70	0.47	0.58	0.49	0.65	0.62	0.62	0.79	0.97	0.85	1.12	1.01	1.01
		80	0.62	0.75	0.62	0.83	0.80	0.81	1.04	1.23	1.01	1.31	1.26	1.31
		90	0.79	0.97	0.85	1.12	1.01	1.03	1.35	1.58	1.20	1.56	1.57	1.66
		100	1.04	1.23	1.01	1.31	1.26	1.27	1.71	1.97	1.56	2.01	1.92	2.04
		110	1.18	1.44	1.14	1.49	1.51	1.53	2.01	2.36	1.76	2.27	2.32	2.48
97	SBC	70	0.33	0.82	0.36	0.67	0.92	0.50	0.33	0.82	0.36	0.67	0.92	0.50
		80	0.48	1.12	0.47	0.87	1.20	0.66	0.48	1.12	0.47	0.87	1.20	0.66
		90	0.62	1.43	0.62	1.15	1.51	0.82	0.62	1.43	0.62	1.15	1.51	0.82
		100	0.83	1.80	0.75	1.39	1.78	1.02	0.83	1.80	0.75	1.39	1.78	1.02
		110	1.07	2.22	1.04	1.84	2.18	1.23	1.07	2.22	1.04	1.84	2.18	1.23

* These factors are for use with the reactions tabulated for Endwall Columns, Bracing, Fascias, and Canopies

Reactions are provided by load cases in order to aid the foundation designer in determining the appropriate load factors and combinations to be used with either Working Stress or Ultimate Strength design methods. Wind load cases are given for each primary wind direction, and the case which produces the largest reaction at a particular column should be used for design.

Anchor bolt embedment lengths and types are not provided by Nucor Building Systems. This information is closely related to the complete foundation design which should be done by a Registered Professional Engineer familiar with the local site conditions and construction practices.

Gable, Clear Span @ 30'-0"

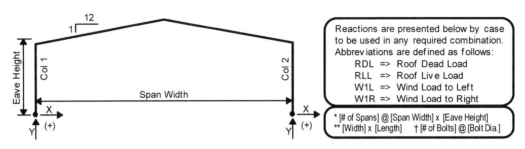

Reactions are presented below by case to be used in any required combination. Abbreviations are defined as follows:
RDL => Roof Dead Load
RLL => Roof Live Load
W1L => Wind Load to Left
W1R => Wind Load to Right

* [# of Spans] @ [Span Width] x [Eave Height]
** [Width] x [Length]   † [# of Bolts] @ [Bolt Dia.]

Frame Description*	Column	Column Base Reactions (kips)								Base-plate Size**	Anchor Bolts†	
		RDL		RLL		W1L		W1R				
		X	Y	X	Y	X	Y	X	Y			
1@30' x 12' RLL = 20 psf	COL 1	0.2	1.3	1.0	7.7	2.6	-1.5	-4.3	-6.0	8"x10"	2 @ 3/4"	
	COL 2	-0.2	1.3	-1.0	7.7	4.3	-6.0	-2.6	-1.5	8"x10"	2 @ 3/4"	
1@30' x 12' RLL = 30 psf	COL 1	0.2	1.3	2.0	11.6	2.4	-1.5	-4.5	-6.0	8"x10"	2 @ 3/4"	
	COL 2	-0.2	1.3	-2.0	11.6	4.5	-6.0	-2.4	-1.5	8"x10"	2 @ 3/4"	
1@30' x 12' RLL = 40 psf	COL 1	0.3	1.4	3.2	15.5	2.3	-1.5	-4.6	-6.0	8"x10"	2 @ 3/4"	
	COL 2	-0.3	1.4	-3.2	15.5	4.6	-6.0	-2.3	-1.5	8"x10"	2 @ 3/4"	
1@30' x 18' RLL = 20 psf	COL 1	0.1	1.6	0.5	7.7	4.1	0.2	-6.3	-8.0	8"x10"	2 @ 3/4"	****
	COL 2	-0.1	1.6	-0.5	7.7	6.3	-8.0	-4.1	0.3	8"x10"	2 @ 3/4"	****
1@30' x 18' RLL = 30 psf	COL 1	0.1	1.5	1.1	11.6	4.0	0.3	-6.5	-8.1	8"x10"	2 @ 3/4"	****
	COL 2	-0.1	1.5	-1.1	11.6	6.5	-8.1	-4.0	0.3	8"x10"	2 @ 3/4"	****
1@30' x 18' RLL = 40 psf	COL 1	0.1	1.5	1.7	15.5	3.9	0.4	-6.5	-8.1	8"x10"	2 @ 3/4"	****
	COL 2	-0.1	1.5	-1.7	15.5	6.5	-8.1	-3.9	0.3	8"x10"	2 @ 3/4"	****
1@30' x 24' RLL = 20 psf	COL 1	0.1	1.6	0.7	7.7	4.5	1.3	-7.4	-9.1	8"x10"	2 @ 3/4"	
	COL 2	-0.1	1.6	-0.7	7.7	7.4	-9.1	-4.5	1.3	8"x10"	2 @ 3/4"	
1@30' x 24' RLL = 30 psf	COL 1	0.1	1.6	1.0	11.6	4.5	1.2	-7.4	-9.0	8"x10"	2 @ 3/4"	
	COL 2	-0.1	1.6	-1.0	11.6	7.4	-9.0	-4.5	1.2	8"x10"	2 @ 3/4"	
1@30' x 24' RLL = 40 psf	COL 1	0.1	1.6	1.3	15.5	4.5	1.3	-7.4	-9.0	8"x10"	2 @ 3/4"	
	COL 2	-0.1	1.6	-1.3	15.5	7.4	-9.0	-4.5	1.3	8"x10"	2 @ 3/4"	
1@30' x 30' RLL = 20 psf	COL 1	0.1	1.7	0.7	7.7	4.6	2.2	-8.1	-9.9	8"x10"	2 @ 3/4"	
	COL 2	-0.1	1.7	-0.7	7.7	8.1	-9.9	-4.6	2.2	8"x10"	2 @ 3/4"	
1@30' x 30' RLL = 30 psf	COL 1	0.1	1.7	1.0	11.6	4.7	2.2	-8.2	-10.0	8"x10"	2 @ 3/4"	
	COL 2	-0.1	1.7	-1.0	11.6	8.2	-10.0	-4.7	2.2	8"x10"	2 @ 3/4"	
1@30' x 30' RLL = 40 psf	COL 1	0.1	1.9	1.5	15.5	7.3	6.9	11.2	-15.5	8"x10"	2 @ 1"	
	COL 2	-0.1	1.9	-1.5	15.5	11.2	-15.4	-7.3	6.9	8"x10"	2 @ 1"	

**** For 18' eave height, tapered beam frames, baseplate length = 12".
**** For 20' eave height, tapered beam frames, baseplate length = 14".

Frame Reaction Tables 267

Gable, Clear Span @ 40'-0"

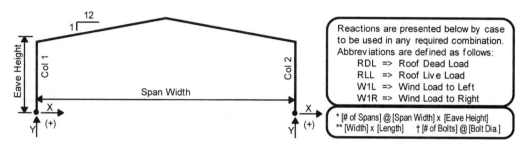

Reactions are presented below by case to be used in any required combination. Abbreviations are defined as follows:
RDL => Roof Dead Load
RLL => Roof Live Load
W1L => Wind Load to Left
W1R => Wind Load to Right

* [# of Spans] @ [Span Width] x [Eave Height]
** [Width] x [Length]   † [# of Bolts] @ [Bolt Dia.]

Frame Description*	Column	Column Base Reactions (kips)								Baseplate Size**	Anchor Bolts†	
		RDL		RLL		W1L		W1R				
		X	Y	X	Y	X	Y	X	Y			
1@40' x 12'	COL 1	0.8	1.8	4.9	10.3	0.6	-3.2	-6.3	-6.7	8"x10"	2 @ 3/4"	
RLL = 20 psf	COL 2	-0.8	1.8	-4.9	10.3	6.3	-6.7	-0.6	-3.2	8"x10"	2 @ 3/4"	
1@40' x 12'	COL 1	0.8	1.9	6.7	15.5	0.8	-3.2	-6.1	-6.7	8"x10"	2 @ 3/4"	
RLL = 30 psf	COL 2	-0.8	1.9	-6.7	15.5	6.1	-6.7	-0.9	-3.2	8"x10"	2 @ 3/4"	
1@40' x 12'	COL 1	0.7	1.9	8.5	20.6	0.9	-3.2	-6.0	-6.7	8"x10"	2 @ 1"	
RLL = 40 psf	COL 2	-0.7	1.9	-8.5	20.6	5.9	-6.7	-0.9	-3.2	8"x10"	2 @ 1"	
1@40' x 18'	COL 1	0.5	1.9	3.3	10.3	2.2	-2.3	-7.4	-7.7	8"x10"	2 @ 3/4"	****
RLL = 20 psf	COL 2	-0.5	1.9	-3.3	10.3	7.3	-7.7	-2.2	-2.3	8"x10"	2 @ 3/4"	****
1@40' x 18'	COL 1	0.5	1.9	4.0	15.5	2.6	-2.3	-7.0	-7.7	8"x10"	2 @ 3/4"	****
RLL = 30 psf	COL 2	-0.5	1.9	-4.0	15.5	7.0	-7.7	-2.6	-2.3	8"x10"	2 @ 3/4"	****
1@40' x 18'	COL 1	0.4	2.0	5.0	20.6	2.7	-2.2	-6.9	-7.7	8"x10"	2 @ 3/4"	****
RLL = 40 psf	COL 2	-0.4	2.0	-5.0	20.6	6.9	-7.7	-2.7	-2.2	8"x10"	2 @ 3/4"	****
1@40' x 24'	COL 1	0.5	2.1	2.7	10.3	4.3	-0.2	-9.9	-10.6	8"x10"	2 @ 1"	
RLL = 20 psf	COL 2	-0.5	2.1	-2.7	10.3	9.9	-10.7	-4.3	-0.2	8"x10"	2 @ 1"	
1@40' x 24'	COL 1	0.4	2.1	3.6	15.5	4.6	-0.2	-9.7	-10.7	8"x10"	2 @ 1"	
RLL = 30 psf	COL 2	-0.4	2.1	-3.6	15.5	9.7	-10.7	-4.6	-0.2	8"x10"	2 @ 1"	
1@40' x 24'	COL 1	0.4	2.2	4.3	20.6	4.8	-0.1	-9.5	-10.7	8"x10"	2 @ 1"	
RLL = 40 psf	COL 2	-0.4	2.2	-4.3	20.6	9.5	-10.7	-4.8	-0.1	8"x10"	2 @ 1"	
1@40' x 30'	COL 1	0.4	2.4	2.4	10.3	6.1	2.6	12.4	-14.0	8"x10"	2 @ 1"	
RLL = 20 psf	COL 2	-0.4	2.4	-2.4	10.3	12.4	-14.0	-6.2	2.6	8"x10"	2 @ 1"	
1@40' x 30'	COL 1	0.4	2.4	3.3	15.5	6.3	2.4	12.1	-13.8	8"x10"	2 @ 1"	
RLL = 30 psf	COL 2	-0.4	2.4	-3.3	15.4	12.1	-13.8	-6.3	2.4	8"x10"	2 @ 1"	
1@40' x 30'	COL 1	0.4	2.5	4.0	20.6	6.5	2.5	12.0	-14.0	8"x10"	2 @ 1"	
RLL = 40 psf	COL 2	-0.4	2.5	-4.0	20.6	12.0	-14.0	-6.5	2.6	8"x10"	2 @ 1"	

**** *For 18' eave height, tapered beam frames, baseplate length = 12".*
**** *For 20' eave height, tapered beam frames, baseplate length = 14".*

Gable, Clear Span @ 50'-0"

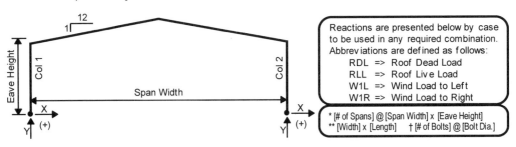

Frame Description*	Column	Column Base Reactions (kips)								Base-plate Size**	Anchor Bolts†	
		RDL		RLL		W1L		W1R				
		X	Y	X	Y	X	Y	X	Y			
1@50' x 12'	COL 1	6.6	11.3	8.8	15.2	-0.7	-5.7	-8.6	-9.0	8"x10"	2 @ 3/4"	
RLL = 20 psf	COL 2	-6.6	11.3	-8.8	15.2	8.6	-9.0	0.7	-5.7	8"x10"	2 @ 3/4"	
1@50' x 12'	COL 1	6.8	11.5	13.1	21.8	-0.7	-5.5	-8.7	-8.9	8"x10"	2 @ 1"	
RLL = 30 psf	COL 2	-6.8	11.4	13.1	21.7	8.7	-8.9	0.7	-5.5	8"x10"	2 @ 1"	
1@50' x 12'	COL 1	7.2	11.6	17.8	28.3	-2.2	-9.7	14.5	-15.0	8"x10"	2 @ 1-1/4"	
RLL = 40 psf	COL 2	-7.2	11.6	17.8	28.3	14.5	-15.0	2.2	-9.7	8"x10"	2 @ 1-1/4"	
1@50' x 18'	COL 1	3.9	11.4	5.3	15.3	2.8	-4.3	-8.9	-10.1	8"x10"	2 @ 1"	****
RLL = 20 psf	COL 2	-3.9	11.4	-5.3	15.3	9.0	-10.1	-2.8	-4.3	8"x10"	2 @ 1"	****
1@50' x 18'	COL 1	3.9	11.6	7.5	21.9	3.0	-4.1	-9.0	-10.0	8"x10"	2 @ 1"	****
RLL = 30 psf	COL 2	-3.9	11.6	-7.5	21.9	9.0	-10.0	-3.0	-4.1	8"x10"	2 @ 1"	****
1@50' x 18'	COL 1	4.1	11.7	10.2	28.5	3.9	-7.7	14.4	-16.8	8"x10"	2 @ 1"	****
RLL = 40 psf	COL 2	-4.1	11.8	-10.2	28.5	14.6	-16.7	-4.1	-7.6	8"x10"	2 @ 1"	****
1@50' x 24'	COL 1	3.5	11.6	4.7	15.5	5.0	-2.7	11.2	-12.4	8"x10"	2 @ 1"	
RLL = 20 psf	COL 2	-3.5	11.6	-4.7	15.5	11.4	-12.3	-5.2	-2.6	8"x10"	2 @ 1"	
1@50' x 24'	COL 1	3.2	11.6	6.0	21.9	5.4	-2.6	11.0	-12.3	8"x10"	2 @ 1"	
RLL = 30 psf	COL 2	-3.2	11.6	-6.0	21.9	11.0	-12.3	-5.4	-2.6	8"x10"	2 @ 1"	
1@50' x 24'	COL 1	4.0	11.9	9.8	28.7	7.0	-5.5	18.4	-20.4	8"x10"	2 @ 1-1/4"	
RLL = 40 psf	COL 2	-4.0	11.9	-9.8	28.7	18.4	-20.4	-7.0	-5.5	8"x10"	2 @ 1-1/4"	
1@50' x 30'	COL 1	3.2	12.0	4.3	15.9	7.3	0.2	13.4	-14.3	8"x10"	2 @ 1"	
RLL = 20 psf	COL 2	-3.2	12.0	-4.3	15.9	13.4	-14.3	-7.3	0.2	8"x10"	2 @ 1"	
1@50' x 30'	COL 1	3.0	12.0	5.8	22.3	7.4	0.1	13.2	-14.2	8"x10"	2 @ 1"	
RLL = 30 psf	COL 2	-3.0	12.0	-5.8	22.3	13.2	-14.2	-7.4	0.1	8"x10"	2 @ 1"	
1@50' x 30'	COL 1	3.2	12.4	8.0	29.1	11.1	-1.1	21.4	-24.0	8"x10"	2 @ 1-1/4"	
RLL = 40 psf	COL 2	-3.2	12.4	-8.0	29.1	21.4	-24.0	11.1	-1.1	8"x10"	2 @ 1-1/4"	

**** *For 18' eave height, tapered beam frames, baseplate length = 12".*
**** *For 20' eave height, tapered beam frames, baseplate length = 14".*

Frame Reaction Tables

Gable, Clear Span @ 60'-0"

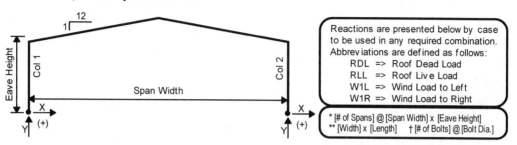

Reactions are presented below by case to be used in any required combination. Abbreviations are defined as follows:
- RDL => Roof Dead Load
- RLL => Roof Live Load
- W1L => Wind Load to Left
- W1R => Wind Load to Right

* [# of Spans] @ [Span Width] x [Eave Height]
** [Width] x [Length]   † [# of Bolts] @ [Bolt Dia.]

Frame Description*	Column	Column Base Reactions (kips)								Base-plate Size**	Anchor Bolts†	
		RDL		RLL		W1L		W1R				
		X	Y	X	Y	X	Y	X	Y			
1@60' x 12'	COL 1	10.4	13.7	14.0	18.3	-3.2	-7.2	11.1	-10.2	8"x10"	2 @ 1"	
RLL = 20 psf	COL 2	-10.4	13.7	-14.0	18.3	11.1	-10.2	3.2	-7.2	8"x10"	2 @ 1"	
1@60' x 12'	COL 1	10.5	14.0	20.1	26.3	-3.0	-6.9	11.0	-9.9	8"x10"	2 @ 1-1/4"	
RLL = 30 psf	COL 2	-10.5	14.0	-20.1	26.3	11.0	-9.9	3.0	-6.9	8"x10"	2 @ 1-1/4"	
1@60' x 12'	COL 1	2.2	3.3	22.7	30.9	-2.5	-6.1	-9.4	-8.8	8"x10"	2 @ 1-1/4"	
RLL = 40 psf	COL 2	-2.2	3.3	-22.6	30.9	9.4	-8.8	2.5	-6.1	8"x10"	2 @ 1-1/4"	
1@60' x 18'	COL 1	7.0	13.7	9.5	18.4	0.7	-6.1	11.2	-11.2	8"x10"	2 @ 1"	****
RLL = 20 psf	COL 2	-7.0	13.7	-9.4	18.3	11.2	-11.2	-0.7	-6.1	8"x10"	2 @ 1"	****
1@60' x 18'	COL 1	6.5	14.2	12.3	26.6	1.3	-5.7	10.6	-10.7	8"x10"	2 @ 1"	****
RLL = 30 psf	COL 2	-6.5	14.2	-12.3	26.5	10.5	-10.7	-1.3	-5.7	8"x10"	2 @ 1"	****
1@60' x 18'	COL 1	7.1	14.1	17.5	34.2	0.6	-10.7	17.9	-18.6	8"x10"	2 @ 1-1/4"	****
RLL = 40 psf	COL 2	-7.1	14.1	-17.5	34.2	17.9	-18.6	-0.6	-10.7	8"x10"	2 @ 1-1/4"	****
1@60' x 24'	COL 1	5.3	13.9	7.1	18.5	3.6	-5.1	12.6	-13.1	8"x10"	2 @ 1"	
RLL = 20 psf	COL 2	-5.3	13.9	-7.1	18.5	12.6	-13.1	-3.6	-5.1	8"x10"	2 @ 1"	
1@60' x 24'	COL 1	5.3	14.1	10.1	26.5	3.7	-4.8	12.5	-12.9	8"x10"	2 @ 1"	
RLL = 30 psf	COL 2	-5.3	14.1	-10.1	26.5	12.5	-12.9	-3.7	-4.8	8"x10"	2 @ 1"	
1@60' x 24'	COL 1	5.5	14.3	13.5	34.4	5.0	-9.2	20.1	-22.0	8"x10"	2 @ 1-1/4"	
RLL = 40 psf	COL 2	-5.5	14.3	-13.5	34.4	20.4	-21.9	-5.3	-9.2	8"x10"	2 @ 1-1/4"	
1@60' x 30'	COL 1	0.9	3.3	5.0	15.5	4.5	-3.0	13.8	-14.1	8"x10"	2 @ 1"	
RLL = 20 psf	COL 2	-0.9	3.3	-5.0	15.5	13.8	-14.1	-4.5	-3.0	8"x10"	2 @ 1"	
1@60' x 30'	COL 1	4.4	14.3	8.5	26.6	6.3	-2.5	14.3	-14.8	8"x10"	2 @ 1"	
RLL = 30 psf	COL 2	-4.4	14.3	-8.5	26.7	14.3	-14.7	-6.3	-2.6	8"x10"	2 @ 1"	
1@60' x 30'	COL 1	0.9	3.6	9.5	30.9	4.6	-3.1	13.5	-14.0	8"x10"	2 @ 1"	
RLL = 40 psf	COL 2	-0.9	3.6	-9.5	30.9	13.5	-14.0	-4.6	-3.1	8"x10"	2 @ 1"	

**** For 18' eave height, tapered beam frames, baseplate length = 12".
**** For 20' eave height, tapered beam frames, baseplate length = 14".

Gable, Clear Span @ 70'-0"

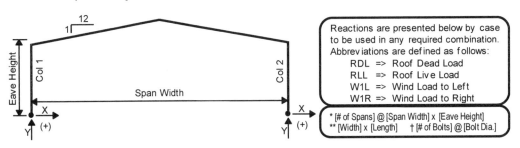

Reactions are presented below by case to be used in any required combination. Abbreviations are defined as follows:
- RDL => Roof Dead Load
- RLL => Roof Live Load
- W1L => Wind Load to Left
- W1R => Wind Load to Right

* [# of Spans] @ [Span Width] x [Eave Height]
** [Width] x [Length]   † [# of Bolts] @ [Bolt Dia.]

Frame Description*	Column	Column Base Reactions (kips)								Base-plate Size**	Anchor Bolts†
		RDL		RLL		W1L		W1R			
		X	Y	X	Y	X	Y	X	Y		
1@70' x 12'	COL 1	3.0	3.5	17.0	18.0	-5.3	-7.5	12.2	-9.9	8"x10"	2 @ 1-1/4"
RLL = 20 psf	COL 2	-3.0	3.5	-17.0	18.0	12.2	-9.9	5.3	-7.5	8"x10"	2 @ 1-1/4"
1@70' x 12'	COL 1	14.5	16.4	27.5	30.8	-5.5	-8.3	13.4	-11.1	8"x10"	4 @ 1"
RLL = 30 psf	COL 2	-14.5	16.4	-27.5	30.8	13.4	-11.1	5.5	-8.3	8"x10"	4 @ 1"
1@70' x 12'	COL 1	3.3	4.0	32.6	36.1	-5.0	-7.5	11.8	-9.9	8"x10"	4 @ 1-1/4"
RLL = 40 psf	COL 2	-3.3	4.0	-32.6	36.1	11.8	-9.9	5.0	-7.5	8"x10"	4 @ 1-1/4"
1@70' x 18'	COL 1	1.8	3.7	10.2	18.0	-0.9	-7.0	11.3	-11.0	8"x10"	2 @ 1"
RLL = 20 psf	COL 2	-1.8	3.7	-10.2	18.0	11.3	-11.0	0.9	-7.0	8"x10"	2 @ 1"
1@70' x 18'	COL 1	1.9	4.0	14.8	27.0	-0.7	-7.0	11.1	-11.0	8"x10"	2 @ 1-1/4"
RLL = 30 psf	COL 2	-1.9	4.0	-14.8	27.0	11.1	-11.0	0.7	-7.0	8"x10"	2 @ 1-1/4"
1@70' x 18'	COL 1	2.1	4.2	20.9	36.1	-1.0	-7.0	11.4	-11.0	8"x10"	2 @ 1-1/4"
RLL = 40 psf	COL 2	-2.1	4.2	-20.9	36.1	11.4	-11.0	1.0	-7.0	8"x10"	2 @ 1-1/4"
1@70' x 24'	COL 1	1.2	3.7	6.8	18.0	2.2	-6.3	12.0	-12.7	8"x10"	2 @ 1"
RLL = 20 psf	COL 2	-1.2	3.7	-6.8	18.0	12.0	-12.7	-2.2	-6.3	8"x10"	2 @ 1"
1@70' x 24'	COL 1	1.5	3.9	12.3	27.0	1.3	-6.3	12.9	-12.7	8"x10"	2 @ 1"
RLL = 30 psf	COL 2	-1.5	3.9	-12.3	27.0	12.9	-12.7	-1.3	-6.3	8"x10"	2 @ 1"
1@70' x 24'	COL 1	1.4	4.4	13.5	36.1	2.2	-6.3	12.1	-12.7	8"x10"	2 @ 1"
RLL = 40 psf	COL 2	-1.4	4.4	-13.5	36.1	12.1	-12.7	-2.2	-6.3	8"x10"	2 @ 1"
1@70' x 30'	COL 1	1.2	3.9	7.0	18.0	3.3	-5.2	14.9	-14.7	8"x10"	2 @ 1"
RLL = 20 psf	COL 2	-1.2	3.9	-7.0	18.0	14.9	-14.7	-3.3	-5.2	8"x10"	2 @ 1"
1@70' x 30'	COL 1	1.2	4.2	9.2	27.0	3.9	-5.1	14.3	-14.8	8"x10"	2 @ 1"
RLL = 30 psf	COL 2	-1.2	4.2	-9.2	27.0	14.4	-14.8	-4.1	-5.1	8"x10"	2 @ 1"
1@70' x 30'	COL 1	1.2	4.5	12.1	36.1	4.0	-5.1	14.3	-14.8	8"x10"	2 @ 1"
RLL = 40 psf	COL 2	-1.2	4.5	-12.1	36.1	14.3	-14.8	-4.0	-5.1	8"x10"	2 @ 1"

Frame Reaction Tables

Gable, Clear Span @ 80'-0"

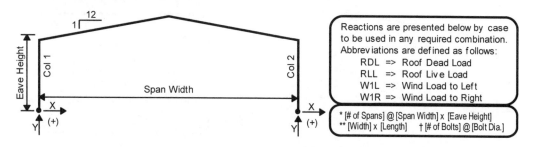

Reactions are presented below by case to be used in any required combination. Abbreviations are defined as follows:
- RDL => Roof Dead Load
- RLL => Roof Live Load
- W1L => Wind Load to Left
- W1R => Wind Load to Right

* [# of Spans] @ [Span Width] x [Eave Height]
** [Width] x [Length]   † [# of Bolts] @ [Bolt Dia.]

Frame Description*	Column	Column Base Reactions (kips)								Base-plate Size**	Anchor Bolts†
		RDL		RLL		W1L		W1R			
		X	Y	X	Y	X	Y	X	Y		
1@80' x 12'	COL 1	4.1	4.2	22.2	20.6	-7.8	-8.8	14.7	-11.1	8"x10"	4 @ 1"
RLL = 20 psf	COL 2	-4.1	4.2	-22.2	20.6	14.7	-11.1	7.8	-8.8	8"x10"	4 @ 1"
1@80' x 12'	COL 1	4.7	4.8	33.1	30.9	-7.8	-8.8	14.6	-11.1	8"x10"	4 @ 1-1/4"
RLL = 30 psf	COL 2	-4.7	4.8	-33.1	30.9	14.6	-11.1	7.8	-8.8	8"x10"	4 @ 1-1/4"
1@80' x 12'	COL 1	4.8	4.9	44.8	41.2	-8.0	-8.8	14.8	-11.1	8"x14"	4 @ 1-1/4"
RLL = 40 psf	COL 2	-4.8	4.9	-44.8	41.2	14.8	-11.1	7.9	-8.8	8"x14"	4 @ 1-1/4"
1@80' x 18'	COL 1	2.6	4.4	14.2	20.6	-3.0	-8.5	13.3	-12.1	8"x10"	2 @ 1-1/4"
RLL = 20 psf	COL 2	-2.6	4.4	-14.2	20.6	13.3	-12.1	3.0	-8.5	8"x10"	2 @ 1-1/4"
1@80' x 18'	COL 1	2.9	4.8	21.8	30.9	-3.1	-8.4	13.6	-12.2	8"x10"	2 @ 1-1/4"
RLL = 30 psf	COL 2	-2.9	4.8	-21.8	30.9	13.6	-12.2	3.1	-8.4	8"x10"	2 @ 1-1/4"
1@80' x 18'	COL 1	3.1	5.0	29.6	41.2	-3.3	-8.4	13.6	-12.2	10"x14"	4 @ 1"
RLL = 40 psf	COL 2	-3.1	5.0	-29.6	41.2	13.6	-12.2	3.3	-8.4	10"x14"	4 @ 1"
1@80' x 24'	COL 1	1.9	4.6	10.2	20.6	0.1	-8.0	13.8	-13.7	8"x10"	2 @ 1"
RLL = 20 psf	COL 2	-1.9	4.6	-10.2	20.6	13.8	-13.7	-0.1	-8.0	8"x10"	2 @ 1"
1@80' x 24'	COL 1	2.1	5.0	16.3	30.9	-0.1	-8.0	14.3	-13.8	8"x10"	2 @ 1-1/4"
RLL = 30 psf	COL 2	-2.1	5.0	-16.3	30.9	14.3	-13.8	0.1	-8.0	8"x10"	2 @ 1-1/4"
1@80' x 24'	COL 1	2.2	5.2	21.7	41.2	-0.1	-8.0	14.3	-13.8	8"x14"	2 @ 1-1/4"
RLL = 40 psf	COL 2	-2.2	5.2	-21.7	41.2	14.3	-13.8	0.1	-8.0	8"x14"	2 @ 1-1/4"
1@80' x 30'	COL 1	1.6	4.5	8.6	20.6	2.5	-7.1	15.9	-15.7	8"x14"	2 @ 1-1/4"
RLL = 20 psf	COL 2	-1.5	4.5	-8.6	20.6	15.8	-15.7	-2.4	-7.1	8"x14"	2 @ 1-1/4"
1@80' x 30'	COL 1	1.7	5.2	12.5	30.9	2.7	-7.1	15.7	-15.7	8"x10"	2 @ 1-1/4"
RLL = 30 psf	COL 2	-1.7	5.2	-12.5	30.9	15.6	-15.7	-2.6	-7.1	8"x10"	2 @ 1-1/4"
1@80' x 30'	COL 1	1.8	5.8	16.5	41.2	2.6	-7.2	15.5	-15.6	8"x14"	2 @ 1-1/4"
RLL = 40 psf	COL 2	-1.8	5.8	-16.5	41.2	15.5	-15.7	-2.6	-7.1	8"x14"	2 @ 1-1/4"

Gable, Clear Span @ 100'-0"

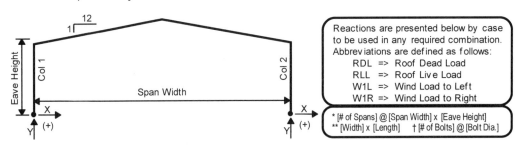

Frame Description*	Column	Column Base Reactions (kips)								Base-plate Size**	Anchor Bolts†
		RDL		RLL		W1L		W1R			
		X	Y	X	Y	X	Y	X	Y		
1@100' x 12' RLL = 20 psf	COL 1	7.0	6.1	32.0	25.8	12.6	-11.3	19.4	-13.6	8"x14"	4 @ 1-1/4"
	COL 2	-7.0	6.1	-32.0	25.8	19.4	-13.6	12.6	-11.3	8"x14"	4 @ 1-1/4"
1@100' x 12' RLL = 30 psf	COL 1	7.9	6.2	52.8	38.6	14.2	-11.3	21.0	-13.6	8"x18"	6 @ 1-1/4"
	COL 2	-7.9	6.2	-52.8	38.6	20.9	-13.6	14.2	-11.3	8"x18"	6 @ 1-1/4"
1@100' x 12' RLL = 40 psf	COL 1	7.8	6.2	52.4	38.6	14.0	-11.3	20.8	-13.6	8"x18"	6 @ 1-1/4"
	COL 2	-7.8	6.2	-52.4	38.6	20.8	-13.6	14.0	-11.3	8"x18"	6 @ 1-1/4"
1@100' x 18' RLL = 20 psf	COL 1	4.5	6.2	22.0	25.8	-6.9	-11.2	17.3	-14.6	8"x14"	4 @ 1"
	COL 2	-4.5	6.2	-22.0	25.8	17.3	-14.6	6.9	-11.2	8"x14"	4 @ 1"
1@100' x 18' RLL = 30 psf	COL 1	5.0	6.8	33.8	38.6	-7.2	-11.2	17.5	-14.6	8"x14"	4 @ 1-1/4"
	COL 2	-5.0	6.8	-33.8	38.6	17.5	-14.5	7.2	-11.2	8"x14"	4 @ 1-1/4"
1@100' x 18' RLL = 40 psf	COL 1	5.8	8.1	44.6	51.5	-7.0	-11.2	17.4	-14.6	12"x18"	4 @ 1-1/4"
	COL 2	-5.8	8.1	-44.6	51.5	17.4	-14.6	7.1	-11.2	12"x18"	4 @ 1-1/4"
1@100' x 24' RLL = 20 psf	COL 1	3.2	6.2	15.9	25.8	-2.8	-11.0	17.0	-16.1	8"x14"	2 @ 1-1/4"
	COL 2	-3.2	6.2	-15.9	25.8	17.0	-16.1	2.8	-11.0	8"x14"	2 @ 1-1/4"
1@100' x 24' RLL = 30 psf	COL 1	3.7	6.9	24.8	38.6	-3.2	-11.1	17.3	-16.1	8"x14"	4 @ 1"
	COL 2	-3.7	6.9	-24.8	38.6	17.3	-16.1	3.2	-11.1	8"x14"	4 @ 1"
1@100' x 24' RLL = 40 psf	COL 1	4.1	8.1	33.6	51.5	-3.4	-11.1	17.4	-16.1	12"x18"	4 @ 1-1/4"
	COL 2	-4.1	8.1	-33.6	51.5	17.4	-16.1	3.3	-11.1	12"x18"	4 @ 1-1/4"
1@100' x 30' RLL = 20 psf	COL 1	3.3	8.4	26.6	51.5	-0.1	-10.6	18.4	-17.9	12"x18"	4 @ 1"
	COL 2	-3.3	8.4	-26.6	51.5	18.4	-17.9	0.1	-10.6	12"x18"	4 @ 1"
1@100' x 30' RLL = 30 psf	COL 1	2.8	7.2	18.9	38.0	0.3	-10.6	18.0	-17.9	8"x14"	2 @ 1-1/4"
	COL 2	-2.8	7.2	-18.9	38.6	18.0	-17.9	-0.3	-10.6	8"x14"	2 @ 1-1/4"
1@100' x 30' RLL = 40 psf	COL 1	3.3	8.4	26.6	51.5	-0.1	-10.6	18.5	-17.9	12"x18"	4 @ 1"
	COL 2	-3.3	8.4	26.6	51.5	18.4	-18.0	0.1	-10.6	12"x18"	4 @ 1"

Frame Reaction Tables

Gable, Clear Span @ 120'-0"

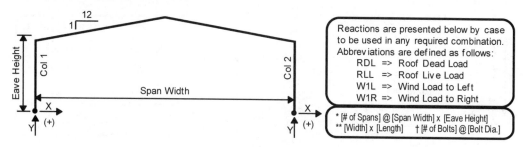

Reactions are presented below by case to be used in any required combination. Abbreviations are defined as follows:
- RDL => Roof Dead Load
- RLL => Roof Live Load
- W1L => Wind Load to Left
- W1R => Wind Load to Right

* [# of Spans] @ [Span Width] x [Eave Height]
** [Width] x [Length]   † [# of Bolts] @ [Bolt Dia.]

Frame Description*	Column	Column Base Reactions (kips)								Base-plate Size**	Anchor Bolts†
		RDL		RLL		W1L		W1R			
		X	Y	X	Y	X	Y	X	Y		
1@120' x 12'	COL 1	12.6	8.6	47.8	30.9	20.3	-13.8	27.1	-16.0	8"x18"	6 @ 1-1/4"
RLL = 20 psf	COL 2	-12.6	8.5	-47.9	30.9	27.1	-16.0	20.3	-13.8	8"x18"	6 @ 1-1/4"
1@120' x 12'	COL 1	13.7	9.3	73.3	46.4	20.8	-13.8	27.6	-16.1	8"x22"	8 @ 1-1/4"
RLL = 30 psf	COL 2	-13.7	9.3	-73.3	46.4	27.6	-16.1	20.8	-13.8	8"x22"	8 @ 1-1/4"
1@120' x 12'	COL 1	14.3	10.0	97.6	61.8	20.8	-13.8	27.5	-16.0	10"x22"	8 @ 1-1/4"
RLL = 40 psf	COL 2	-14.3	10.0	-97.6	61.8	27.5	-16.0	20.8	-13.8	10"x22"	8 @ 1-1/4"
1@120' x 18'	COL 1	7.6	8.2	32.2	30.9	12.1	-13.9	22.4	-17.0	8"x14"	4 @ 1-1/4"
RLL = 20 psf	COL 2	-7.6	8.2	-32.2	30.9	22.4	-17.0	12.1	-13.9	8"x14"	4 @ 1-1/4"
1@120' x 18'	COL 1	8.8	9.8	48.8	46.4	12.2	-13.9	22.5	-17.0	12"x18"	6 @ 1-1/4"
RLL = 30 psf	COL 2	-8.8	9.8	-48.8	46.4	22.5	-17.0	12.2	-13.9	12"x18"	6 @ 1-1/4"
1@120' x 18'	COL 1	8.9	9.8	65.6	61.8	12.3	-13.9	22.6	-17.0	8"x18"	6 @ 1-1/4"
RLL = 40 psf	COL 2	-8.9	9.8	-65.6	61.8	22.6	-17.0	12.3	-13.9	8"x18"	6 @ 1-1/4"
1@120' x 24'	COL 1	5.6	8.2	24.5	30.9	-7.5	-14.0	21.5	-18.6	8"x14"	4 @ 1"
RLL = 20 psf	COL 2	-5.6	8.2	-24.5	30.9	21.5	-18.6	7.5	-14.0	8"x14"	4 @ 1"
1@120' x 24'	COL 1	6.3	9.9	36.7	46.3	-7.4	-14.0	21.5	-18.6	12"x18"	4 @ 1-1/4"
RLL = 30 psf	COL 2	-6.3	9.9	-36.7	46.4	21.5	-18.6	7.4	-14.0	12"x18"	4 @ 1-1/4"
1@120' x 24'	COL 1	6.5	10.1	49.6	61.8	-7.6	-14.0	21.7	-18.6	12"x18"	6 @ 1-1/4"
RLL = 40 psf	COL 2	-6.5	10.1	-49.6	61.8	21.7	-18.6	7.6	-14.0	12"x18"	6 @ 1-1/4"
1@120' x 30'	COL 1	5.1	10.3	39.4	61.8	-3.8	-13.8	22.0	-20.4	12"x18"	4 @ 1-1/4"
RLL = 20 psf	COL 2	-5.1	10.3	-39.4	61.8	21.9	-20.4	3.8	-13.9	12"x18"	4 @ 1-1/4"
1@120' x 30'	COL 1	4.9	10.1	28.8	46.4	-3.6	-13.9	21.6	-20.3	12"x18"	4 @ 1-1/4"
RLL = 30 psf	COL 2	-4.9	10.1	-28.8	46.4	21.6	-20.3	3.6	-13.9	12"x18"	4 @ 1-1/4"
1@120' x 30'	COL 1	5.1	10.3	39.4	61.8	-3.8	-13.8	22.0	-20.4	12"x18"	4 @ 1-1/4"
RLL = 40 psf	COL 2	-5.1	10.3	39.4	61.8	21.9	-20.4	3.8	-13.9	12"x18"	4 @ 1-1/4"

Gable, Two Spans @ 40'-0"

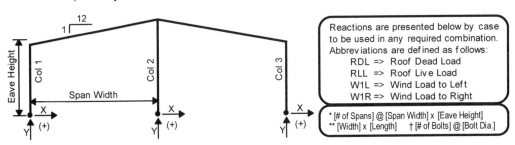

Reactions are presented below by case to be used in any required combination. Abbreviations are defined as follows:
 RDL => Roof Dead Load
 RLL => Roof Live Load
 W1L => Wind Load to Left
 W1R => Wind Load to Right

* [# of Spans] @ [Span Width] x [Eave Height]
** [Width] x [Length]   † [# of Bolts] @ [Bolt Dia.]

Frame Description*	Column	Column Base Reactions (kips)								Base-plate Size**	Anchor Bolts†
		RDL		RLL		W1L		W1R			
		X	Y	X	Y	X	Y	X	Y		
2@40' x 12' RLL = 20 psf	COL 1	0.5	1.6	3.3	9.1	1.5	-3.1	-5.4	-5.4	8"x10"	2 @ 3/4"
	COL 2	0.0	3.9	0.0	22.9	0.0	-11.3	0.0	-11.3	10"x15"	4 @ 3/4"
	COL 3	-0.5	1.6	-3.3	9.1	5.4	-5.4	-1.5	-3.1	8"x10"	2 @ 3/4"
2@40' x 12' RLL = 30 psf	COL 1	0.7	1.7	6.3	14.1	1.0	-3.3	-5.8	-5.6	8"x10"	2 @ 3/4"
	COL 2	0.0	3.9	0.0	33.5	0.0	-11.0	0.0	-11.0	10"x15"	4 @ 3/4"
	COL 3	-0.7	1.7	-6.3	14.1	5.8	-5.6	-1.0	-3.3	8"x10"	2 @ 3/4"
2@40' x 12' RLL = 40 psf	COL 1	0.7	1.8	8.9	19.4	0.9	-3.4	-6.0	-5.7	8"x10"	2 @ 1"
	COL 2	0.0	3.9	0.0	43.6	0.0	-10.8	0.0	-10.8	10"x15"	4 @ 3/4"
	COL 3	-0.7	1.8	-8.9	19.4	6.0	-5.7	-0.9	-3.4	8"x10"	2 @ 1"
2@40' x 18' RLL = 20 psf	COL 1	0.6	1.9	3.4	10.3	2.7	-3.1	-7.7	-6.7	8"x10"	2 @ 1"
	COL 2	0.0	3.5	0.0	20.6	0.0	-10.8	0.0	-10.8	10"x15"	4 @ 3/4"
	COL 3	-0.6	1.9	-3.4	10.3	7.7	-6.7	-2.7	-3.1	8"x10"	2 @ 1"
2@40' x 18' RLL = 30 psf	COL 1	0.5	1.9	4.8	15.0	2.8	-2.9	-7.5	-6.6	8"x10"	2 @ 1"
	COL 2	0.0	3.7	0.0	31.9	0.0	-11.2	0.0	-11.2	10"x15"	4 @ 3/4"
	COL 3	-0.5	1.9	-4.8	14.9	7.5	-6.6	-2.8	-2.9	8"x10"	2 @ 1"
2@40' x 18' RLL = 40 psf	COL 1	0.5	1.9	5.9	19.5	3.0	-2.7	-7.4	-6.4	8"x10"	2 @ 1"
	COL 2	0.0	4.0	0.0	43.5	0.0	-11.5	0.0	-11.5	10"x15"	4 @ 3/4"
	COL 3	-0.5	1.9	-5.9	19.4	7.4	-6.4	-3.0	-2.7	8"x10"	2 @ 1"

Gable, Two Spans @ 40'-0" (Continued)

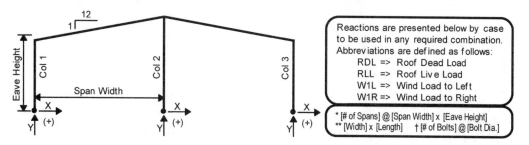

Reactions are presented below by case to be used in any required combination. Abbreviations are defined as follows:
RDL => Roof Dead Load
RLL => Roof Live Load
W1L => Wind Load to Left
W1R => Wind Load to Right

* [# of Spans] @ [Span Width] x [Eave Height]
** [Width] x [Length]   † [# of Bolts] @ [Bolt Dia.]

Frame Description*	Column	Column Base Reactions (kips)								Base-plate Size**	Anchor Bolts†
		RDL		RLL		W1L		W1R			
		X	Y	X	Y	X	Y	X	Y		
2@40' x 24' RLL = 20 psf	COL 1	0.4	2.1	2.4	10.2	4.6	-2.1	-9.3	-7.7	8"x10"	2 @ 1"
	COL 2	0.0	3.6	0.0	20.8	0.0	-11.8	0.0	-12.0	10"x15"	4 @ 3/4"
	COL 3	-0.4	2.1	-2.4	10.2	9.5	-7.8	-4.8	-2.0	8"x10"	2 @ 1"
2@40' x 24' RLL = 30 psf	COL 1	0.4	2.0	3.3	14.7	4.8	-1.8	-9.3	-7.6	8"x10"	2 @ 1"
	COL 2	0.0	3.9	0.0	32.3	0.0	-12.3	0.0	-12.3	10"x15"	4 @ 3/4"
	COL 3	-0.4	2.0	-3.3	14.7	9.3	-7.6	-4.8	-1.8	8"x10"	2 @ 1"
2@40' x 24' RLL = 40 psf	COL 1	0.4	2.0	4.1	19.2	4.8	-1.7	-9.2	-7.4	8"x10"	2 @ 1"
	COL 2	0.0	4.2	0.0	44.0	0.0	-12.6	0.0	-12.6	10"x15"	4 @ 3/4"
	COL 3	-0.4	2.0	-4.1	19.2	9.2	-7.4	-4.9	-1.7	8"x10"	2 @ 1"
2@40' x 30' RLL = 20 psf	COL 1	0.4	2.5	2.3	10.8	6.1	-1.3	11.5	-9.2	8"x10"	2 @ 1"
	COL 2	0.0	3.6	0.0	19.6	0.0	-11.8	0.0	-12.8	10"x15"	4 @ 3/4"
	COL 3	-0.4	2.5	-2.3	10.8	11.9	-9.8	-6.6	-0.7	8"x10"	2 @ 1"
2@40' x 30' RLL = 30 psf	COL 1	0.4	2.5	3.3	15.8	6.3	-1.0	11.5	-9.2	8"x10"	2 @ 1"
	COL 2	0.0	3.8	0.0	30.2	0.0	-12.1	0.0	-13.1	10"x15"	4 @ 3/4"
	COL 3	-0.4	2.5	-3.3	15.8	11.9	-9.7	-6.8	-0.5	8"x10"	2 @ 1"
2@40' x 30' RLL = 40 psf	COL 1	0.4	2.4	4.2	20.7	6.0	-1.2	11.1	-8.7	8"x10"	2 @ 1"
	COL 2	0.0	4.2	0.0	41.1	0.0	-12.0	0.0	-13.8	12"x17"	4 @ 3/4"
	COL 3	-0.4	2.5	-4.2	20.7	12.0	-9.6	-6.9	-0.3	8"x10"	2 @ 1"

Gable, Two Spans @ 50'-0"

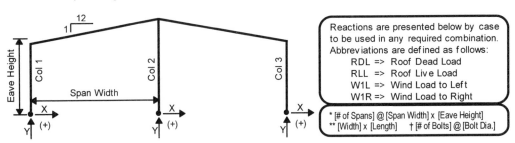

Frame Description*	Column	Column Base Reactions (kips)								Base-plate Size**	Anchor Bolts†
		RDL		RLL		W1L		W1R			
		X	Y	X	Y	X	Y	X	Y		
2@50' x 12' RLL = 20 psf	COL 1	0.9	2.0	5.8	11.6	0.2	-4.4	-6.6	-6.6	8"x10"	2 @ 3/4"
	COL 2	0.0	5.0	0.0	28.3	0.0	-13.9	0.0	-13.9	10"x15"	4 @ 3/4"
	COL 3	-0.9	2.0	-5.8	11.6	6.6	-6.6	-0.2	-4.4	8"x10"	2 @ 3/4"
2@50' x 12' RLL = 30 psf	COL 1	1.2	2.2	10.3	17.9	-0.3	-4.6	-7.1	-6.8	8"x10"	2 @ 1"
	COL 2	0.0	5.1	0.0	41.4	0.0	-13.5	0.0	-13.5	10"x15"	4 @ 3/4"
	COL 3	-1.2	2.2	10.3	17.9	7.1	-6.8	0.3	-4.6	8"x10"	2 @ 1"
2@50' x 12' RLL = 40 psf	COL 1	1.1	2.3	12.7	23.4	-0.1	-4.4	-6.8	-6.7	8"x10"	2 @ 1"
	COL 2	0.0	5.5	0.0	56.2	0.0	-13.8	0.0	-13.8	10"x15"	4 @ 3/4"
	COL 3	-1.1	2.3	12.7	23.4	6.8	-6.6	0.1	-4.4	8"x10"	2 @ 1"
2@50' x 18' RLL = 20 psf	COL 1	0.5	2.0	3.3	11.2	2.6	-3.7	-7.5	-7.0	8"x10"	2 @ 3/4"
	COL 2	0.0	5.1	0.0	29.2	0.0	-15.1	0.0	-15.0	10"x15"	4 @ 3/4"
	COL 3	-0.5	2.0	-3.3	11.2	7.5	-7.0	-2.6	-3.7	8"x10"	2 @ 3/4"
2@50' x 18' RLL = 30 psf	COL 1	0.7	2.2	5.8	17.1	2.5	-3.8	-8.0	-7.2	8"x10"	2 @ 3/4"
	COL 2	0.0	5.5	0.0	43.1	0.0	-14.8	0.0	-14.8	10"x15"	4 @ 3/4"
	COL 3	-0.7	2.2	-5.8	17.1	8.0	-7.1	-2.4	-3.8	8"x10"	2 @ 3/4"
2@50' x 18' RLL = 40 psf	COL 1	0.6	2.2	6.6	22.0	2.8	-3.6	-7.7	-7.0	8"x10"	2 @ 3/4"
	COL 2	0.0	5.9	0.0	59.0	0.0	-15.2	0.0	-15.2	10"x15"	4 @ 3/4"
	COL 3	-0.6	2.2	-6.6	22.0	7.7	-6.9	-2.8	-3.6	8"x10"	2 @ 3/4"

Gable, Two Spans @ 50'-0" (*Continued*)

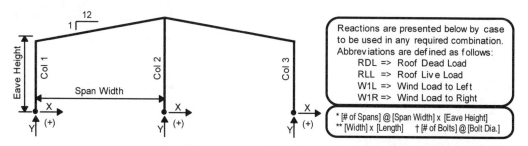

Reactions are presented below by case to be used in any required combination. Abbreviations are defined as follows:
RDL => Roof Dead Load
RLL => Roof Live Load
W1L => Wind Load to Left
W1R => Wind Load to Right

* [# of Spans] @ [Span Width] x [Eave Height]
** [Width] x [Length] † [# of Bolts] @ [Bolt Dia.]

Frame Description*	Column	Column Base Reactions (kips)								Base-plate Size**	Anchor Bolts†
		RDL		RLL		W1L		W1R			
		X	Y	X	Y	X	Y	X	Y		
2@50' x 24' RLL = 20 psf	COL 1	0.5	2.3	3.1	11.8	4.2	-3.4	-9.9	-8.3	8"x10"	2 @ 1"
	COL 2	0.0	5.1	0.0	28.0	0.0	-15.4	0.0	-15.5	10"x15"	4 @ 3/4"
	COL 3	-0.5	2.4	-3.1	11.8	9.9	-8.4	-4.2	-3.3	8"x10"	2 @ 1"
2@50' x 24' RLL = 30 psf	COL 1	0.7	2.6	6.0	18.7	3.7	-3.8	10.4	-8.8	8"x14"	2 @ 1"
	COL 2	0.0	5.0	0.0	39.9	0.0	-14.6	0.0	-14.6	10"x15"	4 @ 3/4"
	COL 3	-0.7	2.6	-6.0	18.7	10.5	-8.8	-3.8	-3.7	8"x14"	2 @ 1"
2@50' x 24' RLL = 40 psf	COL 1	0.5	2.4	6.2	23.1	4.2	-3.2	-9.9	-8.3	8"x14"	2 @ 1"
	COL 2	0.0	5.8	0.0	56.8	0.0	-15.6	0.0	-15.7	12"x17"	4 @ 3/4"
	COL 3	-0.5	2.4	-6.2	23.1	9.9	-8.3	-4.3	-3.2	8"x14"	2 @ 1"
2@50' x 30' RLL = 20 psf	COL 1	0.6	2.9	3.2	12.8	5.7	-3.0	12.5	-10.3	8"x18"	2 @ 1"
	COL 2	0.0	5.0	0.0	26.0	0.0	-15.2	0.0	-15.2	10"x15"	4 @ 3/4"
	COL 3	-0.6	2.9	-3.2	12.8	12.5	-10.3	-5.7	-3.0	8"x18"	2 @ 1"
2@50' x 30' RLL = 30 psf	COL 1	0.6	3.0	5.1	19.3	5.5	-3.2	12.5	-10.2	8"x18"	2 @ 1"
	COL 2	0.0	5.2	0.0	38.7	0.0	-15.0	0.0	-15.3	12"x17"	4 @ 3/4"
	COL 3	-0.6	3.0	-5.1	19.2	12.4	-10.3	-5.4	-3.0	8"x18"	2 @ 1"
2@50' x 30' RLL = 40 psf	COL 1	0.6	2.9	6.1	24.4	5.8	-2.5	12.3	-9.9	8"x10"	2 @ 1"
	COL 2	0.0	5.7	0.0	54.2	0.0	-16.1	0.0	-16.0	12"x17"	4 @ 3/4"
	COL 3	-0.6	2.9	-6.1	24.4	12.4	-9.8	-6.0	-2.6	8"x10"	2 @ 1"

Gable, Two Spans @ 60'-0"

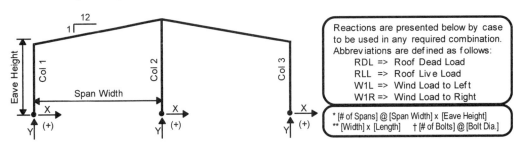

Frame Description*	Column	Column Base Reactions (kips)								Base-plate Size**	Anchor Bolts†
		RDL		RLL		W1L		W1R			
		X	Y	X	Y	X	Y	X	Y		
2@60' x 12' RLL = 20 psf	COL 1	1.8	2.7	10.4	14.8	-2.0	-6.0	-8.8	-8.2	8"x10"	2 @ 1"
	COL 2	0.0	5.8	0.0	32.2	0.0	-15.7	0.0	-15.7	10"x15"	4 @ 3/4"
	COL 3	-1.8	2.7	-10.4	14.8	8.8	-8.2	2.0	-5.9	8"x10"	2 @ 1"
2@60' x 12' RLL = 30 psf	COL 1	1.7	2.8	14.7	21.6	-1.7	-5.7	-8.5	-8.0	8"x10"	2 @ 1"
	COL 2	0.0	6.5	0.0	49.5	0.0	-16.1	0.0	-16.1	10"x15"	4 @ 3/4"
	COL 3	-1.7	2.8	-14.7	21.6	8.5	-8.0	1.7	-5.7	8"x10"	2 @ 1"
2@60' x 12' RLL = 40 psf	COL 1	2.2	3.3	23.0	30.7	-2.6	-6.2	-9.3	-8.4	8"x10"	2 @ 1-1/4"
	COL 2	0.0	6.6	0.0	62.3	0.0	-15.2	0.0	-15.2	10"x15"	4 @ 3/4"
	COL 3	-2.2	3.3	-23.0	30.7	9.3	-8.4	2.6	-6.2	8"x10"	2 @ 1-1/4"
2@60' x 18' RLL = 20 psf	COL 1	1.1	2.7	6.6	14.3	0.9	-5.3	-9.1	-8.3	8"x10"	2 @ 1"
	COL 2	0.0	6.1	0.0	33.1	0.0	-16.3	0.0	-16.3	10"x15"	4 @ 3/4"
	COL 3	-1.1	2.7	-6.6	14.3	9.0	-8.3	-0.9	-5.3	8"x10"	2 @ 1"
2@60' x 18' RLL = 30 psf	COL 1	0.9	2.7	7.6	20.2	1.6	-4.8	-8.3	-7.8	8"x10"	2 @ 1"
	COL 2	0.0	6.9	0.0	52.3	0.0	-17.2	0.0	-17.2	10"x15"	4 @ 3/4"
	COL 3	-0.9	2.7	-7.6	20.2	8.3	-7.8	-1.7	-4.8	8"x10"	2 @ 1"
2@60' x 18' RLL = 40 psf	COL 1	1.2	3.2	12.9	28.9	0.9	-5.3	-8.9	-8.3	8"x10"	2 @ 1"
	COL 2	0.0	7.1	0.0	65.9	0.0	-16.2	0.0	-16.2	12"x17"	4 @ 3/4"
	COL 3	-1.2	3.2	-12.9	28.8	8.9	-8.3	-0.9	-5.3	8"x10"	2 @ 1"

Frame Reaction Tables

Gable, Two Spans @ 60'-0" (Continued)

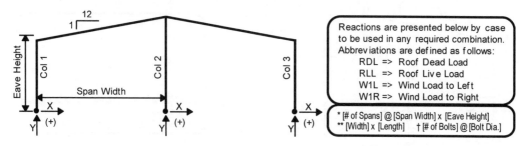

Reactions are presented below by case to be used in any required combination. Abbreviations are defined as follows:
- RDL => Roof Dead Load
- RLL => Roof Live Load
- W1L => Wind Load to Left
- W1R => Wind Load to Right

* [# of Spans] @ [Span Width] x [Eave Height]
** [Width] x [Length]   † [# of Bolts] @ [Bolt Dia.]

Frame Description*	Column	Column Base Reactions (kips)								Base-plate Size**	Anchor Bolts†
		RDL		RLL		W1L		W1R			
		X	Y	X	Y	X	Y	X	Y		
2@60' x 24' RLL = 20 psf	COL 1	0.9	2.9	5.3	14.7	2.7	-5.4	10.8	-9.6	8"x14"	2 @ 1"
	COL 2	0.0	6.1	0.0	32.4	0.0	-17.3	0.0	-17.9	10"x15"	4 @ 3/4"
	COL 3	-0.9	3.0	-5.3	14.7	11.3	-9.9	-3.3	-5.1	8"x14"	2 @ 1"
2@60' x 24' RLL = 30 psf	COL 1	0.7	2.8	6.2	20.5	3.5	-4.7	-10.2	-9.1	8"x10"	2 @ 1"
	COL 2	0.0	7.1	0.0	51.7	0.0	-18.7	0.0	-18.9	12"x17"	4 @ 3/4"
	COL 3	-0.7	2.9	-6.2	20.5	10.4	-9.2	-3.7	-4.6	8"x10"	2 @ 1"
2@60' x 24' RLL = 40 psf	COL 1	0.9	3.3	9.5	28.6	3.1	-5.0	10.6	-9.4	8"x10"	2 @ 1"
	COL 2	0.0	7.3	0.0	66.5	0.0	-18.1	0.0	-18.1	12"x17"	4 @ 3/4"
	COL 3	-0.9	3.3	-9.5	28.6	10.6	-9.4	-3.1	-5.0	8"x10"	2 @ 1"
2@60' x 30' RLL = 20 psf	COL 1	0.9	3.5	4.7	15.3	4.9	-4.8	13.4	-11.4	8"x18"	2 @ 1"
	COL 2	0.0	6.1	0.0	31.2	0.0	-18.0	0.0	-18.0	10"x15"	4 @ 3/4"
	COL 3	-0.9	3.5	-4.7	15.3	13.4	-11.4	-4.9	-4.8	8"x18"	2 @ 1"
2@60' x 30' RLL = 30 psf	COL 1	0.8	3.5	6.9	22.6	4.9	-4.7	13.2	-11.2	8"x18"	2 @ 1"
	COL 2	0.0	6.6	0.0	47.5	0.0	-18.3	0.0	-18.3	12"x17"	4 @ 3/4"
	COL 3	-0.8	3.5	-6.9	22.6	13.2	-11.2	-4.9	-4.7	8"x18"	2 @ 1"
2@60' x 30' RLL = 40 psf	COL 1	0.8	3.5	8.7	29.6	5.0	-4.6	13.1	-11.1	8"x18"	2 @ 1"
	COL 2	0.0	6.9	0.0	64.4	0.0	-18.6	0.0	-18.6	12"x17"	4 @ 3/4"
	COL 3	-0.8	3.5	-8.7	29.6	13.1	-11.1	-5.1	-4.5	8"x18"	2 @ 1"

Gable, Three Spans @ 40'-0"

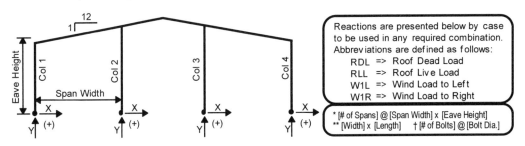

Frame Description*	Column	Column Base Reactions (kips)								Base-plate Size**	Anchor Bolts†
		RDL		RLL		W1L		W1R			
		X	Y	X	Y	X	Y	X	Y		
3@40' x 12' RLL = 20 psf	COL 1	0.6	1.7	4.0	10.0	1.0	-3.4	-5.6	-6.0	8"x10"	2 @ 3/4"
	COL 2	0.0	3.5	0.0	21.0	0.0	-10.9	0.0	-9.6	10"x15"	4 @ 3/4"
	COL 3	0.0	3.5	0.0	20.7	0.0	-9.5	0.0	-10.8	10"x15"	4 @ 3/4"
	COL 4	-0.6	1.8	-4.0	10.1	5.7	-6.1	-1.2	-3.4	8"x10"	2 @ 3/4"
3@40' x 12' RLL = 30 psf	COL 1	0.8	1.9	7.3	15.4	0.6	-3.5	-6.1	-6.2	8"x10"	2 @ 3/4"
	COL 2	0.0	3.6	0.0	31.0	0.0	-10.8	0.0	-9.3	10"x15"	4 @ 3/4"
	COL 3	0.0	3.5	0.0	30.8	0.0	-9.3	0.0	-10.8	10"x15"	4 @ 3/4"
	COL 4	-0.8	1.9	-7.3	15.5	6.1	-6.3	-0.6	-3.5	8"x10"	2 @ 3/4"
3@40' x 12' RLL = 40 psf	COL 1	0.9	1.9	10.0	20.7	0.6	-3.5	-6.2	-6.3	8"x10"	2 @ 1"
	COL 2	0.0	3.7	0.0	41.2	0.0	-10.8	0.0	-9.3	10"x15"	4 @ 3/4"
	COL 3	0.0	3.7	0.0	41.1	0.0	-9.2	0.0	-10.8	10"x15"	4 @ 3/4"
	COL 4	-0.9	1.9	-10.0	20.7	6.2	-6.3	-0.6	-3.5	8"x10"	2 @ 1"
3@40' x 18' RLL = 20 psf	COL 1	0.6	2.0	3.7	10.6	2.5	-2.9	-7.8	-7.3	8"x10"	2 @ 1"
	COL 2	0.0	3.4	0.0	20.3	0.0	-12.1	0.0	-8.6	10"x15"	4 @ 3/4"
	COL 3	0.0	3.4	0.0	20.3	0.0	-8.6	0.0	-12.1	10"x15"	4 @ 3/4"
	COL 4	-0.6	2.0	-3.7	10.6	7.8	-7.3	-2.5	-2.9	8"x10"	2 @ 1"
3@40' x 18' RLL = 30 psf	COL 1	0.6	2.0	5.2	15.6	2.6	-2.7	-7.7	-7.2	8"x10"	2 @ 1"
	COL 2	0.0	3.6	0.0	30.7	0.0	-12.4	0.0	-8.6	10"x15"	4 @ 3/4"
	COL 3	0.0	3.6	0.0	30.7	0.0	-8.6	0.0	-12.4	10"x15"	4 @ 3/4"
	COL 4	-0.6	2.0	-5.3	15.6	7.7	-7.2	-2.6	-2.7	8"x10"	2 @ 1"
3@40' x 18' RLL = 40 psf	COL 1	0.6	2.0	6.8	20.6	2.6	-2.6	-7.7	-7.2	8"x10"	2 @ 1"
	COL 2	0.0	3.8	0.0	41.3	0.0	-12.6	0.0	-8.5	10"x15"	4 @ 3/4"
	COL 3	0.0	3.8	0.0	41.1	0.0	-8.5	0.0	-12.5	10"x15"	4 @ 3/4"
	COL 4	-0.6	2.0	-6.8	20.6	7.7	-7.2	-2.6	-2.7	8"x10"	2 @ 1"

Gable, Three Spans @ 40'-0" (Continued)

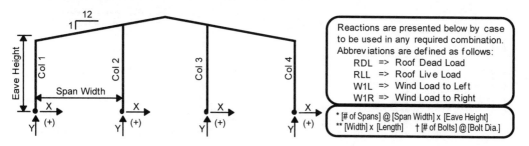

Reactions are presented below by case to be used in any required combination. Abbreviations are defined as follows:
- RDL => Roof Dead Load
- RLL => Roof Live Load
- W1L => Wind Load to Left
- W1R => Wind Load to Right

* [# of Spans] @ [Span Width] x [Eave Height]
** [Width] x [Length]   † [# of Bolts] @ [Bolt Dia.]

Frame Description*	Column	Column Base Reactions (kips)								Base-plate Size**	Anchor Bolts†
		RDL		RLL		W1L		W1R			
		X	Y	X	Y	X	Y	X	Y		
3@40' x 24' RLL = 20 psf	COL 1	0.4	2.1	2.6	10.4	4.4	-1.7	-9.4	-8.5	8"x10"	2 @ 1"
	COL 2	0.0	3.5	0.0	20.5	0.0	-14.6	0.0	-7.8	10"x15"	4 @ 3/4"
	COL 3	0.0	3.5	0.0	20.5	0.0	-7.4	0.0	-14.9	10"x15"	4 @ 3/4"
	COL 4	-0.4	2.2	-2.6	10.4	9.9	-8.8	-4.8	-1.4	8"x10"	2 @ 1"
3@40' x 24' RLL = 30 psf	COL 1	0.4	2.1	3.6	15.1	4.4	-1.5	-9.2	-8.3	8"x10"	2 @ 1"
	COL 2	0.0	3.8	0.0	31.4	0.0	-15.0	0.0	-7.9	10"x15"	4 @ 3/4"
	COL 3	0.0	3.7	0.0	30.9	0.0	-7.4	0.0	-15.1	10"x15"	4 @ 3/4"
	COL 4	-0.4	2.1	-3.6	15.3	9.7	-8.7	-4.9	-1.3	8"x10"	2 @ 1"
3@40' x 24' RLL = 40 psf	COL 1	0.4	2.1	4.6	20.1	4.7	-1.2	-9.4	-8.6	8"x10"	2 @ 1"
	COL 2	0.0	3.9	0.0	41.7	0.0	-15.5	0.0	-7.3	10"x15"	4 @ 3/4"
	COL 3	0.0	3.9	0.0	41.7	0.0	-7.3	0.0	-15.5	10"x15"	4 @ 3/4"
	COL 4	-0.4	2.1	-4.6	20.1	9.4	-8.6	-4.7	-1.2	8"x10"	2 @ 1"
3@40' x 30' RLL = 20 psf	COL 1	0.4	2.5	2.3	10.6	6.1	-0.4	11.6	-10.2	8"x10"	2 @ 1"
	COL 2	0.0	3.7	0.0	20.3	0.0	-17.0	0.0	-6.6	10"x15"	4 @ 3/4"
	COL 3	0.0	3.7	0.0	20.3	0.0	-6.3	0.0	-17.4	10"x15"	4 @ 3/4"
	COL 4	-0.4	2.5	-2.3	10.6	11.9	-10.5	-6.5	-0.1	8"x10"	2 @ 1"
3@40' x 30' RLL = 30 psf	COL 1	0.4	2.4	3.3	15.6	6.2	-0.2	11.6	-10.2	8"x10"	2 @ 1"
	COL 2	0.0	3.8	0.0	30.8	0.0	-17.6	0.0	-6.3	10"x15"	4 @ 3/4"
	COL 3	0.0	3.8	0.0	30.6	0.0	-5.7	0.0	-18.0	10"x15"	4 @ 3/4"
	COL 4	-0.4	2.5	-3.3	15.7	12.1	-10.7	-6.8	0.3	8"x10"	2 @ 1"
3@40' x 30' RLL = 40 psf	COL 1	0.4	2.4	4.3	20.7	6.3	0.1	11.5	-10.3	8"x10"	2 @ 1"
	COL 2	0.0	4.1	0.0	41.1	0.0	-18.0	0.0	-5.9	12"x17"	4 @ 3/4"
	COL 3	0.0	4.1	0.0	41.1	0.0	-5.5	0.0	-18.5	12"x17"	4 @ 3/4"
	COL 4	-0.4	2.5	-4.2	20.7	11.9	-10.7	-6.7	0.5	8"x10"	2 @ 1"

Gable, Three Spans @ 50'-0"

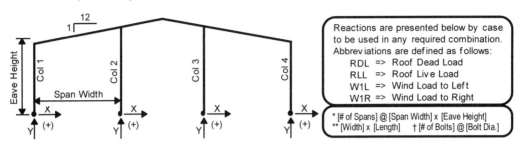

Reactions are presented below by case to be used in any required combination. Abbreviations are defined as follows:
- RDL => Roof Dead Load
- RLL => Roof Live Load
- W1L => Wind Load to Left
- W1R => Wind Load to Right

* [# of Spans] @ [Span Width] x [Eave Height]
** [Width] x [Length]   † [# of Bolts] @ [Bolt Dia.]

Frame Description*	Column	Column Base Reactions (kips)								Base-plate Size**	Anchor Bolts†
		RDL		RLL		W1L		W1R			
		X	Y	X	Y	X	Y	X	Y		
3@50' x 12' RLL = 20 psf	COL 1	1.3	2.3	7.8	13.1	-0.8	-5.0	-7.5	-7.4	8"x10"	2 @ 1"
	COL 2	0.0	4.4	0.0	25.5	0.0	-12.5	0.0	-12.3	10"x15"	4 @ 3/4"
	COL 3	0.0	4.4	0.0	25.5	0.0	-12.4	0.0	-12.5	10"x15"	4 @ 3/4"
	COL 4	-1.3	2.3	-7.8	13.1	7.5	-7.4	0.8	-5.0	8"x10"	2 @ 1"
3@50' x 12' RLL = 30 psf	COL 1	1.6	2.6	12.9	19.9	-1.2	-5.1	-8.0	-7.6	8"x10"	2 @ 1"
	COL 2	0.0	4.6	0.0	38.0	0.0	-12.5	0.0	-12.1	10"x15"	4 @ 3/4"
	COL 3	0.0	4.6	0.0	38.0	0.0	-12.1	0.0	-12.5	10"x15"	4 @ 3/4"
	COL 4	-1.6	2.6	-12.9	19.9	7.9	-7.6	1.2	-5.1	8"x10"	2 @ 1"
3@50' x 12' RLL = 40 psf	COL 1	1.5	2.7	15.9	26.3	-0.9	-5.1	-7.6	-7.5	8"x10"	2 @ 1-1/4"
	COL 2	0.0	5.0	0.0	51.0	0.0	-12.5	0.0	-12.3	10"x15"	4 @ 3/4"
	COL 3	0.0	5.0	0.0	51.0	0.0	-12.3	0.0	-12.5	10"x15"	4 @ 3/4"
	COL 4	-1.5	2.7	-15.9	26.3	7.6	-7.5	0.9	-5.1	8"x10"	2 @ 1-1/4"
3@50' x 18' RLL = 20 psf	COL 1	0.8	2.3	4.5	12.5	2.1	-4.1	-8.2	-8.0	8"x10"	2 @ 1"
	COL 2	0.0	4.6	0.0	26.1	0.0	-14.4	0.0	-12.1	10"x15"	4 @ 3/4"
	COL 3	0.0	4.6	0.0	26.1	0.0	-12.1	0.0	-14.4	10"x15"	4 @ 3/4"
	COL 4	-0.8	2.3	-4.5	12.5	8.3	-8.0	-2.1	-4.1	8"x10"	2 @ 1"
3@50' x 18' RLL = 30 psf	COL 1	0.9	2.5	7.3	18.9	1.8	-4.1	-8.4	-8.1	8"x10"	2 @ 1"
	COL 2	0.0	4.9	0.0	39.0	0.0	-14.5	0.0	-11.8	10"x15"	4 @ 3/4"
	COL 3	0.0	4.9	0.0	39.1	0.0	-11.8	0.0	-14.6	10"x15"	4 @ 3/4"
	COL 4	-0.9	2.5	-7.3	18.9	8.4	-8.1	-1.8	-4.1	8"x10"	2 @ 1"
3@50' x 18' RLL = 40 psf	COL 1	0.9	2.7	9.5	25.2	1.9	-4.2	-8.4	-8.1	8"x10"	2 @ 1"
	COL 2	0.0	5.2	0.0	52.0	0.0	-14.4	0.0	-11.9	10"x15"	4 @ 3/4"
	COL 3	0.0	5.2	0.0	52.0	0.0	-11.9	0.0	-14.4	10"x15"	4 @ 3/4"
	COL 4	-0.9	2.7	-9.5	25.2	8.4	-8.1	-1.9	-4.1	8"x10"	2 @ 1"

Gable, Three Spans @ 50'-0" (Continued)

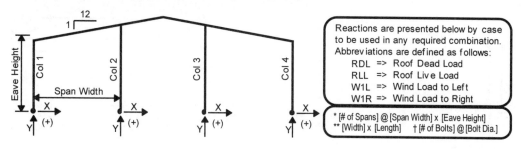

Frame Description*	Column	Column Base Reactions (kips)								Base-plate Size**	Anchor Bolts†
		RDL		RLL		W1L		W1R			
		X	Y	X	Y	X	Y	X	Y		
3@50' x 24' RLL = 20 psf	COL 1	0.7	2.6	4.0	12.9	3.7	-3.5	10.4	-9.6	8"x14"	2 @ 1"
	COL 2	0.0	4.6	0.0	25.7	0.0	-16.4	0.0	-11.2	10"x15"	4 @ 3/4"
	COL 3	0.0	4.6	0.0	25.7	0.0	-11.2	0.0	-16.4	10"x15"	4 @ 3/4"
	COL 4	-0.7	2.6	-4.0	12.9	10.4	-9.6	-3.7	-3.5	8"x14"	2 @ 1"
3@50' x 24' RLL = 30 psf	COL 1	0.7	2.6	5.6	19.1	3.8	-3.3	-10.2	-9.5	8"x14"	2 @ 1"
	COL 2	0.0	5.0	0.0	38.9	0.0	-16.8	0.0	-11.1	10"x15"	4 @ 3/4"
	COL 3	0.0	5.0	0.0	38.9	0.0	-11.1	0.0	-16.8	10"x15"	4 @ 3/4"
	COL 4	-0.7	2.6	-5.6	19.1	10.2	-9.5	-3.8	-3.3	8"x14"	2 @ 1"
3@50' x 24' RLL = 40 psf	COL 1	0.7	2.7	7.2	25.2	3.9	-3.3	-10.2	-9.5	8"x14"	2 @ 1"
	COL 2	0.0	5.3	0.0	52.0	0.0	-16.8	0.0	-11.2	12"x17"	4 @ 3/4"
	COL 3	0.0	5.3	0.0	52.0	0.0	-11.2	0.0	-16.8	12"x17"	4 @ 3/4"
	COL 4	-0.7	2.7	-7.2	25.2	10.2	-9.5	-3.9	-3.3	8"x14"	2 @ 1"
3@50' x 30' RLL = 20 psf	COL 1	0.7	3.0	3.8	13.4	5.3	-2.7	12.8	-11.5	8"x14"	2 @ 1"
	COL 2	0.0	4.6	0.0	25.1	0.0	-18.7	0.0	-9.9	10"x15"	4 @ 3/4"
	COL 3	0.0	4.7	0.0	25.4	0.0	-10.1	0.0	-18.8	10"x15"	4 @ 3/4"
	COL 4	-0.7	3.0	-3.8	13.3	12.7	-11.4	-5.2	-2.7	8"x14"	2 @ 1"
3@50' x 30' RLL = 30 psf	COL 1	0.7	3.0	5.5	19.8	5.4	-2.3	12.8	-11.6	8"x14"	2 @ 1"
	COL 2	0.0	5.2	0.0	38.1	0.0	-19.6	0.0	-9.3	12"x17"	4 @ 3/4"
	COL 3	0.0	5.1	0.0	38.1	0.0	-9.4	0.0	-19.5	12"x17"	4 @ 3/4"
	COL 4	-0.7	3.0	-5.5	19.8	12.7	-11.5	-5.3	-2.4	8"x14"	2 @ 1"
3@50' x 30' RLL = 40 psf	COL 1	0.7	3.0	7.3	26.2	5.4	-2.2	12.7	-11.6	8"x14"	2 @ 1"
	COL 2	0.0	5.3	0.0	51.0	0.0	-19.7	0.0	-9.2	12"x17"	4 @ 3/4"
	COL 3	0.0	5.3	0.0	51.0	0.0	-9.2	0.0	-19.7	12"x17"	4 @ 3/4"
	COL 4	-0.7	3.0	-7.3	26.2	12.7	-11.6	-5.4	-2.2	8"x14"	2 @ 1"

Gable, Three Spans @ 60'-0"

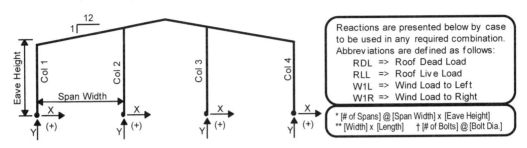

Frame Description*	Column	Column Base Reactions (kips)								Base-plate Size**	Anchor Bolts†
		RDL		RLL		W1L		W1R			
		X	Y	X	Y	X	Y	X	Y		
3@60' x 12' RLL = 20 psf	COL 1	2.4	3.1	13.9	16.6	-3.8	-6.8	10.5	-9.1	8"x10"	2 @ 1"
	COL 2	0.0	5.4	0.0	29.8	0.0	-14.1	0.0	-14.8	10"x15"	4 @ 3/4"
	COL 3	0.0	5.4	0.0	29.8	0.0	-14.8	0.0	-14.1	10"x15"	4 @ 3/4"
	COL 4	-2.4	3.1	13.9	16.6	10.4	-9.1	3.8	-6.8	8"x10"	2 @ 1"
3@60' x 12' RLL = 30 psf	COL 1	2.4	3.3	19.9	24.7	-3.5	-6.7	-10.1	-9.0	8"x10"	2 @ 1-1/4"
	COL 2	0.0	5.8	0.0	44.9	0.0	-14.2	0.0	-14.9	10"x15"	4 @ 3/4"
	COL 3	0.0	5.8	0.0	44.7	0.0	-14.8	0.0	-14.1	10"x15"	4 @ 3/4"
	COL 4	-2.4	3.4	19.9	24.7	10.1	-9.0	3.5	-6.8	8"x10"	2 @ 1-1/4"
3@60' x 12' RLL = 40 psf	COL 1	2.5	3.6	25.9	32.7	-3.4	-6.7	-10.0	-9.0	8"x14"	4 @ 1"
	COL 2	0.0	6.3	0.0	60.0	0.0	-14.2	0.0	-14.9	10"x15"	4 @ 3/4"
	COL 3	0.0	6.3	0.0	60.1	0.0	-14.9	0.0	-14.2	10"x15"	4 @ 3/4"
	COL 4	-2.5	3.6	25.9	32.6	10.0	-9.0	3.4	-6.7	8"x14"	4 @ 1"
3@60' x 18' RLL = 20 psf	COL 1	1.3	3.0	7.9	15.6	0.2	-5.9	-10.0	-9.5	8"x10"	2 @ 1"
	COL 2	0.0	5.6	0.0	30.7	0.0	-16.2	0.0	-14.8	10"x15"	4 @ 3/4"
	COL 3	0.0	5.6	0.0	30.7	0.0	-14.8	0.0	-16.2	10"x15"	4 @ 3/4"
	COL 4	-1.3	3.0	-7.9	15.6	10.0	-9.5	-0.2	-5.9	8"x10"	2 @ 1"
3@60' x 18' RLL = 30 psf	COL 1	1.3	3.3	10.1	22.9	0.9	-5.7	-9.4	-9.2	8"x10"	2 @ 1"
	COL 2	0.0	6.1	0.0	46.7	0.0	-16.2	0.0	-15.2	10"x15"	4 @ 3/4"
	COL 3	0.0	6.1	0.0	46.6	0.0	-15.2	0.0	-16.1	10"x15"	4 @ 3/4"
	COL 4	-1.3	3.3	-10.1	22.9	9.3	-9.2	-0.8	-5.7	8"x10"	2 @ 1"
3@60' x 18' RLL = 40 psf	COL 1	1.4	3.5	14.3	30.5	0.7	-5.7	-9.6	-9.3	8"x10"	2 @ 1"
	COL 2	0.0	6.4	0.0	62.2	0.0	-16.2	0.0	-15.1	10"x15"	4 @ 3/4"
	COL 3	0.0	6.4	0.0	62.2	0.0	-15.1	0.0	-16.2	10"x15"	4 @ 3/4"
	COL 4	-1.4	3.5	14.3	30.5	9.6	-9.3	-0.7	-5.7	8"x10"	2 @ 1"

Frame Reaction Tables

Gable, Three Spans @ 60'-0" (Continued)

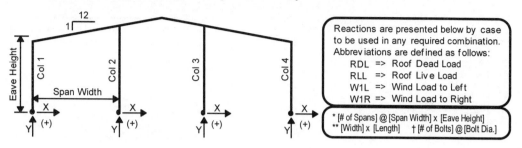

Reactions are presented below by case to be used in any required combination. Abbreviations are defined as follows:
RDL => Roof Dead Load
RLL => Roof Live Load
W1L => Wind Load to Left
W1R => Wind Load to Right

* [# of Spans] @ [Span Width] x [Eave Height]
** [Width] x [Length]   † [# of Bolts] @ [Bolt Dia.]

Frame Description*	Column	Column Base Reactions (kips)								Base-plate Size**	Anchor Bolts†
		RDL		RLL		W1L		W1R			
		X	Y	X	Y	X	Y	X	Y		
3@60' x 24' RLL = 20 psf	COL 1	1.0	3.1	6.2	15.7	2.3	-5.3	11.5	-10.9	8"x14"	2 @ 1"
	COL 2	0.0	5.7	0.0	30.7	0.0	-18.5	0.0	-14.2	10"x15"	4 @ 3/4"
	COL 3	0.0	5.7	0.0	30.7	0.0	-14.2	0.0	-18.6	10"x15"	4 @ 3/4"
	COL 4	-1.0	3.1	-6.2	15.7	11.5	-10.9	-2.3	-5.2	8"x14"	2 @ 1"
3@60' x 24' RLL = 30 psf	COL 1	1.0	3.2	8.5	23.1	2.7	-5.1	11.2	-10.7	8"x14"	2 @ 1"
	COL 2	0.0	6.0	0.0	46.3	0.0	-18.6	0.0	-14.4	10"x15"	4 @ 3/4"
	COL 3	0.0	6.1	0.0	46.7	0.0	-14.5	0.0	-18.7	10"x15"	4 @ 3/4"
	COL 4	-1.0	3.2	-8.5	22.9	11.2	-10.7	-2.7	-5.0	8"x14"	2 @ 1"
3@60' x 24' RLL = 40 psf	COL 1	1.0	3.4	11.0	30.5	2.8	-5.0	11.1	-10.6	8"x14"	2 @ 1"
	COL 2	0.0	6.6	0.0	62.2	0.0	-18.7	0.0	-14.5	12"x17"	4 @ 3/4"
	COL 3	0.0	6.6	0.0	62.3	0.0	-14.6	0.0	-18.7	12"x17"	4 @ 3/4"
	COL 4	-1.0	3.4	-11.0	30.4	11.1	-10.6	-2.8	-5.0	8"x14"	2 @ 1"
3@60' x 30' RLL = 20 psf	COL 1	1.0	3.5	5.7	16.1	4.0	-4.6	13.8	-12.7	8"x18"	2 @ 1"
	COL 2	0.0	5.7	0.0	30.3	0.0	-20.9	0.0	-13.2	10"x15"	4 @ 3/4"
	COL 3	0.0	5.7	0.0	30.1	0.0	-13.0	0.0	-20.9	10"x15"	4 @ 3/4"
	COL 4	-1.0	3.6	-5.7	16.2	13.9	-12.8	-4.2	-4.6	8"x18"	2 @ 1"
3@60' x 30' RLL = 30 psf	COL 1	1.0	3.6	8.2	23.9	4.3	-4.4	13.8	-12.7	8"x18"	2 @ 1"
	COL 2	0.0	6.2	0.0	45.6	0.0	-21.3	0.0	-12.9	12"x17"	4 @ 3/4"
	COL 3	0.0	6.2	0.0	45.6	0.0	-12.8	0.0	-21.4	12"x17"	4 @ 3/4"
	COL 4	-1.0	3.6	-8.2	23.9	13.8	-12.8	-4.3	-4.3	8"x18"	2 @ 1"
3@60' x 30' RLL = 40 psf	COL 1	1.0	3.7	10.3	31.4	4.6	-4.1	13.8	-12.8	8"x18"	2 @ 1"
	COL 2	0.0	6.6	0.0	61.2	0.0	-21.7	0.0	-12.7	12"x17"	4 @ 3/4"
	COL 3	0.0	6.6	0.0	61.3	0.0	-12.9	0.0	-21.5	12"x17"	4 @ 3/4"
	COL 4	-1.0	3.7	-10.3	31.4	13.5	-12.6	-4.3	-4.3	8"x18"	2 @ 1"

Gable, Four Spans @ 40'-0"

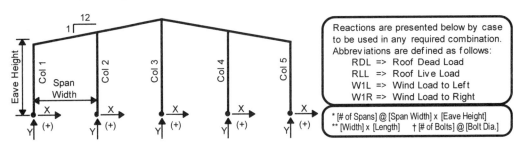

Reactions are presented below by case to be used in any required combination. Abbreviations are defined as follows:
RDL => Roof Dead Load
RLL => Roof Live Load
W1L => Wind Load to Left
W1R => Wind Load to Right

* [# of Spans] @ [Span Width] x [Eave Height]
** [Width] x [Length] † [# of Bolts] @ [Bolt Dia.]

Frame Description*	Column	Column Base Reactions (kips)								Base-plate Size**	Anchor Bolts†
		RDL		RLL		W1L		W1R			
		X	Y	X	Y	X	Y	X	Y		
4@40' x 12' RLL = 20 psf	COL 1	0.6	1.7	3.7	9.7	1.2	-3.4	-5.5	-5.7	8"x10"	2 @ 3/4"
	COL 2	0.0	3.7	0.0	21.9	0.0	-10.6	0.0	-10.8	10"x15"	4 @ 3/4"
	COL 3	0.0	3.2	0.0	19.2	0.0	-9.3	0.0	-9.3	10"x15"	4 @ 3/4"
	COL 4	0.0	3.7	0.0	21.9	0.0	-10.8	0.0	-10.6	10"x15"	4 @ 3/4"
	COL 5	-0.6	1.7	-3.7	9.7	5.5	-5.7	-1.2	-3.4	8"x10"	2 @ 3/4"
4@40' x 12' RLL = 30 psf	COL 1	0.8	1.8	6.8	14.9	0.7	-3.5	-5.9	-5.9	8"x10"	2 @ 3/4"
	COL 2	0.0	3.8	0.0	32.6	0.0	-10.6	0.0	-10.6	10"x15"	4 @ 3/4"
	COL 3	0.0	3.3	0.0	28.6	0.0	-9.2	0.0	-9.2	10"x15"	4 @ 3/4"
	COL 4	0.0	3.8	0.0	32.6	0.0	-10.6	0.0	-10.6	10"x15"	4 @ 3/4"
	COL 5	-0.8	1.8	-6.8	14.9	5.9	-5.9	-0.8	-3.5	8"x10"	2 @ 3/4"
4@40' x 12' RLL = 40 psf	COL 1	0.8	1.8	8.9	19.6	0.8	-3.4	-5.8	-5.8	8"x10"	2 @ 1"
	COL 2	0.0	4.0	0.0	44.1	0.0	-10.7	0.0	-10.8	10"x15"	4 @ 3/4"
	COL 3	0.0	3.3	0.0	37.6	0.0	-9.1	0.0	-9.1	10"x15"	4 @ 3/4"
	COL 4	0.0	4.0	0.0	43.9	0.0	-10.8	0.0	-10.7	10"x15"	4 @ 3/4"
	COL 5	-0.8	1.8	-8.9	19.7	5.9	-5.8	-0.8	-3.5	8"x10"	2 @ 1"
4@40' x 18' RLL = 20 psf	COL 1	0.6	1.9	3.4	10.2	2.6	-2.9	-7.6	-6.9	8"x10"	2 @ 1"
	COL 2	0.0	3.6	0.0	21.0	0.0	-11.5	0.0	-9.9	10"x15"	4 @ 3/4"
	COL 3	0.0	3.4	0.0	19.9	0.0	-10.0	0.0	-10.0	10"x15"	4 @ 3/4"
	COL 4	0.0	3.6	0.0	21.0	0.0	-9.9	0.0	-11.5	10"x15"	4 @ 3/4"
	COL 5	-0.6	1.9	-3.4	10.2	7.6	-6.9	-2.6	-2.9	8"x10"	2 @ 1"
4@40' x 18' RLL = 30 psf	COL 1	0.5	1.9	4.9	15.3	2.7	-2.9	-7.6	-6.8	8"x10"	2 @ 1"
	COL 2	0.0	3.7	0.0	31.7	0.0	-11.6	0.0	-9.9	10"x15"	4 @ 3/4"
	COL 3	0.0	3.6	0.0	29.7	0.0	-9.9	0.0	-9.9	10"x15"	4 @ 3/4"
	COL 4	0.0	3.7	0.0	31.7	0.0	-9.9	0.0	-11.6	10"x15"	4 @ 3/4"
	COL 5	-0.5	1.9	-4.9	15.3	7.6	-6.8	-2.7	-2.9	8"x10"	2 @ 1"
4@40' x 18' RLL = 40 psf	COL 1	0.5	1.9	6.1	19.7	2.8	-2.7	-7.4	-6.7	8"x10"	2 @ 1"
	COL 2	0.0	4.0	0.0	43.5	0.0	-12.0	0.0	-10.2	10"x15"	4 @ 3/4"
	COL 3	0.0	3.6	0.0	38.5	0.0	-9.6	0.0	-9.7	10"x15"	4 @ 3/4"
	COL 4	0.0	4.0	0.0	43.5	0.0	-10.2	0.0	-12.0	10"x15"	4 @ 3/4"
	COL 5	-0.5	1.9	-6.1	19.7	7.4	-6.7	-2.8	-2.7	8"x10"	2 @ 1"

Frame Reaction Tables 287

Gable, Four Spans @ 40'-0" (Continued)

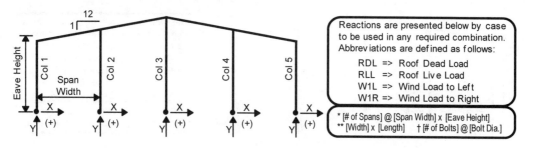

Reactions are presented below by case to be used in any required combination. Abbreviations are defined as follows:

RDL => Roof Dead Load
RLL => Roof Live Load
W1L => Wind Load to Left
W1R => Wind Load to Right

* [# of Spans] @ [Span Width] x [Eave Height]
** [Width] x [Length]   † [# of Bolts] @ [Bolt Dia.]

Frame Description*	Column	Column Base Reactions (kips)								Base-plate Size**	Anchor Bolts†
		RDL		RLL		W1L		W1R			
		X	Y	X	Y	X	Y	X	Y		
4@40' x 24' RLL = 20 psf	COL 1	0.4	2.1	2.5	10.3	4.5	-1.9	-9.6	-8.3	8"x14"	2 @ 1"
	COL 2	0.0	3.7	0.0	20.8	0.0	-13.3	0.0	-9.2	10"x15"	4 @ 3/4"
	COL 3	0.0	3.6	0.0	20.3	0.0	-10.8	0.0	-10.8	10"x15"	4 @ 3/4"
	COL 4	0.0	3.7	0.0	20.8	0.0	-9.2	0.0	-13.3	10"x15"	4 @ 3/4"
	COL 5	-0.4	2.1	-2.5	10.3	9.6	-8.3	-4.5	-1.9	8"x14"	2 @ 1"
4@40' x 24' RLL = 30 psf	COL 1	0.4	2.1	3.8	15.5	4.5	-1.9	-9.6	-8.3	8"x14"	2 @ 1"
	COL 2	0.0	3.8	0.0	31.2	0.0	-13.4	0.0	-9.2	10"x15"	4 @ 3/4"
	COL 3	0.0	3.7	0.0	30.3	0.0	-10.7	0.0	-10.7	10"x15"	4 @ 3/4"
	COL 4	0.0	3.8	0.0	31.2	0.0	-9.1	0.0	-13.4	10"x15"	4 @ 3/4"
	COL 5	-0.4	2.1	-3.8	15.5	9.6	-8.3	-4.5	-1.9	8"x14"	2 @ 1"
4@40' x 24' RLL = 40 psf	COL 1	0.4	2.1	4.8	20.3	4.6	-1.8	-9.5	-8.2	8"x14"	2 @ 1"
	COL 2	0.0	4.0	0.0	42.0	0.0	-13.5	0.0	-9.2	10"x15"	4 @ 3/4"
	COL 3	0.0	3.8	0.0	40.3	0.0	-10.7	0.0	-10.7	10"x15"	4 @ 3/4"
	COL 4	0.0	4.0	0.0	41.9	0.0	-9.2	0.0	-13.5	10"x15"	4 @ 3/4"
	COL 5	-0.4	2.1	-4.8	20.3	9.5	-8.2	-4.6	-1.8	8"x14"	2 @ 1"
4@40' x 30' RLL = 20 psf	COL 1	0.4	2.5	2.3	10.7	6.1	-0.9	11.7	-9.9	8"x14"	2 @ 1"
	COL 2	0.0	3.7	0.0	20.1	0.0	-15.0	0.0	-8.1	10"x15"	4 @ 3/4"
	COL 3	0.0	3.8	0.0	20.9	0.0	-11.7	0.0	-11.7	10"x15"	4 @ 3/4"
	COL 4	0.0	3.7	0.0	20.1	0.0	-8.0	0.0	-15.2	10"x15"	4 @ 3/4"
	COL 5	-0.4	2.5	-2.3	10.7	12.0	-10.1	-6.4	-0.7	8"x14"	2 @ 1"
4@40' x 30' RLL = 30 psf	COL 1	0.4	2.5	3.4	15.9	6.2	-0.8	11.7	-9.9	8"x14"	2 @ 1"
	COL 2	0.0	3.8	0.0	30.2	0.0	-15.2	0.0	-8.1	10"x15"	4 @ 3/4"
	COL 3	0.0	4.1	0.0	31.3	0.0	-11.6	0.0	-11.7	12"x17"	4 @ 3/4"
	COL 4	0.0	3.8	0.0	30.2	0.0	-7.9	0.0	-15.3	10"x15"	4 @ 3/4"
	COL 5	-0.4	2.5	-3.4	15.9	12.0	-10.1	-6.4	-0.6	8"x14"	2 @ 1"
4@40' x 30' RLL = 40 psf	COL 1	0.4	2.6	-0.4	16.2	6.6	-0.3	11.9	-10.1	8"x14"	2 @ 1"
	COL 2	0.0	4.3	0.0	46.6	0.0	-15.6	0.0	-8.0	12"x17"	4 @ 3/4"
	COL 3	0.0	4.2	0.0	37.0	0.0	-11.6	0.0	-11.9	12"x17"	4 @ 3/4"
	COL 4	0.0	3.6	0.0	-2.9	0.0	-8.4	0.0	-14.7	10"x15"	4 @ 3/4"
	COL 5	-0.4	2.6	0.4	-14.5	11.5	-9.7	-6.2	-1.0	8"x14"	2 @ 1"

Gable, Four Spans @ 50'-0"

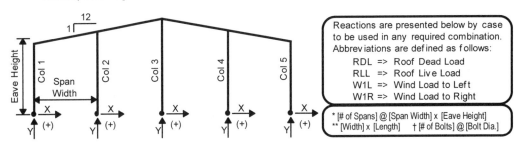

Reactions are presented below by case to be used in any required combination. Abbreviations are defined as follows:
- RDL => Roof Dead Load
- RLL => Roof Live Load
- W1L => Wind Load to Left
- W1R => Wind Load to Right

* [# of Spans] @ [Span Width] x [Eave Height]
** [Width] x [Length]   † [# of Bolts] @ [Bolt Dia.]

Frame Description*	Column	\multicolumn{8}{c	}{Column Base Reactions (kips)}	Base-plate Size**	Anchor Bolts†
		RDL X	RDL Y	RLL X	RLL Y	W1L X	W1L Y	W1R X	W1R Y		
4@50' x 12' RLL = 20 psf	COL 1	1.0	2.1	6.3	12.1	-0.1	-4.7	-6.8	-6.8	8"x10"	2 @ 1"
	COL 2	0.0	4.9	0.0	27.8	0.0	-13.0	0.0	-14.1	10"x15"	4 @ 3/4"
	COL 3	0.0	4.0	0.0	23.2	0.0	-11.2	0.0	-11.2	10"x15"	4 @ 3/4"
	COL 4	0.0	4.9	0.0	27.8	0.0	-14.1	0.0	-13.0	10"x15"	4 @ 3/4"
	COL 5	-1.0	2.1	-6.3	12.1	6.8	-6.8	0.1	-4.7	8"x10"	2 @ 1"
4@50' x 12' RLL = 30 psf	COL 1	1.3	2.4	11.2	18.7	-0.7	-4.9	-7.3	-7.0	8"x10"	2 @ 1"
	COL 2	0.0	5.0	0.0	40.9	0.0	-12.8	0.0	-13.8	10"x15"	4 @ 3/4"
	COL 3	0.0	4.1	0.0	35.2	0.0	-11.3	0.0	-11.3	10"x15"	4 @ 3/4"
	COL 4	0.0	5.0	0.0	40.9	0.0	-13.8	0.0	-12.8	10"x15"	4 @ 3/4"
	COL 5	-1.3	2.4	-11.2	18.7	7.3	-7.0	0.7	-4.9	8"x10"	2 @ 1"
4@50' x 12' RLL = 40 psf	COL 1	1.2	2.4	13.6	24.4	-0.4	-4.7	-7.0	-6.9	8"x10"	2 @ 1"
	COL 2	0.0	5.4	0.0	55.4	0.0	-13.0	0.0	-14.0	10"x15"	4 @ 3/4"
	COL 3	0.0	4.4	0.0	46.4	0.0	-11.2	0.0	-11.2	10"x15"	4 @ 3/4"
	COL 4	0.0	5.4	0.0	55.4	0.0	-14.0	0.0	-13.0	10"x15"	4 @ 3/4"
	COL 5	-1.2	2.4	-13.6	24.4	7.0	-6.9	0.4	-4.7	8"x10"	2 @ 1"
4@50' x 18' RLL = 20 psf	COL 1	0.6	2.2	3.6	11.7	2.3	-4.0	-7.5	-7.3	8"x10"	2 @ 1"
	COL 2	0.0	4.9	0.0	27.8	0.0	-14.1	0.0	-14.1	10"x15"	4 @ 3/4"
	COL 3	0.0	4.2	0.0	23.9	0.0	-11.9	0.0	-11.9	10"x15"	4 @ 3/4"
	COL 4	0.0	4.9	0.0	27.8	0.0	-14.0	0.0	-14.2	10"x15"	4 @ 3/4"
	COL 5	-0.6	2.2	-3.6	11.8	7.7	-7.4	-2.5	-4.0	8"x10"	2 @ 1"
4@50' x 18' RLL = 30 psf	COL 1	0.8	2.3	6.7	17.9	1.9	-4.0	-8.1	-7.6	8"x10"	2 @ 1"
	COL 2	0.0	5.3	0.0	41.7	0.0	-14.4	0.0	-13.8	10"x15"	4 @ 3/4"
	COL 3	0.0	4.3	0.0	35.2	0.0	-11.7	0.0	-11.7	10"x15"	4 @ 3/4"
	COL 4	0.0	5.3	0.0	41.7	0.0	-13.8	0.0	-14.4	10"x15"	4 @ 3/4"
	COL 5	-0.8	2.3	-6.7	17.9	8.1	-7.6	-1.9	-4.0	8"x10"	2 @ 1"
4@50' x 18' RLL = 40 psf	COL 1	0.7	2.5	8.2	23.7	2.1	-4.0	-7.9	-7.5	8"x10"	2 @ 1"
	COL 2	0.0	5.5	0.0	55.6	0.0	-14.3	0.0	-13.9	10"x15"	4 @ 3/4"
	COL 3	0.0	4.5	0.0	47.4	0.0	-11.8	0.0	-11.8	10"x15"	4 @ 3/4"
	COL 4	0.0	5.5	0.0	55.7	0.0	-13.9	0.0	-14.4	10"x15"	4 @ 3/4"
	COL 5	-0.7	2.5	-8.2	23.7	7.9	-7.5	-2.1	-4.0	8"x10"	2 @ 1"

Gable, Four Spans @ 50'-0" (Continued)

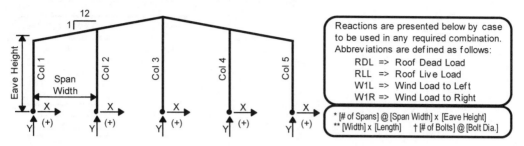

Reactions are presented below by case to be used in any required combination. Abbreviations are defined as follows:
RDL => Roof Dead Load
RLL => Roof Live Load
W1L => Wind Load to Left
W1R => Wind Load to Right

* [# of Spans] @ [Span Width] x [Eave Height]
** [Width] x [Length]   † [# of Bolts] @ [Bolt Dia.]

Frame Description*	Column	Column Base Reactions (kips)								Base-plate Size**	Anchor Bolts†
		RDL		RLL		W1L		W1R			
		X	Y	X	Y	X	Y	X	Y		
4@50' x 24' RLL = 20 psf	COL 1	0.6	2.5	3.1	12.2	4.2	-3.6	-9.8	-8.7	8"x14"	2 @ 1"
	COL 2	0.0	4.9	0.0	26.9	0.0	-15.3	0.0	-13.6	10"x15"	4 @ 3/4"
	COL 3	0.0	4.4	0.0	24.8	0.0	-13.2	0.0	-13.2	10"x15"	4 @ 3/4"
	COL 4	0.0	4.9	0.0	26.9	0.0	-13.6	0.0	-15.2	10"x15"	4 @ 3/4"
	COL 5	-0.6	2.5	-3.1	12.2	9.8	-8.7	-4.2	-3.6	8"x14"	2 @ 1"
4@50' x 24' RLL = 30 psf	COL 1	0.5	2.5	4.4	17.9	4.2	-3.5	-9.7	-8.5	8"x14"	2 @ 1"
	COL 2	0.0	5.3	0.0	41.0	0.0	-15.6	0.0	-13.8	10"x15"	4 @ 3/4"
	COL 3	0.0	4.5	0.0	36.6	0.0	-12.9	0.0	-12.9	10"x15"	4 @ 3/4"
	COL 4	0.0	5.3	0.0	41.0	0.0	-13.8	0.0	-15.6	10"x15"	4 @ 3/4"
	COL 5	-0.5	2.5	-4.4	18.0	9.7	-8.5	-4.2	-3.5	8"x14"	2 @ 1"
4@50' x 24' RLL = 40 psf	COL 1	0.5	2.6	5.5	23.6	4.4	-3.3	-9.6	-8.5	8"x14"	2 @ 1"
	COL 2	0.0	5.8	0.0	55.2	0.0	-15.8	0.0	-13.9	12"x17"	4 @ 3/4"
	COL 3	0.0	4.9	0.0	48.5	0.0	-12.8	0.0	-12.8	12"x17"	4 @ 3/4"
	COL 4	0.0	5.8	0.0	55.1	0.0	-13.9	0.0	-15.8	12"x17"	4 @ 3/4"
	COL 5	-0.5	2.6	-5.5	23.6	9.6	-8.5	-4.4	-3.3	8"x14"	2 @ 1"
4@50' x 30' RLL = 20 psf	COL 1	0.7	3.1	3.6	13.1	5.3	-3.1	12.5	-10.7	8"x14"	2 @ 1"
	COL 2	0.0	4.8	0.0	25.9	0.0	-16.9	0.0	-12.3	10"x15"	4 @ 3/4"
	COL 3	0.0	4.8	0.0	25.2	0.0	-14.0	0.0	-14.1	12"x17"	4 @ 3/4"
	COL 4	0.0	4.8	0.0	25.6	0.0	-12.0	0.0	-17.0	10"x15"	4 @ 3/4"
	COL 5	-0.7	3.1	-3.6	13.2	12.9	-11.0	-5.7	-2.9	8"x14"	2 @ 1"
4@50' x 30' RLL = 30 psf	COL 1	0.6	3.0	5.1	19.2	5.4	-2.8	12.5	-10.6	8"x14"	2 @ 1"
	COL 2	0.0	5.3	0.0	39.4	0.0	-17.3	0.0	-12.5	12"x17"	4 @ 3/4"
	COL 3	0.0	5.0	0.0	37.4	0.0	-13.9	0.0	-13.9	12"x17"	4 @ 3/4"
	COL 4	0.0	5.3	0.0	39.0	0.0	-12.2	0.0	-17.3	12"x17"	4 @ 3/4"
	COL 5	-0.6	3.0	-5.1	19.4	12.6	-10.8	-5.5	-2.8	8"x14"	2 @ 1"
4@50' x 30' RLL = 40 psf	COL 1	0.6	2.9	6.9	25.6	5.4	-2.6	12.6	-10.8	8"x14"	2 @ 1"
	COL 2	0.0	5.5	0.0	52.8	0.0	-17.7	0.0	-12.2	12"x17"	4 @ 3/4"
	COL 3	0.0	5.2	0.0	49.3	0.0	-13.7	0.0	-13.7	12"x17"	4 @ 3/4"
	COL 4	0.0	5.5	0.0	52.6	0.0	-12.1	0.0	-17.7	12"x17"	4 @ 3/4"
	COL 5	-0.6	2.9	-6.9	25.7	12.6	-10.9	-5.5	-2.6	8"x14"	2 @ 1"

Gable, Four Spans @ 60'-0"

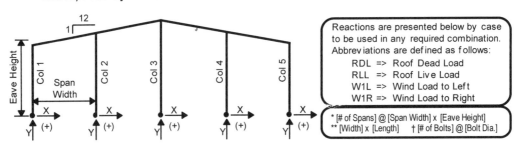

Reactions are presented below by case to be used in any required combination. Abbreviations are defined as follows:
- RDL => Roof Dead Load
- RLL => Roof Live Load
- W1L => Wind Load to Left
- W1R => Wind Load to Right

* [# of Spans] @ [Span Width] x [Eave Height]
** [Width] x [Length]   † [# of Bolts] @ [Bolt Dia.]

Frame Description*	Column	Column Base Reactions (kips)								Base-plate Size**	Anchor Bolts†
		RDL		RLL		W1L		W1R			
		X	Y	X	Y	X	Y	X	Y		
4@60' x 12' RLL = 20 psf	COL 1	1.8	2.7	10.8	15.1	-2.4	-6.2	-8.9	-8.2	8"x10"	2 @ 1"
	COL 2	0.0	6.1	0.0	33.1	0.0	-15.2	0.0	-17.1	10"x15"	4 @ 3/4"
	COL 3	0.0	4.9	0.0	27.2	0.0	-13.1	0.0	-13.0	10"x15"	4 @ 3/4"
	COL 4	0.0	6.1	0.0	33.1	0.0	-17.1	0.0	-15.2	10"x15"	4 @ 3/4"
	COL 5	-1.8	2.7	10.8	15.1	8.9	-8.2	2.4	-6.2	8"x10"	2 @ 1"
4@60' x 12' RLL = 30 psf	COL 1	1.8	2.9	14.8	22.1	-1.9	-6.0	-8.5	-8.0	8"x10"	2 @ 1-1/4"
	COL 2	0.0	6.6	0.0	50.0	0.0	-15.3	0.0	-17.2	10"x15"	4 @ 3/4"
	COL 3	0.0	5.2	0.0	41.2	0.0	-13.2	0.0	-13.2	10"x15"	4 @ 3/4"
	COL 4	0.0	6.5	0.0	50.0	0.0	-17.2	0.0	-15.3	10"x15"	4 @ 3/4"
	COL 5	-1.8	2.9	14.8	22.1	8.5	-8.0	1.9	-6.0	8"x10"	2 @ 1-1/4"
4@60' x 12' RLL = 40 psf	COL 1	2.1	3.2	21.8	30.3	-2.4	-6.2	-8.9	-8.3	8"x10"	2 @ 1-1/4"
	COL 2	0.0	6.7	0.0	65.4	0.0	-14.9	0.0	-16.9	10"x15"	4 @ 3/4"
	COL 3	0.0	5.5	0.0	55.7	0.0	-13.4	0.0	-13.4	10"x15"	4 @ 3/4"
	COL 4	0.0	6.7	0.0	65.4	0.0	-16.9	0.0	-14.9	10"x15"	4 @ 3/4"
	COL 5	-2.1	3.3	21.8	30.3	8.9	-8.2	2.4	-6.2	8"x10"	2 @ 1-1/4"
4@60' x 18' RLL = 20 psf	COL 1	1.2	2.8	7.2	14.8	0.6	-5.4	-9.4	-8.6	8"x10"	2 @ 1"
	COL 2	0.0	6.2	0.0	33.1	0.0	-15.9	0.0	-16.4	10"x15"	4 @ 3/4"
	COL 3	0.0	5.2	0.0	27.9	0.0	-13.4	0.0	-13.4	10"x15"	4 @ 3/4"
	COL 4	0.0	6.1	0.0	33.0	0.0	-16.3	0.0	-16.0	10"x15"	4 @ 3/4"
	COL 5	-1.2	2.8	-7.2	14.8	9.6	-8.7	-0.8	-5.3	8"x10"	2 @ 1"
4@60' x 18' RLL = 30 psf	COL 1	1.1	2.9	8.8	21.2	1.3	-5.1	-8.8	-8.3	8"x10"	2 @ 1"
	COL 2	0.0	6.7	0.0	50.6	0.0	-16.2	0.0	-16.7	10"x15"	4 @ 3/4"
	COL 3	0.0	5.3	0.0	41.7	0.0	-13.4	0.0	-13.4	10"x15"	4 @ 3/4"
	COL 4	0.0	6.7	0.0	50.6	0.0	-16.7	0.0	-16.2	10"x15"	4 @ 3/4"
	COL 5	-1.1	2.9	-8.8	21.2	8.8	-8.3	-1.3	-5.1	8"x10"	2 @ 1"
4@60' x 18' RLL = 40 psf	COL 1	1.2	3.2	12.5	28.9	1.1	-5.2	-9.0	-8.4	8"x10"	2 @ 1"
	COL 2	0.0	7.1	0.0	66.5	0.0	-16.0	0.0	-16.5	12"x17"	4 @ 3/4"
	COL 3	0.0	5.7	0.0	56.4	0.0	-13.6	0.0	-13.6	12"x17"	4 @ 3/4"
	COL 4	0.0	7.0	0.0	66.5	0.0	-16.5	0.0	-16.0	12"x17"	4 @ 3/4"
	COL 5	-1.2	3.2	12.5	28.9	8.9	-8.4	-1.1	-5.2	8"x10"	2 @ 1"

Frame Reaction Tables

Gable, Four Spans @ 60'-0" (Continued)

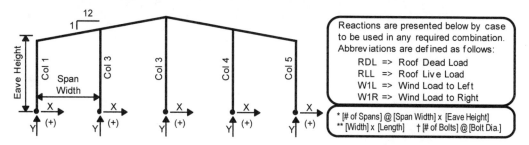

Reactions are presented below by case to be used in any required combination. Abbreviations are defined as follows:

RDL => Roof Dead Load
RLL => Roof Live Load
W1L => Wind Load to Left
W1R => Wind Load to Right

* [# of Spans] @ [Span Width] x [Eave Height]
** [Width] x [Length]   † [# of Bolts] @ [Bolt Dia.]

Frame Description*	Column	Column Base Reactions (kips)								Base-plate Size**	Anchor Bolts†
		RDL		RLL		W1L		W1R			
		X	Y	X	Y	X	Y	X	Y		
4@60' x 24' RLL = 20 psf	COL 1	0.9	3.0	5.0	14.8	3.1	-5.2	10.8	-9.9	8"x14"	2 @ 1"
	COL 2	0.0	6.2	0.0	32.6	0.0	-17.8	0.0	-17.1	10"x15"	4 @ 3/4"
	COL 3	0.0	5.4	0.0	28.9	0.0	-15.2	0.0	-15.2	10"x15"	4 @ 3/4"
	COL 4	0.0	6.2	0.0	32.6	0.0	-17.1	0.0	-17.8	10"x15"	4 @ 3/4"
	COL 5	-0.9	3.0	-5.0	14.8	10.8	-9.9	-3.1	-5.2	8"x14"	2 @ 1"
4@60' x 24' RLL = 30 psf	COL 1	0.9	3.1	7.1	21.8	3.2	-5.0	10.6	-9.8	8"x14"	2 @ 1"
	COL 2	0.0	6.8	0.0	49.5	0.0	-18.1	0.0	-17.3	12"x17"	4 @ 3/4"
	COL 3	0.0	5.8	0.0	42.8	0.0	-15.0	0.0	-15.0	12"x17"	4 @ 3/4"
	COL 4	0.0	6.8	0.0	49.5	0.0	-17.3	0.0	-18.1	12"x17"	4 @ 3/4"
	COL 5	-0.9	3.1	-7.1	21.8	10.7	-9.8	-3.2	-5.0	8"x14"	2 @ 1"
4@60' x 24' RLL = 40 psf	COL 1	0.9	3.2	9.3	28.9	3.2	-5.0	10.6	-9.7	8"x14"	2 @ 1"
	COL 2	0.0	7.3	0.0	66.2	0.0	-18.1	0.0	-17.3	12"x17"	4 @ 3/4"
	COL 3	0.0	6.2	0.0	57.1	0.0	-15.0	0.0	-15.0	12"x17"	4 @ 3/4"
	COL 4	0.0	7.3	0.0	66.2	0.0	-17.3	0.0	-18.1	12"x17"	4 @ 3/4"
	COL 5	-0.9	3.2	-9.3	28.9	10.6	-9.7	-3.2	-5.0	8"x14"	2 @ 1"
4@60' x 30' RLL = 20 psf	COL 1	0.8	3.3	4.3	15.1	4.9	-4.6	12.9	-11.4	8"x14"	2 @ 1"
	COL 2	0.0	6.3	0.0	31.7	0.0	-19.4	0.0	-16.4	12"x17"	4 @ 3/4"
	COL 3	0.0	5.8	0.0	29.9	0.0	-16.6	0.0	-16.6	12"x17"	4 @ 3/4"
	COL 4	0.0	6.3	0.0	31.7	0.0	-16.4	0.0	-19.4	12"x17"	4 @ 3/4"
	COL 5	-0.8	3.3	-4.3	15.1	12.9	-11.4	-4.9	-4.6	8"x14"	2 @ 1"
4@60' x 30' RLL = 30 psf	COL 1	0.8	3.5	6.3	22.5	5.1	-4.4	12.9	-11.5	8"x14"	2 @ 1"
	COL 2	0.0	6.6	0.0	47.7	0.0	-19.6	0.0	-16.3	12"x17"	4 @ 3/4"
	COL 3	0.0	6.1	0.0	45.0	0.0	-16.6	0.0	-16.7	12"x17"	4 @ 3/4"
	COL 4	0.0	6.6	0.0	47.6	0.0	-16.6	0.0	-19.3	12"x17"	4 @ 3/4"
	COL 5	-0.8	3.4	-6.3	22.5	12.7	-11.3	-4.9	-4.6	8"x14"	2 @ 1"
4@60' x 30' RLL = 40 psf	COL 1	0.7	3.5	7.6	29.4	5.4	-4.2	12.6	-11.2	8"x14"	2 @ 1"
	COL 2	0.0	7.1	0.0	64.5	0.0	-19.8	0.0	-16.7	12"x17"	4 @ 3/4"
	COL 3	0.0	6.3	0.0	59.5	0.0	-16.5	0.0	-16.4	12"x17"	4 @ 3/4"
	COL 4	0.0	7.1	0.0	64.5	0.0	-16.6	0.0	-19.9	12"x17"	4 @ 3/4"
	COL 5	-0.7	3.4	-7.6	29.4	12.6	-11.2	-5.3	-4.2	8"x14"	2 @ 1"

Index

A

ACI:
  ACI 117, 189, 190, 234, 245, 258, 261
  ACI 318 Appendix D, 196, 197, 201, 202, 204, 208, 213, 217, 218, 224, 235, 236, 240, 243, 260
  ACI 349, 243, 247, 249, 250, 255, 256, 261
  ACI 351.2R-10, 192, 234, 239, 261
  ACI 360R, 103, 104, 111
  ACI Design Handbook, 73, 74, 77, 134
Adjacent splices, 94
AISC:
  Specification, 5, 34, 237, 240, 253
  Steel Construction Manual, 186, 188, 190, 234, 261
  Steel Design Guide 1, 239, 243, 248, 262
Allowable bearing capacity of soil, 35
Anchor, 9, 52, 54, 56, 70, 92, 171, 172, 174, 185–234, 235–246, 250, 251, 253, 255–262. *See also* Anchor bolt
  cast-in, 201, 204, 207, 211–214, 229, 231
  column, 170, 185, 235, 236
  combined tension and shear, 194, 228
  embedded bolts, 185, 194
  headed bolt, 193, 194, 201, 204, 212, 220
  hooked, 186, 215, 216, 220
  post-installed, 193–197, 201, 207, 216, 228, 229
  pullout, 197–202, 212
  reduced diameter, 197, 240
  strength, pryout, 197, 207, 215, 227–231
  stud, headed, 193, 194, 215, 216, 220, 231
Anchor bolt, 172, 188, 197, 228, 233, 237, 243, 244
Anchor rod, 186, 188, 189, 192, 193, 201, 234, 239, 240, 243, 251, 262. *See also* Anchor bolt

Anchorage to concrete, 193, 194
Applied bending moment, 122, 123
ASCE:
  ASCE 32, 168
  ASCE/SEI 7, 36, 77, 117, 140
ASTM:
  D1557, 21
  F1554, 193, 200, 201, 211, 216
Atterberg limits, 19–22, 29

B

Blow count, 17, 19, 23, 26
BOCA National Building Code, 218

C

Cantilevered retaining wall, 82, 112–115
Cap plate, 236, 257
Centroid, 66, 67, 116, 125, 160, 198
Column:
  fixed-base, 11, 41, 45, 47, 122, 140
  footing, 21, 30, 35, 39, 59–64, 69–71, 80, 83–85, 93, 100, 101, 130–134
  interior, 83
  pedestals, 52–56, 62, 93–96, 110, 126, 173, 233–236, 247, 248, 257, 259. *See also* Column, pier
  pier, 39, 52, 55, 83, 91–93, 97, 98, 101, 109, 129, 133, 135, 172, 210
  pin-base, 41, 45, 121, 127, 148
Column reactions:
  estimating, 43, 44
  horizontal, 3, 40, 41, 51, 53, 101–108, 112–114, 121, 143, 175, 177, 182, 188, 191, 192, 198, 214, 224, 233–237, 242, 244
  lateral. *See* Column reactions, horizontal
  vertical, 50, 88, 143, 146, 177, 260

Index

Compressibility, 15, 28, 29
Concrete:
    bearing stress for, 249
    breakout prism, 198–204, 207, 208, 216, 224
    breakout strength, 197, 198, 202, 204, 207, 209, 211–217, 220–222, 224, 227–231
    crushing, 201, 239
    embedments, 54, 172, 188, 224, 234–236, 242, 243, 257, 262
    masonry units, 61, 79, 81
    placement tolerances, 177, 188, 237, 243
    plain, 60, 61, 65, 68, 70, 84, 165
    pryout, 197, 215, 227
    shear strength, 68, 247, 256
    side-face blowout, 197, 200–202, 212
    tensile capacity, 198
    uncracked, 210, 207, 211, 212
Corrosion resistance, 108
CRSI Design Handbook, 57, 60, 65, 77, 112, 116–118, 121, 127, 140, 149, 172–174, 183

D

Designing:
    for moment, 64
    for shear, 62
Development length, 73, 76, 91, 94, 95, 101, 102, 105, 107, 108, 135–137, 157–162, 207, 208, 224, 225
Differential settlement, 5, 21, 31, 32, 34, 63, 83, 180
Dowels, 55, 81, 82, 97, 104, 108–110, 126, 139, 177, 179, 207, 208, 213, 214, 233
Drilled piers, 173, 175
Ductility, 165, 199

E

Edge:
    distance, 54, 194–197, 202–204, 207, 211, 216, 217, 220–224, 228–231
    exterior, 98, 143, 57, 160, 179, 204
    reinforcement, 227
Elongation, 85, 89, 96, 99–101, 106, 176
Embedment length, 74, 136, 158, 160, 197, 198, 201, 202, 209
Endwall framing, 8
    nonexpandable, 8

F

Factor:
    cracked concrete, 224
    edge effect, 223
    of safety, 29, 34, 41, 42, 48, 119, 120, 129, 131, 151, 153–155
    shallow concrete, 224
FHWA-RD-75-130, 183
Floor slab, 53, 82, 99, 102, 104, 142, 178, 180–182
Footing:
    minimum width, 59, 83
    trench, 164
Force:
    horizontal, 88, 95, 102, 105–108, 120, 129, 131, 150, 153, 155, 175, 218, 219, 222, 234–239, 241–243, 249, 253, 254, 257, 261
    resultant, 125, 130, 132, 151, 152, 156, 165
Foundation:
    deep, 35, 39, 49, 51, 53, 79, 165, 173, 177–182
    moment-resisting, 11, 51, 80, 112–118, 120–122, 126, 127, 132, 133, 138, 140, 143, 165, 183
    shallow, 32, 35, 49, 52, 53, 142, 170, 172, 180
Frame-and-purlin buildings, 1, 9
Frames:
    portal, 11, 12, 244
    primary, 1–4, 7–9, 11, 96, 107, 186–188
    single-span rigid, 3–5, 41, 44, 48, 99, 107, 127, 148, 251
Frost line, 49, 51, 79, 142, 168–171

G

Girt:
    bypass, 8, 9, 143, 145, 149, 165, 173, 180, 204, 205, 228, 233, 235
    flush, 18, 9, 71, 99, 143, 145, 191, 204, 205, 228, 233
Groundwater pressure, 116
Grout:
    minimum clearance, 245
    nonshrink, 96, 97, 190–193, 238, 243–245, 250, 251, 255, 256, 258
    pad, 52, 54, 191, 216, 239, 242, 243
Grouting, 52, 190, 192, 234, 238, 243–245

H

Hairpin, 89, 224

I

International Building Code (IBC), 23, 34, 36, 37, 41, 47, 57, 59, 77, 80, 87, 110, 117, 118, 120, 140, 168, 172, 173, 183, 185, 193, 234, 243, 261

K

Kern limit, 123–126, 130, 143, 152, 156, 165

L

Lateral sliding, 42, 112, 114, 118–121, 126, 131, 147, 153, 155, 183
    coefficient of friction, 118, 119
Liquidity index, 20
Load:
    dead, 31, 37–39, 41, 49, 61, 63, 64, 69, 71, 75, 85, 86, 120, 121, 133, 143, 145, 156, 162, 167, 183, 256
    factor, 61, 62, 122, 135, 138, 157, 161, 162, 201
    nominal, 42, 61, 120
    roof live, 31, 38, 48, 63, 107, 149, 150, 195, 232
Load combinations, 31, 34, 38, 41, 42, 47, 48, 61–63, 65, 69, 71, 118, 120, 126, 128, 150, 195

M

Metal Building Manufacturers Association (MBMA), 3, 14, 42, 57, 234, 262
Metal Building System Manual, 3, 14, 42, 57, 188, 234, 262
Modulus of vertical subgrade reaction, 29, 30
Mohr's failure envelope, 116
Moment:
    fixity, of, 122, 140
    negative, 139, 162
    positive, 139, 162
    restoring, 130–132, 147, 151, 155

N

NAVFAC:
    DM-7.1, 16, 17, 19, 21, 24, 27, 36
    DM-7.2, 39, 57, 77, 120, 122, 140, 167, 172, 176, 183
Newman lug, 242, 243, 250, 251
Nuts, 187, 190–192, 243, 251, 255
    leveling, 190, 243, 251, 255

O

One-third stress increase, 34, 41, 42, 120
OSHA, 185–187

Oversized holes, 186, 188–193, 214, 235, 237–239, 243, 250, 251, 257–260
Overturning, 41, 42, 53, 113–115, 119–123, 127, 130–133, 141, 144, 148–156, 162, 165, 179, 183

P

PCA Notes on ACI 318, 77, 87, 201, 210, 234
Piles, 176
    battered, 182, 183
    single-pile system, 50, 177
Plasticity index, 18–23
Post-tensioned, 97–99, 176
Pressure, soil:
    active, 115–118, 121, 131
    passive, 117–121, 131–133, 144, 153, 156, 174, 175
Pressure wedge method, 125, 126

Q

Quonset hut, 1, 13, 14

R

Recessed column base, 242, 254–256
Redundancy, 50, 51, 92, 99, 100
Reinforcement:
    distribution in rectangular footings, 68
    minimum, 54–56, 68, 73, 76, 83, 92, 135, 161, 252
    spacing, 69
    supplemental, 232, 233, 235–237, 260
    surface, 208
    tension, 70, 95, 213, 214
    at top of footings, 70
    transverse, 73, 100, 136, 137, 159
    vertical pier, 54, 70, 209, 210, 213, 233, 236
Reliability, 9, 49, 50, 51, 112, 148, 167
Retaining wall, 118–122, 127, 133, 175
Rotation, 127–130, 147, 148, 237

S

Seismic design category, 24, 34, 81, 210
Seismic ties, 60, 181
Settlement, 5, 15, 21–35, 61, 63, 83, 97, 110, 180
    differential, 5, 21–34, 63, 83, 180, 254
Shaft, 173–176, 214, 220
Shear:
    lug, 243
    two-way action, 65–68, 75
    wide-beam action, 65–67, 70, 73, 75

Shims, 192, 238, 258–260
Shrinkage:
   drying, 90, 103
   limit, 19, 20
Sidewalls, 2, 7, 8
Slab:
   downturned, 142
   on grade, 32, 35, 51, 53, 64, 69, 71, 79, 81–83, 89–91, 97–100, 104, 108, 111, 113, 115, 128, 150, 167–172, 180, 182, 259
   with haunch, 14, 51, 79, 80, 143–150, 153–156, 161, 165, 167
   structural, 103–105, 171, 178–182
Sliding, 41, 42, 102, 113, 115, 119–123, 127, 131–134, 144, 148, 154, 156, 183
Soil:
   anchors, 39, 124, 175, 176
   borings, 25, 26, 28, 35
   coarse-grained, 16, 17, 20
   cohesionless, 17, 29, 30, 119, 166, 181
   expansive, 22, 23, 32, 98, 111, 180
   fill, structural, 21
   friction, 99, 115, 120, 131, 144, 154, 156
   highly organic, 16, 18, 23, 25
   lateral pressure, 19, 22, 33, 116–118, 128
   mixtures, 18, 20
   pressure:
      factored, 65, 86
      trapezoidal, 124, 127, 144, 147, 149, 153, 156
   presumptive load-bearing values, 33, 34, 119, 176
Special inspection, 194, 196
Splices, mechanical, 90, 91
Standard penetration test, 26
Steel, high-strength, 89, 99, 186, 187, 198
Stemwalls, 84
Stirrups, 93–96, 163, 176, 227
Strength design, 61–65, 193–197
Support, lateral, 8, 83, 173–176, 182, 183
Survivability, 50, 51, 90

T

Template, 187, 188
Tensile crack, 103, 106,
Tension-tie members, 89, 90
Threaded parts, 240
Tie rod, 51, 89–105, 167
TMS 602, 245, 262
Toe, 113, 114
Truncated prism, 203

U

U.S. Department of Agriculture, 25

V

Versatility, 50, 51, 90, 149, 167

W

Wall:
   footing, 14, 21, 35, 49, 61, 79, 82–85, 133
   loadbearing, 85, 177
   nonstructural finishes, 32, 33, 97, 115, 127, 170, 175, 182, 237
Washers, 186–193, 214, 238–243, 250, 251
   hillside, 9, 10
Waterproofing, 97, 98
Weldability, 186
Welded-wire fabric, 103
Wind post, 11